Dictionary of Mathematics

Everyday Handbooks

Dictionary of Mathematics

T. ALARIC MILLINGTON
WILLIAM MILLINGTON

BARNES & NOBLE, INC. • NEW YORK
Publishers • Booksellers • Since 1873

First American Edition 1966

Reprinted 1971 by Barnes & Noble, Inc.
through special arrangement with
A. S. Barnes and Co., Inc.
Cranbury, New Jersey 08512

Library of Congress Catalog Card Number: 76-149830

SBN 389 00320 4

Printed in the United States of America

Contents

Preface

by

DR. HYMAN LEVY, M.A., D.SC., F.R.S.E.

Emeritus Professor, University of London. Formerly Head of the Department of Mathematics, Imperial College of Science

The generation that has grown up during the past twenty years has entered a society that is very different indeed from anything that preceded it. There is hardly an aspect of social life that has not been drastically affected by scientific discovery and technological application. The obstacles of distance to personal communication, visual and aural, between individuals have been all but removed. Machines have been devised that can multiply the speed of ordinary computation a million-fold, assemble data in factories, in a certain sense draw general conclusions from them, and immediately apply the latter so that large sectors of factory production can proceed automatically, with the minimum of personal guidance and supervision. In fact, that part of our thought and action that was mechanical, can be delegated to a machine.

What all this can mean for our thinking, our creative activity, and our way of life generally, the future will show. Behind it all, however, lies a network of mathematical thought, of abstract ideas about points, lines, sets, and groups, represented by concrete symbols on paper, so that the study of the relations between these ideas, and the outcome of their combination, may be carried through by an examination of the various relations that may be established between the concrete symbols. Mathematicians have opened up a fascinating world, which, in a figurative sense, is a symbolic reflection or model of the world around us, and our thinking about it. Thus mathematics supplements experimental science, working out the logical conclusions of its theories, and helping to probe the secrets of the universe.

The recent output of books on and about mathematics is evidence of a growing realization by the layman of the cultural importance of the subject, and with this, the possibilities that its pursuit opens up for the careers of the oncoming generation. This is the immediate justification for the production of a Dictionary of Mathematics that will not only enlighten the enquiring layman, but will open the door of understanding to the student who is anxious to know, with accuracy and precision, the background of ideas with which he will have to become familiar.

There are two kinds of dictionary. One is a mere index to aid translation from one language to another. The second confines itself entirely to one language and seeks to express the various shades of meaning attached to a word, in terms of other words found in the same dictionary. Sometimes of course the words are so basic—*and*, for example—that one finds the same word is used again in the explanation! A language is a social growth, and it is inevitable that the same word, used in different contexts, will vary in meaning. A mathematical dictionary does not belong to either of these categories. A word like *equation* has a unique meaning in the sense that it is permissible to use it in very definite mathematical circumstances only. It would never be used metaphorically, for example, as one might use the word *page* in the expression '*a page of history*'. The reason is that mathematical methods, and the notation which describes these, are deliberate creations designed to communicate ideas as precisely as possible, and divorced as far as possible from the kind of feeling and emotion that is inevitable in ordinary speech. It is concerned with logic and not with poetry; although many of its theorems, in the simplicity with which their deeper meaning is revealed, are by no means devoid of aesthetic appeal.

For all these reasons the writing of a mathematical dictionary is not only an important social task at the present moment, but a vital educational one. It can be performed adequately only by those who have had both direct experience of the art and the techniques of teaching, but who have also grasped the newer ideas that are fermenting the world of mathematics at the present time, and revolutionizing its teaching methods. It is therefore fortunate that the Millingtons, father and son, with their wealth of teaching experience, extending over two rapidly changing generations, in institutions ranging from primary school to college of education and university, should have combined together to the necessary task of producing a dictionary, which is necessarily part way towards a mathematical encyclopaedia writ small. The outcome is excellent and can be recommended to pupil, parent and teacher alike.

Introduction

The world today is experiencing revolutions of many kinds—scientific, technological, psychological, economic, social. In the field of education, traditional beliefs which have been accepted as truths for generations are being questioned more and more. In no subject is this more evident than in mathematics. Students of all ages, and their teachers, are meeting newer mathematics as well as learning the established branches of the subject. The need for clear thinking and clarity of expression has never been greater. It was in this climate that this book was created.

We do not believe that the acquisition of mathematical knowledge is the same thing as learning to think mathematically. But we do believe that the development of sound concepts demands clear thinking and the use of correct language. We have tried to convey concepts, where possible, rather than over-simplified technical definitions. Adequate cross-references are provided for readers who wish to widen their study; and, where appropriate, reference has been made to the historical development of mathematics. The language of such modern theories as sets, groups, rings, fields, vectors, logic and motion geometry is explained, and the traditional branches of mathematics, pure and applied, are given a modern interpretation.

The dictionary is intended for several kinds of reader. It is a basic reference book for all students of mathematics in schools and colleges. It is hoped that teachers and student-teachers will find it a useful source-book; and that those no longer actively working in the field of mathematics will find the book stimulating to read.

We wish to record our gratitude to Professor Hyman Levy for giving so generously of his time in painstakingly reading the typescript and offering constructive suggestions.

Hampstead Garden Suburb
1966

WILLIAM MILLINGTON
T. ALARIC MILLINGTON

How to use the Dictionary

Cross-references are given in two ways. If a term used in one definition is defined elsewhere such that further elucidation can be obtained by reference to it, the term is printed in **bold face type**. Additional cross-references are given at the end of a definition in **bold face type**.

Supplementary definitions, which appear as part of a main definition, are printed in **bold face type** and indented under the main definition. These are not defined elsewhere.

Italics have been used to emphasize certain words and symbols.

For example: in the definition of ACCELERATION, **curvature** and **vector sum** are defined separately. Supplementary definitions, **absolute, average** and **instantaneous acceleration**, are not defined elsewhere. Additional information is found in the definitions of **Angular Acceleration**; **Equations of Motion**; **Relative Acceleration**; **Velocity.**

A

Å. **Angstrom.**

A.G.P. **Arithmetico-Geometric Progression.**

A.P. **Arithmetic Progression.**

A.U. **Astronomical Unit.**

ABACUS. The simplest and earliest type of calculating device. Originally, it consisted of a set of grooves in the sand in which pebbles were placed to represent units, tens, hundreds, according to the groove used. It was first made transportable by using a sand board (*abaq*, in old Semitic). The *counting frame* was extensively used by the Egyptians, Greeks and Romans, as computation in their number notations was cumbersome. The ease of operation on the abacus, in fact, removed any need to improve the notations. The abacus eventually spread throughout the civilized world in different forms. The Roman one consisted of a metal plate with two sets of parallel grooves, the lower containing four pebbles (the word *calculus* from which calculation is derived, is Latin for pebble), the upper set one pebble, equivalent in value to five times the value of one in the corresponding groove in the lower set. The Japanese abacus, having beads on wires, is a development of this idea. The Chinese abacus and the Bulgarian abacus have sets of two beads and five beads.

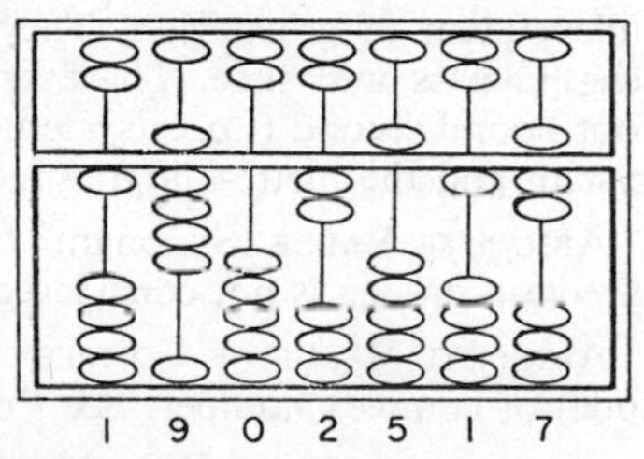

ABELIAN GROUP. See **Commutative (Abelian) Group.**

ABOVE PAR. See **Premium.**

ABSCISSA. The x-coordinate of a point in a plane. See **Cartesian Coordinates.**

ABSOLUTE. Independent; not relative.

ABSOLUTE CONSTANT. See **Constant.**

ABSOLUTE CONVERGENCE. The sum of an infinite **series**, with terms $a_1, a_2, a_3, \ldots$ converges absolutely if $|a_1| + |a_2| + |a_3| + \ldots$ converges, e.g. the **alternating series** $1 - \frac{1}{2} + \frac{1}{4} - \frac{1}{8} + \ldots$ has absolute convergence because $1 + (\frac{1}{2}) + (\frac{1}{2})^2 + (\frac{1}{2})^3 + \ldots$ is the **geometric progression** which converges to the limit 2. See **Absolute Value; Convergence, Divergence of Series.**

ABSOLUTE ERROR. In theory the difference between the measured value of a quantity and its true value. The true value can never be found by measurement, which is an **approximation** within limits determined by the accuracy of the instruments and the measurer. In practice absolute error is taken as the **deviation** of a measurement from the average of a set of measurements of the same quantity.

ABSOLUTE INEQUALITY. See **Inequality.**

ABSOLUTE NUMBER. A non-**literal** constant.

ABSOLUTE SCALE. Synonym of **Kelvin Scale** of temperature.

ABSOLUTE TERM. The term in an **expression** which does not contain an unknown quantity. In the quadratic equation representing all **conic sections,**

$$ax^2+2hxy+by^2+2gx+2fy+c=0$$

c is the absolute term. It is sometimes referred to as the **constant term.**

ABSOLUTE UNITS. Units of **force, work, energy** and **power** which are independent of gravitation. They are based on the **fundamental units** of **length, mass** and **time.** There are three possible systems of units: the foot-pound-second (f.p.s.) system, the centimetre-gramme-second (c.g.s.) system and the metre-kilogramme-second (m.k.s.) system.

ABSOLUTE VALUE. Synonym of **numerical value.** The value when direction or sign is not considered.

ABSOLUTE VALUE OF COMPLEX NUMBER. Synonym of modulus (magnitude) of complex number. See **Polar Form of Complex Number.**

ABSOLUTE VALUE OF REAL NUMBER. The **numerical value** when direction or sign is not considered. The absolute value of both a and $-a$ is a, written as $|a|=|-a|=a$. There are four laws: $|0|=0$; when $x\neq 0$, $|x|>0$; $|x \cdot y|=|x|\cdot|y|$; $|x+y|\leqslant|x|+|y|$.

ABSOLUTE VALUE OF VECTOR. The **numerical value** of the length of the line segment representing a **vector quantity**. The absolute value of $x\mathbf{i}+y\mathbf{j}+z\mathbf{k}$ is $\surd(x^2+y^2+z^2)$.

ABSOLUTE VELOCITY. See **Velocity.**

ABSOLUTE ZERO. Zero on the **Kelvin Scale** or absolute scale of temperature.

ABSTRACT. ABSTRACTION. The process of developing a concept. Concrete experience is generalized and idealized to form a logical mental structure or a mathematical model. Thus, many experiences with three objects produce the concept of the number 3. Many experiences with boundaries produce the concepts of curve and surface. One level of abstraction can lead to another. Thus the expression $a+b$ is a mathematical model for the addition of any two numbers, which may be two **cardinal numbers** which are themselves abstractions from a concrete level of experience.

ABSTRACT MATHEMATICS. Synonym of pure **mathematics.**

ABSTRACT NUMBER, SYMBOL, WORD. A number, symbol, word dissociated from the concrete or any specific cases.

ABSTRACT SPACE. A formalized mathematical **system** which has geometric qualities.

ABUNDANT NUMBER. See **Perfect Number.**

ACCELERATION. The rate of change of velocity with respect to time. When the displacement is along a line (*rectilinear motion*) the acceleration of a point with displacement x is the derivative dv/dt or $\ddot{x}$ (linear or rectilinear acceleration). When the displacement is along a curve (*curvilinear motion*) its acceleration has two components, one along the tangent, one along the normal to the path and directed towards the centre of **curvature.** These are dv/dt (*tangential component*) and v^2/ρ (*normal or centripetal component*) where ρ is the radius of curvature at the point. The resultant acceleration is the **vector sum** of these two components.

absolute acceleration. The acceleration of a moving point with respect to a fixed point. The acceleration relative to a fixed system of coordinates.

average acceleration. The difference in velocities of a point at the beginning and end of a time interval, divided by the time interval, $\Delta v/\Delta t$. If the acceleration is uniform during this interval the average acceleration is the actual acceleration.

instantaneous acceleration. The limit of the average acceleration when the time interval, Δt, approaches zero: dv/dt.

See **Angular Acceleration**; **Equations of Motion**; **Relative Acceleration, Velocity.**

ACCELERATION DUE TO GRAVITY. The **acceleration** of a body which is moving under the **force of gravity** only, and ideally in a vacuum. Its value, symbol g, varies according to the distance of the body from the centre of the earth. By international agreement standard gravity $g_0 = 980.665$ cm/s^2 or 32.173 9 ft/s^2.

ACCLIVITY. See **Angular Levelling.**

ACCOUNTANCY. The principles and practices of keeping accounts.

ACCOUNTS. Records of financial transactions.

ACCUMULATED VALUE. Synonym of **amount** in **simple interest** and in **compound interest.**

ACCUMULATION POINT. The limit of a sequence of points of a set. Synonym of cluster point, limit point. Thus the sequence $\{1, \frac{1}{2}, \frac{1}{4}, \frac{1}{8}, \ldots\}$ has zero as an accumulation point. See **Bound of Set**; **Crowd About**; **Limit of Sequence.**

ACCURACY. The numerical measure of the closeness of an **approximation** to the truth. See **Correct to *n* Decimal Places**; **Significant Figures.**

ACCURATE. Exact. Without error.

ACCURATE TO *n* DECIMAL PLACES. Synonym of **correct to *n* decimal places.**

ACNODE. Synonym of **isolated point.**

ACRE. British unit of area: 4 840 yd². Equivalent to 4 047 m² or 40·47 ares. See **Area, British Units; Area, Metric Units.**

ACTION AND REACTION. Any force exerted on a body in pushing, pulling, lifting, supporting, etc., is resisted by another force exerted by the body being pushed, etc. Not until Newton's time was the relationship between these opposing forces, called action and reaction, expressed mathematically. Newton's third law of motion states: Action and reaction are equal and opposite. This implies (1) If a body A exerts a force on body B, then B exerts an equal force on A in the opposite direction; (2) All the internal forces of a body may be grouped into actions and reactions which have no effect on the motion of a body; (3) Every change in the velocity of a body is caused by external forces applied to the body.

ACUTE ANGLE. An **angle (plane angle)** less than a right angle (90° or $\pi/2$ radians).

ACUTE (ACUTE-ANGLED) TRIANGLE. A **triangle** with three **acute angles.**

ADDEND. Any one of a set of numbers to be added, e.g., 1, 2, 3 and 4 are the addends of the sum $1+2+3+4=10$.

ADDITION. A general term for the **binary operation** equivalent to that of combining two equal or unequal groups into a whole; the reverse operation of subtraction. The name of the combination is variously referred to as aggregate, sum, total. Fundamentally, the operation is an abstraction from the concrete experience of combining groups of real objects and is expressed in terms of **natural numbers**. The concept of addition is developed by extending the number concept to embrace **fractions** (rational numbers). It is further developed by the **algebraic addition** of **directed numbers,** and further still by its application to the field of **complex numbers**. The addition of numbers obeys the **commutative law,** $a+b=b+a$, and the **associative law,** $a+(b+c)=(a+b)+c$.

The term addition can be applied to fields outside traditional arithmetic: two sets of things can often be combined in an additive sense as in union (sum) in **set theory.**

ADDITION OF SETS. See **Set Theory.**

ADDITION, SUBTRACTION FORMULAE.

$$(1)\ \sin(A\pm B)=\sin A\cos B\pm\cos A\sin B,$$
$$(2)\ \cos(A\pm B)=\cos A\cos B\mp\sin A\sin B,$$
$$(3)\ \tan(A\pm B)=(\tan A\pm\tan B)/(1\mp\tan A\tan B).$$

If $B = A$,

(4) $\sin 2A = 2 \sin A \cos A$,
(5) $\cos 2A = \cos^2 A - \sin^2 A$,
(6) $\tan 2A = 2 \tan A/(1 - \tan^2 A)$.

If $B = 2A$,

(7) $\sin 3A = 3 \sin A - 4 \sin^3 A$,
(8) $\cos 3A = 4 \cos^3 A - 3 \cos A$,
(9) $\tan 3A = (3 \tan A - \tan^3 A)/(1 - 3 \tan^2 A)$.

ADDITION, SUBTRACTION OF COMPLEX NUMBERS. See **Complex Numbers**.

ADDITION, SUBTRACTION OF VECTORS. See **Composition of Vectors**.

ADDITIVE FUNCTION. A function of the form $f(x+y)$ defined as being equal to $f(x)+f(y)$.

AD INFINITUM. Continuing without end in some specified manner. Usually indicated by . . . or $\rightarrow\infty$, written after the last term of a sequence or the last written factor of a product. See **Infinite**; **Infinity**.

ADJACENT ANGLES. Two angles which do not overlap displayed by three **rays** in a plane with a common initial point. If the angles are supplementary, two of the rays form a straight line.

ADJACENT SIDES. Any pair of sides of a **polygon** that meet at a **vertex**.

ADJUNCTION. The process of enlarging a given set to include a new element with the intention of extending the properties of the given set. The **field of real numbers** $[R^*]$ does not include roots of quadratic equations such as $x^2+1=0$, since $x^2+1>0$ for all real x. The adjunction of a new number $i = \sqrt{-1}$, $[R^*(i)]$, and a redefining of addition and multiplication to accommodate the new number produces the field of **complex numbers**, c.

AFFINE GEOMETRY. The study of properties of geometric **configurations** which are unaltered by *parallel* (*cylindrical*) **geometric projection.** It includes the study of **affinities** and can be treated by the methods of analysis. Branch of **projective geometry** in which centres of projection are at infinity.

AFFINITIES. The group of **transformations** which are combinations of **isometries, shear transformations** and **similarities**.

AGGREGATE. A term used to emphasize the wholeness of a collection as opposed to the parts constituting it. A sum of totals as in the aggregate of a set of scores. A totality such as the system or set of rational numbers. Such an aggregate may be closed or ordered as are sets in **set theory**.

AGGREGATION. The process of indicating terms to be treated collectively by using signs of aggregation. Terms may be enclosed between brackets [ ], braces { }, parentheses ( ) or they may be placed above or below a vinculum, ———.

For example: $a - 3[2b + \{c + 2(d - \overline{e - f})\}]$, $\dfrac{x^3 - y^3}{x - y}$, $\sqrt{x - 5}$.

Agonic Line. A line on a map joining places of zero **magnetic declination**. At such places a compass needle points to geographic and magnetic north.

Air Navigation (Avigation). The navigation of aircraft can involve various techniques: visual observation of landmarks; dead reckoning; astronavigation (celestial observation); radio and radar navigational aids. Dead reckoning requires knowledge of the speed and direction of the wind before an aircraft can be headed in the correct direction to counteract its effect. The air speed as registered by the air speed indicator, and the wind velocity can be combined in a **triangle of vectors** to determine the ground speed. Alternatively, the triangle of vectors may be used to determine the direction the aircraft is headed, and the air speed necessary to counteract wind effect, for a given direction of flight and a given ground speed. $\overrightarrow{AB}$ is the direction in which the aircraft is headed; $\overrightarrow{AC}$ is the actual direction of flight.

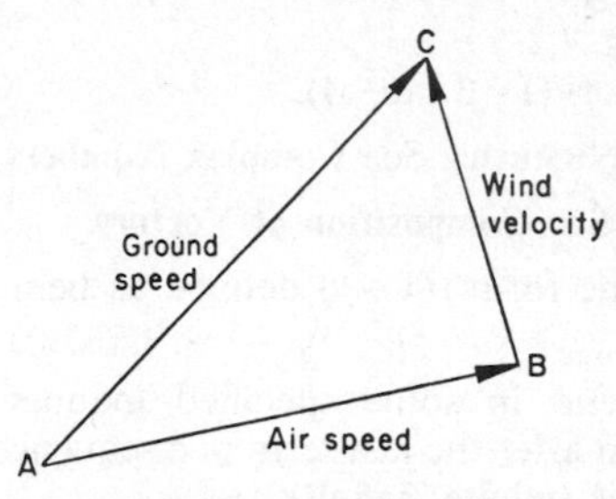

Astronavigation involves the determination of the position of the aircraft in terms of latitude and longitude by means of observations on the sun or the stars. The course is chosen by reference to a navigational map and periodical observations are made to correct the course. A knowledge of **magnetic declination** is required when compass directions are used.

Air Pressure. See **Atmospheric Pressure**.

Air Resistance. The total **force** exerted by the atmosphere, assumed to be stationary, to the passage through it of any body. It is equal to the total force exerted by the atmosphere (the wind force) on the same body at rest, if wind speed is equal to original speed of body and direction of wind is opposite to the direction in which the body was moving.

Air Speed. See **Air Navigation (Avigation)**.

Algebra. The methods of reasoning about numbers by employing letters to represent them and signs to represent their relationships. The use of symbols first occurred in the works of the early Egyptians and Hindus; the papyrus of Ahmes dates from about 1700 B.C. The Greek, Diophantus, contributed to the subject around A.D. 250, but true algebra was not developed until the seventeenth century. The origin of the word is not known with certainty but in A.D. 825 an Arabian mathematician wrote a work entitled Al-jebr-w'al-muqabâlah. See **Operations in Ordinary Algebra**; **Signs of Operation**.

Algebra, Boolean. See **Boolean Algebra**.

ALGEBRA OF PROPOSITIONS. The **Boolean algebra** of logic which treats mathematically the validity of propositions involving phrase connectives: and; or; not; if; then; if and only if, then.

conditional. If P then Q, written $P \rightarrow Q$, $P \supset Q$.

conjunction. P and Q, written $P \wedge Q$.

denial (negation). Not P, written P', $-P$, $\sim P$, $\bar{P}$.

disjunction. P or Q (or both), written $P \vee Q$.

equivalence or biconditional. If P, and only if P, then Q, written P iff Q, $P \longleftrightarrow Q$, $P \equiv Q$, $P \sim Q$.

The operational rules are **isomorphic** to those of **algebra of sets** when *union* (sum) is equivalent to the connective *or*, and *intersection* (product) to *and*. If *or* is interpreted as meaning P or Q (but not both), the systems are isomorphic if the symmetric **difference of two sets** is interpreted as addition and not union. See **Truth Set**.

ALGEBRA OF SETS. The **Boolean algebra** of **set theory**. For a set of sets $\{A, B, C, \ldots\}$, subject to two **binary operations**, union ($\cup$) and intersection ($\cap$), with ϕ as the null set and I as the universal set, the following operational rules apply:

closure. $A \cup B$ is a set; $A \cap B$ is a set.

associative law. $A \cup (B \cup C) = (A \cup B) \cup C$; $A \cap (B \cap C) = (A \cap B) \cap C$.

commutative law. $A \cup B = B \cup A$; $A \cap B = B \cap A$.

distributive law. $A \cap (B \cup C) = (A \cap B) \cup (A \cap C)$; $A \cup (B \cap C) = (A \cup B) \cap (A \cup C)$.

idempotent law. $A \cup A = A$; $A \cap A = A$.

bounds. $A \cup \phi = A$; $A \cup I = I$; $A \cap \phi = \phi$; $A \cap I = A$.

complements. $A \cup A' = I$; $A \cap A' = \phi$; $(A \cup B)' = A' \cap B'$; $(A \cap B)' = A' \cup B'$. These are **De Morgan's Laws.**

involution. $(A')' = A$; $\phi' = I$; $I' = \phi$.

The operational rules are **isomorphic** to those of **algebra of propositions** when *union* (sum) is equivalent to the connective *or* and *intersection* (product) to *and*. If *or* is interpreted as meaning P or Q (but not both), as in the case of P, Q not being equivalent in value, then sum is interpreted as the symmetric **difference of two sets.** See **Cartesian Product**; **Separation (2)**; **Truth Set.**

ALGEBRAIC ADDITION. The addition of directed numbers according to the following rules of signs. The quantities in brackets [ ] are the sums or differences of **numerical values** of a and b.

$$
\begin{array}{ll}
(+a)+(+b) \equiv +[a+b], & \\
\left\{\begin{array}{l}(+a)+(-b) \equiv +[a-b], \\ (+a)+(-b) \equiv -[b-a],\end{array}\right. & \begin{array}{l}a>b, \\ a<b,\end{array} \\
(-a)+(-b) \equiv -[a+b], & \\
\left\{\begin{array}{l}(-a)+(+b) \equiv -[a-b], \\ (-a)+(+b) \equiv +[b-a],\end{array}\right. & \begin{array}{l}a>b, \\ a<b.\end{array}
\end{array}
$$

See **Operations in Ordinary Algebra.**

ALGEBRAIC CURVE. A graphical representation of an algebraic **relation.** A set of points having coordinates which satisfy an **algebraic equation.** See **Coordinates of Point.**

ALGEBRAIC DIVISION. The division of **directed numbers** according to the following rules of signs. The quantities in brackets [] are quotients of **numerical values** of a and b.

$$\frac{+a}{+b} \equiv \frac{-a}{-b} \equiv +\left[\frac{a}{b}\right]; \quad \frac{+a}{-b} \equiv \frac{-a}{+b} \equiv -\left[\frac{a}{b}\right].$$

See **Operations in Ordinary Algebra.**

ALGEBRAIC EQUATION. An equation containing only **algebraic expressions** and **signs of operation.**

ALGEBRAIC EXPRESSION. A symbol or a collection of symbols which may be associated with **signs of operation.** Those parts of the expression which are connected by plus or minus signs are called its terms. See **Multinominal Expression; Polynominal Expression.**

ALGEBRAIC FUNCTION. A **function** containing only **algebraic expressions** and **signs of operation.**

ALGEBRAIC MULTIPLICATION. The multiplication of **directed numbers** according to the following rules of signs. The quantities in brackets [] are the products of **numerical values** of a and b.

$$(+a)\times(+b) \equiv (-a)\times(-b) \equiv +[ab]; \quad (+a)\times(-b) \equiv (-a)\times(+b) \equiv -[ab].$$

See **Operations in Ordinary Algebra.**

ALGEBRAIC NUMBERS. Real or complex **numbers** which are solutions of **polynominal expressions** having rational coefficients.

ALGEBRAIC OPERATION. One of the fundamental operations of arithmetic: **addition, multiplication** and **involution,** and the reverse operations, **subtraction, division** and **evolution.** See **Operations in Ordinary Algebra.**

ALGEBRAIC SUBTRACTION. The subtraction of **directed numbers** according to the following rules of signs. The quantities in brackets [] are differences or sums of **numerical values** of a and b.

$$\begin{cases}(+a)-(+b) \equiv +[a-b], & a>b,\\ (+a)-(+b) \equiv -[b-a], & a<b,\end{cases}$$
$$(+a)-(-b) \equiv +[a+b],$$
$$\begin{cases}(-a)-(-b) \equiv -[a-b], & a>b,\\ (-a)-(-b) \equiv +[b-a], & a<b,\end{cases}$$
$$(-a)-(+b) \equiv -[a+b].$$

See **Operations in Ordinary Algebra.**

ALGORITHM. Any standardized procedure for the solution of a particular type of problem. See **Euclid's Algorithm.**

ALIQUOT PART. Any named portion of a quantity which is a **unit fraction** of the quantity, e.g., one furlong is an aliquot part of one mile ($\frac{1}{8}$).

ALTERNATE ANGLES. See **Transversal**.

ALTERNATE SEGMENT. Of two **segments of circle**, with a common separating chord, the larger is the major segment and the smaller is the minor segment. The term *alternate* refers to the segment other than the one under immediate consideration. See **Alternate Segment Theorem**.

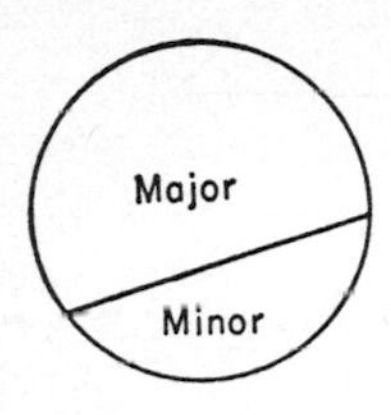

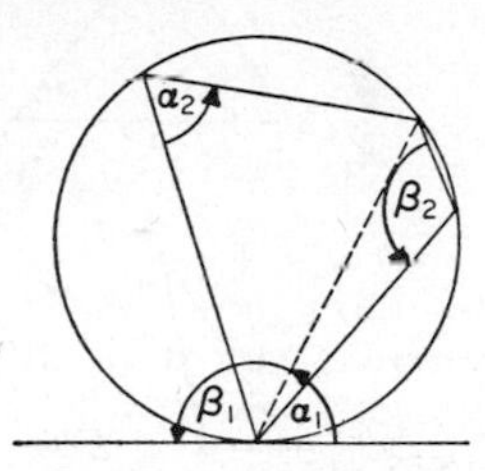

ALTERNATE SEGMENT THEOREM. The acute or obtuse angle between a tangent to a circle and a chord passing through the point of contact is equal to the acute angle in the major segment or the obtuse angle in the minor segment respectively, i.e., either is equal to the angle in the **alternate segment**.

ALTERNATING GROUP. See **Group Theory**.

ALTERNATING SERIES. An infinite **series** with terms alternately positive and negative.

ALTERNATION. (1) In proportion, if $a/b = c/d$, the deduction that $a/c = b/d$. (2) In **algebra of propositions**, a synonym of disjunction.

ALTITUDE ABOVE SEA-LEVEL. The vertical distance of a point above mean sea-level.

ALTITUDE OF CELESTIAL BODY. The spherical coordinate of a celestial body measured in the plane of the circle passing through the body and crossing the horizon at right angles. See **Spherical Coordinates in Astronomy**.

ALTITUDE OF PLANE RECTILINEAR FIGURE. A line segment and its length drawn from a vertex perpendicular to the opposite side (called the base), which may be extended. If the figure has n sides there will be $n-2$ altitudes through each vertex.

ALTITUDE OF SOLID. A line segment and its length drawn from a vertex to the opposite face (called the base), which may be extended.

ALTITUDE OF TRIANGLE. A special case of **altitude of plane rectilinear figure** when $n=3$. The three altitudes are **concurrent**.

AMBIGUOUS CASE. If two sides and one angle of a triangle are given, four possible triangles can be drawn according to the relative lengths of the sides and the relative positions of the sides and the angle.

1. cBa; two sides and the included angle—1 possible triangle.
2. bcB; $b>c$—1 possible triangle.

3, 4. bcB; $b<c$—2 possible triangles, referred to as the *ambiguous case.*

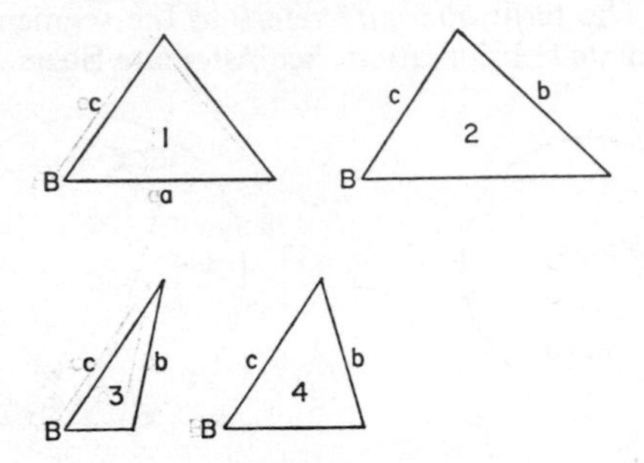

AMICABLE NUMBERS. Pairs of **numbers**, each of which is the sum of unity and all the factors of the other. They were known to the Hindus and given this name by the Pythagoreans. The Greeks knew 220 and 284. Euler published a list of 64 of which two were false. It is a reasonable assumption that there is an infinite number of such pairs.

AMOUNT. (1) Quantity. (2) Principal plus interest. See **Simple Interest**; **Compound Interest.**

AMPERE (amp, A). Unit of electric current. The international ampere is the electric current which, passed through a solution of silver nitrate, deposits silver at the rate of 0.001 118 grammes per second. Other units of electric current are milliamp (mA) $=10^{-3}$ A and microamp (μA) $=10^{-6}$ A. See **Ampere-hour.**

AMPERE-HOUR. The practical unit of electric current, defined as the amount of current which flows through a conductor in one hour when the current is one **ampere.** It is equivalent to 3 600 **coulombs.**

AMPLITUDE IN OSCILLATORY MOTION. The maximum value of the displacement of a particle or a body in such a motion. If a coiled spring hangs from a support and its end is pulled downwards, then released, the end will rise and fall rhythmically about its normal position. The amplitude of the motion is one half of the difference in length between the length of the spring when most compressed and its length when most extended. See **Simple Harmonic Motion.**

AMPLITUDE IN POLAR COORDINATES. See **Polar Coordinates in Plane.**

AMPLITUDE OF COMPLEX NUMBER. See **Polar Form of Complex Number.**

ANALOGY. The equality of **ratios, i.e., proportion.** In trigonometry, the analogy

$$\frac{\tan \frac{1}{2}(B-C)}{\cot \frac{1}{2}A} = \frac{\sin \frac{1}{2}(B-C)}{\sin \frac{1}{2}(B+C)}$$

is attributed to Napier.

ANALYSIS. Those parts of mathematics which exclude (*a*) the plane and solid geometry so exhaustively treated by the Greeks, and (*b*) the algebra (including arithmetic) associated with computation. It stems from the work of the mathematicians of the seventeenth century and is based on the theories of **continuous functions, convergence** and **limits**. A few of the many subjects included in analysis are **analytic geometry, calculus, exponential functions, logarithmic functions, trigonometric functions, series.**

ANALYSIS SITUS. The study of those properties of spaces which are independent of shape or size: continuity, proximity, and the **topological properties** of figures which can be expressed as sets of points. The study also considers those shapes which are topologically equivalent (homeomorphic), that is, those shapes which can be placed in **one-one correspondence** under a two-way **mapping. See Topological Transformation**; **Topology.**

ANALYTIC GEOMETRY (COORDINATE GEOMETRY). Until the middle of the seventeenth century, **geometry** and **algebra** had developed as two separate branches of **mathematics**. In the sixteenth century geometry differed very little from that of Euclid and Apollonius. In 1637 René Descartes published *La Géométrie* in which he applied algebra to the study of theorems and the solution of problems.

Points are defined by means of a **frame of reference** and coordinates. Analytic or coordinate geometry consists of drawing graphs of equations with two or three variables and, inversely, finding equations for loci on a plane or in space. See **Cartesian Coordinates**; **Coordinates of Point.**

ANCHOR RING. See **Torus.**

ANGLE (PLANE ANGLE). When a straight line rotates in a plane about a fixed point in the line, the angle is the measure of the amount of rotation. It is also the measure of the difference in direction of the **initial and terminal lines.** It is a convention that an anti-clockwise rotation generates a positive angle, a clockwise rotation a negative one. Though the term angle is often used for the **vertex** (corner) of a polygon, the concept of angle is strictly one of rotation.

The unit angle is one revolution. This occurs when the terminal line lies along the initial one. The Babylonians divided this angle into six equal parts, the parts into 60 equal parts, these again into 60 and these again into 60. Their scheme of measuring rotation was:

1 revolution = 6 equilateral triangle angles,
= 360 degrees,
= 21,600 first minute parts (minutes),
= 1,296,000 second minute parts (seconds).

The difference between the horizontal direction (horizon) and the vertical direction (plumbline) at any point is a quarter of a revolution. Any angle of this size is called a right angle.

The table of angular measure is:

60 seconds = 1 minute
60 minutes = 1 degree
90 degrees = 1 right angle
4 right angles or 360 degrees = 1 perigon or revolution.

Two right angles are referred to as a straight angle. In **analysis** the unit angle is the **radian. See Polyhedral Angle; Rotation; Solid Angle; Spherical Angle.**

ANGLE AT CENTRE. The angle between any two radii of a circle. The angle α is said to be subtended at the centre by the arc a. The size of the angle is proportional to the length of the arc and can range from 0° to 360° (0 to 2π **radians**).

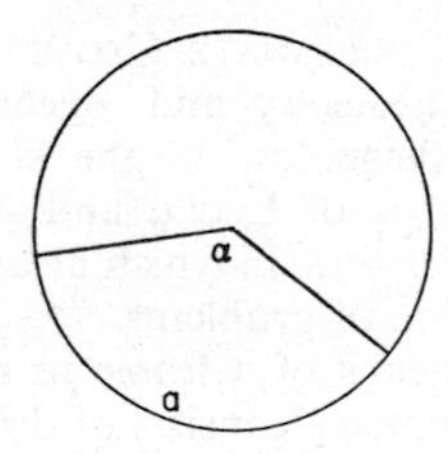

ANGLE AT CIRCUMFERENCE. The angle between any two chords of a circle which have a common point on the circumference. The angle β is said to be subtended at the circumference by the arc b. The size of the angle is determined by the length of the arc, and can range from 0° to 180° (0 to π **radians**). See **Angle in Segment**.

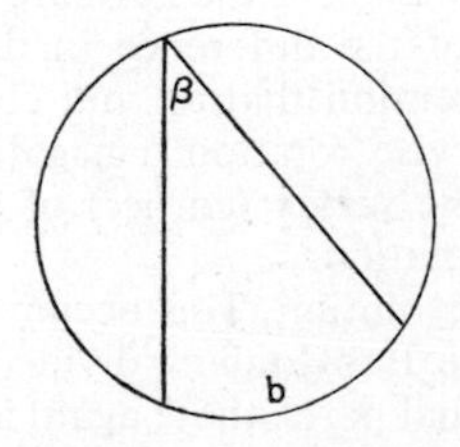

ANGLE BETWEEN CHORD AND TANGENT. See **Alternate Segment Theorem.**

ANGLE BETWEEN TWO CURVES. The **angle between two lines** which are tangents to the curves at the point of intersection.

ANGLE BETWEEN TWO LINES. The **angle (plane angle)** between two lines L_1 and L_2 can be expressed in terms of the **angles of inclination** of L_1 and L_2 with respect to a third line. If this is the x-axis, it follows, since $\theta = \theta_1 - \theta_2$,

$$\theta = \tan^{-1} \{(\tan \theta_1 - \tan \theta_2)/(1 + \tan \theta_1 \tan \theta_2)\}$$

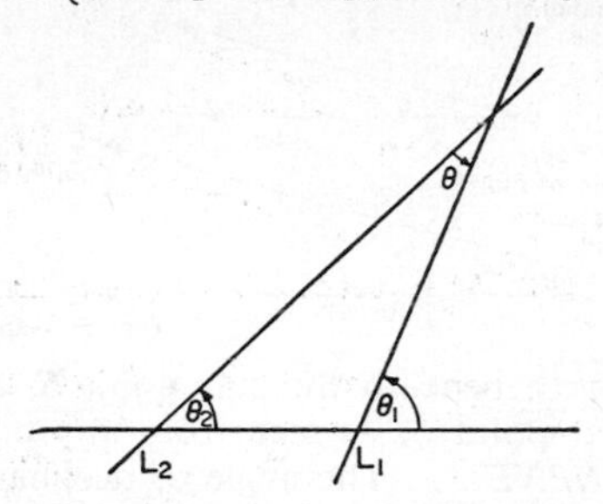

ANGLE BETWEEN TWO PLANES. See **Angle of Inclination.**

ANGLE IN SEGMENT. Any interior angle of an **inscribed triangle of circle,** the angle α described as being subtended at the circumference by the chord c. See **Angle at Circumference.**

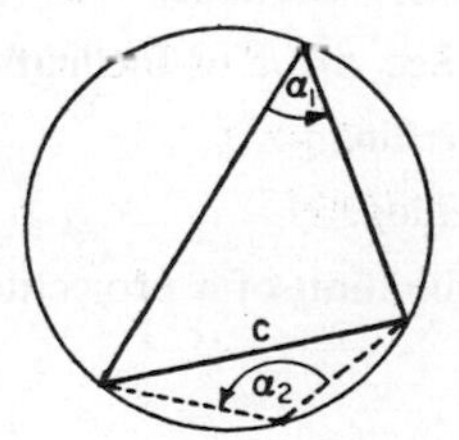

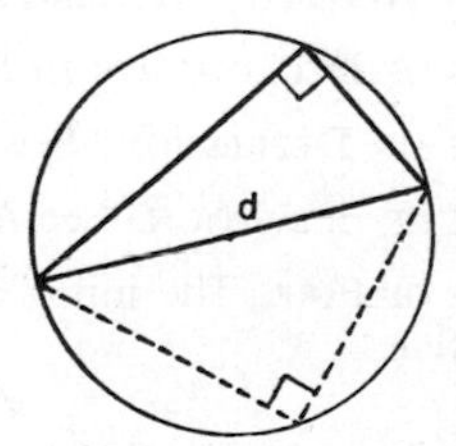

ANGLE IN SEMICIRCLE. The **angle in segment** when the chord is a diameter. It is a right angle or $\pi/2$ **radians.**

ANGLE OF ATTACK. The angle between wind direction and a surface exposed to it. The resultant force acting on the surface depends on this angle and consists of two **components,** the *drag* in the direction of the wind and the *lift*, at right angles.

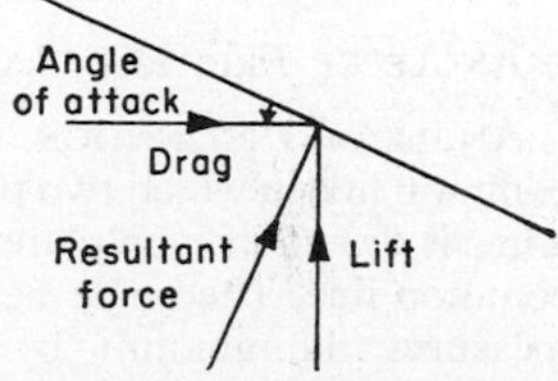

ANGLE OF DECLINATION IN ASTRONOMY. One of the **spherical coordinates in astronomy** used to fix the position of a point on the **celestial sphere.** A **great circle** is passed through the point and the celestial pole to determine a point on the celestial equator. The angle between the lines joining the observer to this point and the one on the celestial sphere is the positive angle of declination in the northern hemisphere and the negative angle of declination in the southern. The angle ranges from 0° at points on the celestial equator to 90° at points above the poles of the earth. In the diagram, the number of degrees in the angle θ measures both (*a*) the latitude

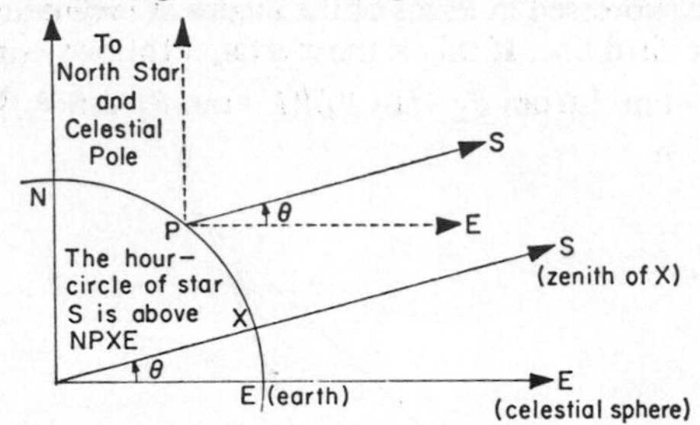

of the point X on earth beneath the star point S, and (*b*) the positive declination of the star point S, as seen from any point P, on the great circle quadrant . . . *NPXE* . . . The angle of declination of a star at the **zenith** is the same as the angle of latitude of the place of observation.

Angle of Declination in Navigation. The compass correction that has to be made in the direction of the horizontal component of the earth's magnetic field at any place in order to obtain the meridian direction of that place. See **Magnetic Declination; Meridian of Longitude.**

Angle of Declination in Surveying. See **Angle of Inclination.**

Angle of Depression. See **Angular Levelling.**

Angle of Elevation. See **Angular Levelling.**

Angle of Fire. The initial angle of inclination of a projectile in **parabolic motion.**

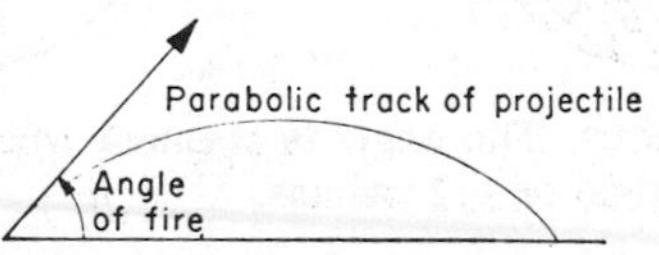

Angle of Friction. See **Friction.**

Angle of Inclination. From a point on the common line in which two plane surfaces intersect a line is drawn in each plane at right angles to the common line. The angle between these two lines measures the amount by which one plane is inclined to the other. The angle of inclination of roofs, hillsides, roads, railway lines, etc., is always given with respect to a theoretical horizontal plane tangential to the earth. An angle measured below the plane of the horizon is called an *angle of*

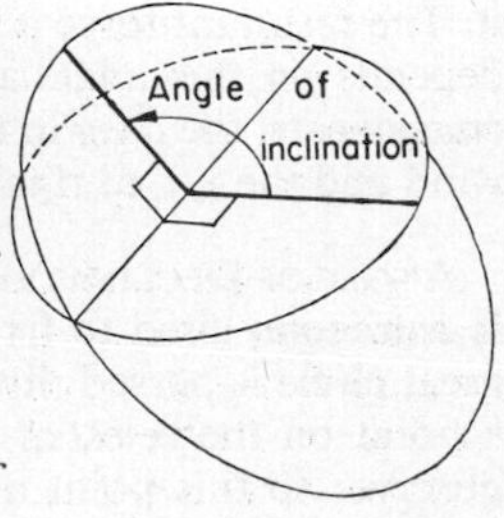

declination. Angles of inclination and declination are **dihedral angles.** See **Angular Levelling**; **Gradient**; **Pitch.**

ANGLE OF INTERSECTION OF GREAT CIRCLES. The **dihedral angle of inclination** between the plane of one **great circle** and the plane of the other.

ANGLE OF REPOSE. The maximum angle with the horizontal at which any dry, loose material such as sand or coal will remain in equilibrium. It is an important factor in the design of bunkers, dams and embankments.

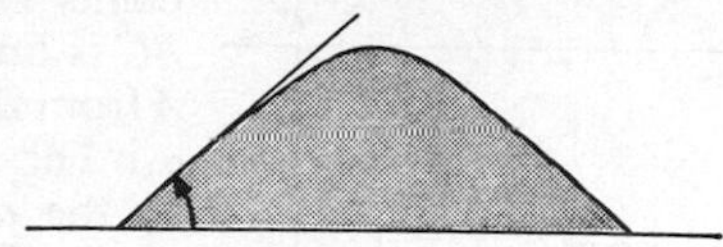

ANGLE OF ROTATION. Fundamentally, the concept of **angle.** The term is used specifically when the angle is a variable as in **uniform circular motion.** It is also the angle between the **initial and terminal lines** in **polar coordinates in plane.** See **Rotation.**

ANGLE PROPERTIES OF CIRCLE. (1) The **angle at centre** is double the **angle at circumference** subtended by the same or an equal arc. (2) **Angles in same segment** or in equal segments are equal. (3) The **angle in semicircle** is a **right angle.** (4) The opposite angles of a **cyclic quadrilateral** are **supplementary angles.** (5) The angle between a chord and a tangent is equal to the angle in the alternate segment—this is the **alternate segment theorem.**

ANGLES IN SAME SEGMENT. Angles subtended at the circumference of a circle by a single chord, the angles being on the same side of the chord. See **Angle in Segment**; **Angle Properties of Circle.**

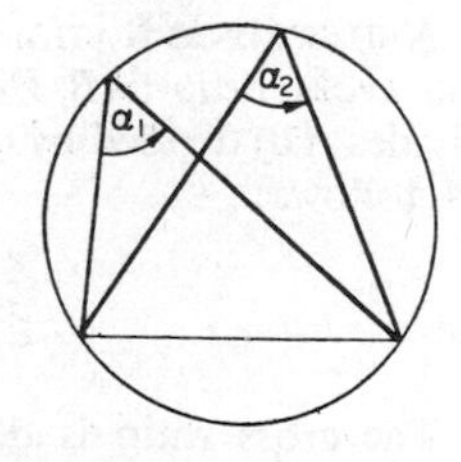

ANGLES, SPHERICAL. See **Spherical Angle.**

ANGSTROM, Å. The unit of length named after the spectroscopist Ångstrom who first used it. It is 10^{-10} **metres** (0.000 000 01 cm). It is used for giving **wave lengths** of light and thickness of films of oil, etc. (The wave length of yellow light from a sodium flame is 5 890 Å.)

ANGULAR ACCELERATION. The rate of change (first **derivative**) of **angular velocity**, with respect to time, written $d\omega/dt$, $\dot{\omega}$, or $d^2\theta/dt^2$, $\ddot{\theta}$, the second derivative of the **angular distance**, with respect to time.

ANGULAR DISPLACEMENT. Synonym of **angular distance.**

ANGULAR DISTANCE. The **angle of rotation** θ = angle POX, of some ray $\overrightarrow{OP}$ with respect to some initial direction $\overrightarrow{OX}$. The angular distance is positive when the rotation is in an anti-clockwise direction.

ANGULAR DISTANCE BETWEEN TWO POINTS. The angle between two rays from a point of reference passing through the two points.

ANGULAR DISTANCE IN ASTRONOMY. The distance between two heavenly bodies measured by the angle they subtend at the point of observation.

ANGULAR LEVELLING. The determination of differences in **elevation** of points on the earth's surface. In the figure, *BC* is horizontal, *AC* vertical. The *aclivity* of *A* from *B* is measured by the *angle of elevation*, α_1. The *declivity* of *B* from *A* is measured by the *angle of depression*, α_2. The vertical and horizontal distances are related by the expressions: $h = d \tan \alpha_1 = d \tan \alpha_2$. See **Levelling.**

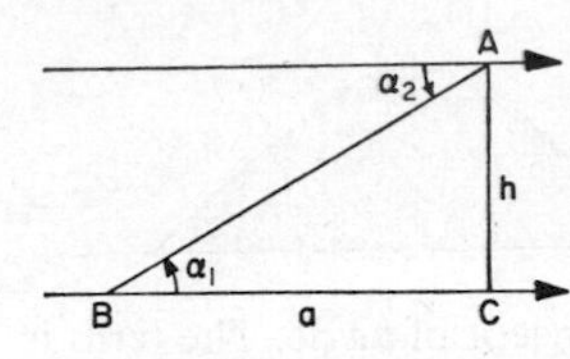

ANGULAR MEASURE. See **Angle (Plane Angle); Angle of Rotation.**

ANGULAR MINUTE, SECOND. See **Angle (Plane Angle).**

ANGULAR MOMENTUM. See **Moment of Momentum.**

ANGULAR MOTION. Synonym of **rotation.** See **Angular Acceleration; Angular Distance; Angular Velocity.**

ANGULAR VELOCITY. The rate of change (first **derivative**) of **angular distance** with respect to time, written $d\theta/dt$, $\dot{\theta}$, or ω.

ANHARMONIC RATIO. If *A*, *B*, *C*, *D* are distinct points on a straight line, the cross ratio (*AB*, *DC*) is defined as the quotient (ratio in which *C* divides *AB*) divided by (ratio in which *D* divides *AB*). This may be reduced as follows:

$$\frac{AC}{CB} \div \frac{AD}{DB} = \frac{AC \cdot DB}{CB \cdot AD} = \frac{AC \cdot BD}{BC \cdot AD}$$

The cross ratio is defined as an anharmonic ratio if (*AB*, *CD*) is *not* equal to -1. (If $\{AC \cdot BD\}/\{BC \cdot AD\} = -1$, the ratio is harmonic and *C* and *D* are said to divide *AB* harmonically.) See **Harmonic Section.**

ANNUAL VARIATION. Yearly changes in **magnetic declination.**

ANNULAR. Ringed. Shaped like a ring. See **Annulus.**

ANNULAR ECLIPSE. An eclipse of the sun when the disc of the moon leaves a ring of light.

ANNULAR SPACE. The space between an inner and an outer cylinder if the two cylinders have coincident axes.

ANNULUS. The portion of a plane bounded by the circumferences of two concentric circles in the plane. The area is $\pi\,(r_1^2 - r_2^2)$, where r_1 and r_2 are the radii of the circles.

ANNUM. See **Per Annum.**

ANOMALISTIC YEAR. The time taken by the earth to travel round its orbit from the time when it is nearest to the sun to the next time at which it is nearest: 365.259 64 **mean solar days,** or 365 days 6 hours 13 minutes 53 seconds. The major axis of the ellipse of the earth's orbit moves through an angle of 11.25 seconds annually about the centre in a sense which adds 25 minutes 7 seconds to the **tropical year** to give this period from **perihelion** to perihelion. See **Sidereal Year.**

ANSWER. From the Anglo-Saxon, to swear against, this word means to make a reply to a question or charge under oath. Hence the answers to mathematical questions must be correct. Correct answers are not necessarily exact ones since all answers should be reasonable and correct within certain limits determined by the data in the question and the methods employed in the solution. No measurement, for example, is exact, since instrumental and human **errors** are involved. See **Approximation.**

ANTARCTIC CAP. See **Arctic, Antarctic Caps.**

ANTARCTIC CIRCLE. See **Arctic, Antarctic Circles.**

ANTECEDENT, CONSEQUENT IN LOGIC. In the conditional proposition: If P then Q, P is the *antecedent* and Q the *consequent*. See **Algebra of Propositions.**

ANTECEDENT, CONSEQUENT IN RATIO AND PROPORTION. In the ratio $A : B$, the *antecedent* is A, and the *consequent*, B. In the proportion $A : B = C : D$, the antecedents are A and C, and the consequents, B and D.

ANTE MERIDIEM (A.M.), POST MERIDIEM (P.M.). Wherever the twelve-hour clock is used, the day is divided into two parts. From midnight to noon is referred to as A.M. (a.m.), from noon to midnight, P.M. (p.m.). See **Apparent Solar Time; Mean Solar Day.**

ANTI-DERIVATIVE. If $f(x)$ is a given function, $F(x)$ is the anti-derivative of $f(x)$ if $F'(x) = f(x)$. The term is synonymous with integral when **integration** is seen as the reverse operation of differentiation.

ANTILOGARITHMS. In the exponential equation $B^L = N$, N is the antilogarithm of L to the base B. See **Logarithms.**

ANTINODES IN PHYSICS. See **Standing (Stationary) Waves.**

ANTIPARALLEL. Two lines L_1 and L_2 are antiparallel if they cut two given lines λ_1 and λ_2 to form two pairs of equal angles in opposite order.

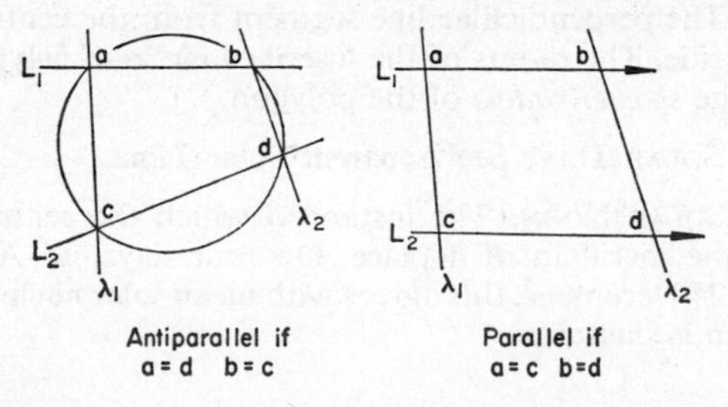

Antiparallel if
a = d b = c

Parallel if
a = c b = d

The opposite sides of a **cyclic quadrilateral** are antiparallel. When the corresponding angles are equal, L_1 and L_2 are parallel.

ANTIPODAL POINTS. The opposite ends of any diameter of a **sphere.** The word is derived from the Greek word *pous*, a foot. A pair of points are called *antipodes.*

ANTIPODAL TRIANGLES. If A, B and C are the vertices of one triangle drawn on a sphere, and AA', BB' and CC' are diameters of the sphere, ABC and $A'B'C'$ are antipodal triangles.

APEX. The highest point of a plane figure or solid with respect to a line or a plane chosen as base. The **vertex** of a triangle opposite the side chosen as base.

APOGEE. The point on the orbit of the moon, or of any planet, or of the apparent orbit of the sun, which is farthest from the earth. See **Perigee.**

APHELION. The point on the orbit of the earth or of any heavenly body which is farthest from the sun. See **Perihelion.**

APOLLONIUS'S THEOREM. In the triangle ABC with median AM_a, $b^2+c^2 = 2(a/2)^2+2(m_a)^2$.

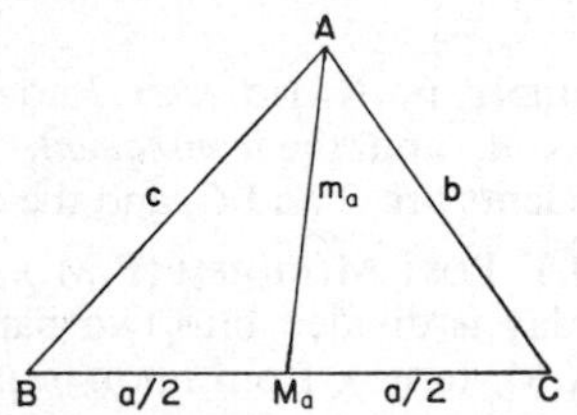

APOTHECARIES' FLUID MEASURE.

60 minims = 1 fluid dram	16 fluid ounces = 1 pint
8 fluid drams = 1 fluid ounce	8 pints = 1 gallon

See **Fluid Ounce.**

APOTHECARIES' WEIGHT. A system of measure used in the preparation and dispensing of drugs. See **Avoirdupois; Troy Weight.**

20 grains = 1 scruple (scr)	8 dr = 1 ounce (oz)
3 scr = 1 drachm (dram, dr)	12 oz = 1 pound (lb)

APOTHEM. The perpendicular line segment from the centre of a regular polygon to a side. The radius of the **inscribed circle of polygon** and sometimes called the *shorter radius* of the polygon.

APPARENT SOLAR DAY. See **Apparent Solar Time.**

APPARENT SOLAR NOON. The instant at which the centre of the sun's disc crosses the meridian of a place. On four days, 15 April, 14 June, 1 September, 24 December, this agrees with **mean solar noon**. The deviation from the mean is **sinusoidal.**

APPARENT SOLAR TIME. The interval between successive **transits** of the sun is not the same throughout the year. Thus, the *apparent solar day* is not a constant. The time kept by the apparent sun, as indicated on a sundial, is known as *apparent time*. Clock-time, known as *mean solar time*, is based on the **mean solar day.**

APPLIED MATHEMATICS. One of the traditional aspects of **mathematics** when contrasted with **pure mathematics.** It is that aspect concerned with the solutions of physical problems where empiricism is eventually supported by logically deduced solutions based on broadening concepts of energy, mass and time.

APPROXIMATION. A number which is accepted as an estimate of another number; and the method of finding the substitute number. All measurements are approximations. See **Answer**; **Correct to *n* Decimal Places**; **Significant Figures.**

APPROXIMATION SIGN. When one expression is an approximation for another, the equal sign is replaced by $\simeq$ or $\approx$ or $\risingdotseq$. Thus, $\pi \simeq 3.141\,6$, $\sqrt{2} \approx 1.414\,2$, $e \risingdotseq 2.718\,3$.

APSE, APSIDAL DISTANCE. The point on a central orbit at which the line of force is normal to the curve; the length of the radius vector at such a point.

ARABIC NUMERALS. See **Hindu-Arabic Numerals.**

ARBITRARY CONSTANT. See **Constant.**

ARC. Part of a **curve** as distinct from the whole.

ARC HYPERBOLIC FUNCTION. Synonym of **inverse hyperbolic function.**

ARC TRIGONOMETRIC FUNCTION. Synonym of **inverse trigonometric function.**

ARCHIMEDES' CONSTANT. See **Pi, Π, π.**

ARCHIMEDES' PRINCIPLE. Archimedes (287–212 B.C.) discovered that when a body was partly or wholly immersed in a liquid there was an apparent loss in weight. This led him to enunciate the following principle: the resultant upward thrust on the closed surface of a body immersed in liquid is a vertical force equal to the weight of the liquid displaced and acting through the centre of gravity of the space originally occupied by the displaced liquid. The principle also applies to immersion in gases and leads to the concept that the true weight of a body is its weight *in vacuo*. The principle is used in determining the **specific gravity** or relative density of bodies.

ARCHIMEDES' SPIRAL. The **spiral** whose **polar equation** is $r = a\theta$.

ARCTIC, ANTARCTIC CAPS. The two areas bounded by the **Arctic, Antarctic circles** respectively. Each covers about 4 per cent of the earth's surface.

ARCTIC, ANTARCTIC CIRCLES. The **parallels of latitude** 66½°N and 66½°S respectively. The boundaries of the **Arctic, Antarctic caps**. Each circle is about 10,000 miles long.

ARE. Unit of **area** in the **metric system**; 100 square metres. It is equivalent to 119.6 square yards. See **Area, Metric Units.**

AREA. The size of a surface. The amount of surface is expressed in numbers of square units or equivalent square units. If the surface is rectangular the product of the pair of numbers representing the length and the breadth is the number of units of area of the rectangle. **Rectilinear figures** can be sub-divided into rectangles: but plane areas bounded by curves and the areas of the surfaces of solids like cones and spheres cannot be calculated without recourse to the **Calculus.**

AREA, BRITISH UNITS.

1 in² (6.451 6 cm²)
144 in² = 1 ft² (929 cm²)
9 ft² = 1 yd²
484 yd² = 1 square chain
4 840 yd² = 1 acre (40.47 ares)
640 acres = 1 square mile (2.59 Km²)

AREA, METRIC UNITS.

1 cm² (0.155 in²)
10,000 cm² = 1 centare or 1 m²
100 m² = 1 are (119.60 yd²)
100 ares = 1 hectare (2.471 1 acres)
100 hectares = 1 Km² (0.386 1 square mile)

AREA OF REVOLUTION. See **Solid of Revolution.**

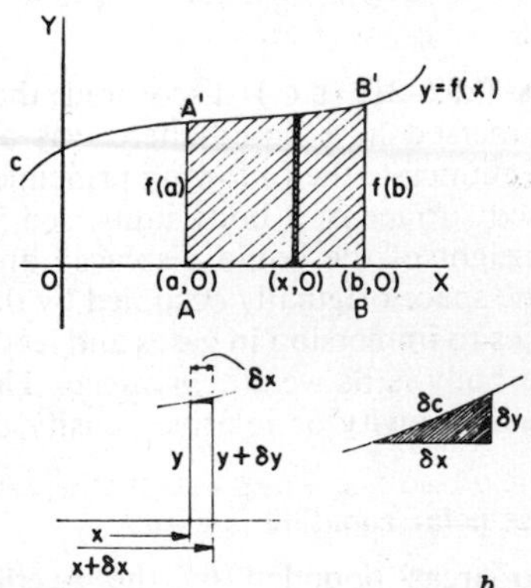

AREA UNDER CURVE. Given a curve c, representing a relation $y=f(x)$, in respect to axes OX, OY, the area under the curve is bounded by the ordinates AA', BB', with lengths $f(a)$ and $f(b)$ respectively, the x-axis, OX and the segment $A'B'$ of c. The points A and B are chosen arbitrarily and $f(a)$ and $f(b)$ may be positive or negative. The area is defined as positive or negative according to whether it is above or below the x-axis. The area is found by summation of the areas of elements of widths δx and height $f(x)$. Thus, total area $= \sum_a^b f(x)\delta x$. In the limit as $\delta x \to 0$, the small area next to the curve, $\frac{1}{2}\,\delta x\ \ \delta y \rightarrow 0$, and the total area is the definite **integral of function,** $\int_a^b f(x)dx$.

Areal Coordinates. The **coordinates of a point** P, in the plane of a fundamental triangle ABC are the three ratios of areas: $\triangle PBC/\triangle BCA$, $\triangle PCA/\triangle CAB$, $\triangle PAB/\triangle ABC$.

Argand Diagram. Synonym of complex plane. The representation of **complex numbers** by **rectangular coordinates** in a plane. The complex number $x+iy$ is associated with the point (x, y). All real numbers are represented along the x-axis, the *real axis*, and all pure imaginary numbers along the y-axis, the *imaginary axis*. The Argand diagram permits a geometric interpretation of relations and operations involving complex numbers. A complex number can be expressed in polar form when polar coordinates are used. See **Polar Form of Complex Number.**

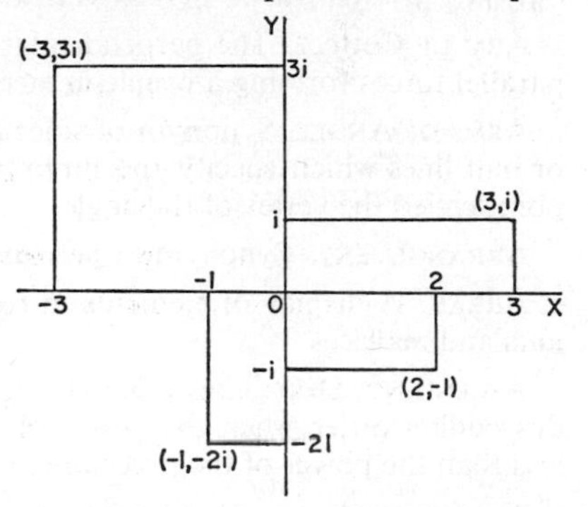

Argument of Complex Number. See **Polar Form of Complex Number.**

Arithmetic. The numerical aspect of **mathematics.** All the concepts which develop from an individual's need to accommodate and assimilate in a physical environment. The awareness of size and order as expressed in **cardinal numbers** and **ordinal numbers** respectively. The knowledge of notation and operation in mental and written **calculation** and **computation.** The use of numeral concepts and computational techniques involved in **measure, mensuration** and the solution of numerical problems.

Arithmetic Average. See **Arithmetic Means.**

Arithmetic Means. The terms of an **arithmetic progression** lying between the first and the last are called the arithmetic means of the first and the last terms. To insert n means between numbers a and b implies having $n+2$ numbers in arithmetic progression. Hence the common difference d is $(b-a)/(n+1)$. The n means are therefore $a+(b-a)/(n+1)$, $a+2(b-a)/(n+1) \ldots a+n(b-a)/(n+1)$. If four means are needed between 7 and 19, $(19-7)/5$ being 2.4, they are: 9.4, 11.8, 14.2 and 16.6. When $n=1$, the arithmetic mean of a and b is found to be $a+(b-a)/2$ or $\frac{1}{2}(a+b)$, and is referred to as the *arithmetic mean.* This is half the sum of the given terms.

The arithmetic average (mean) of n different quantities, $a_1, a_2, a_3, \ldots a_n$, is the one nth part of their sum $(a_1+a_2+a_3 \ \ldots \ a_n)/n$. See **Harmonic, Geometric, Arithmetic Means.**

ARITHMETIC PROGRESSION (A.P.). A **series** in which the difference between any term and the preceding one is a constant called the *common difference*. If a is the first term, d the common difference, n the number of terms, then the rth term is $a+(r-1)d$ and the sum to n terms is S_n, equal to $\frac{1}{2}n\{2a+(n-1)d\}$. See **Progression.**

ARITHMETICO-GEOMETRIC PROGRESSION (A.G.P.). A series in which the rth term is the product of the rth term of an A.P. and the rth term of a G.P. The rth term is therefore $\{a+(r-1)d\}x^{r-1}$ and the sum to n terms, $S_n=[a-\{a+(n-1)d\}x^n]/(1-x)+d\cdot x(1-x^{n-1})/(1-x)^2$.

ARM OF COUPLE. The perpendicular and shortest distance between the parallel forces forming a **couple in mechanics.**

ARMS OF ANGLE. Synonym of sides of **angle (plane angle).** The two rays or half lines which specify the limits of the angle at their common end point called the vertex of the angle.

ARRANGEMENT. Synonym of **permutation.**

ARRAY. A display of elements in rows and columns, used in **determinants** and **matrices.**

ASCENDING, DESCENDING ORDER. A **sequence** or **series** is in ascending or descending order when the **power** of any term is respectively greater or less than the power of the preceding term. See **Power of Quantity.**

ASSOCIATIVE LAW. This law states: $(a\odot b)\odot c=a\odot(b\odot c)=[a\odot b\odot c]$. In these quantities, a, b, c are undefined elements of a set and $\odot$ represents an undefined **binary operation** in a set. Examples of operations that satisfy this law are **addition, multiplication, translation** and **rotation.**

ASTROID. A **hypocycloid** of four **cusps.** It is the **envelope** of a line segment of fixed length with its end points on rectangular coordinate axes.

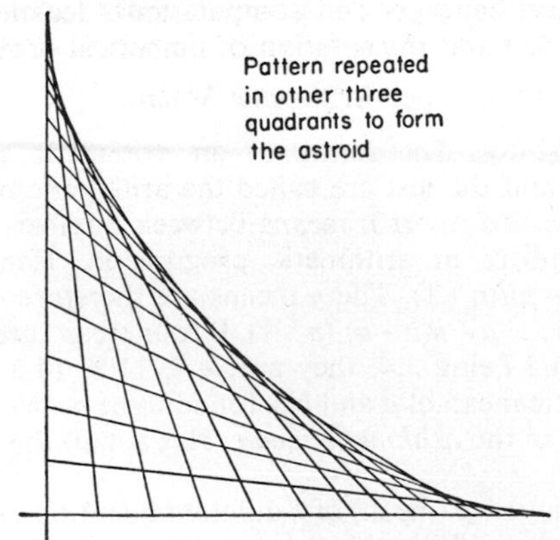

ASTRONAVIGATION. Navigation based on celestial observation. See **Air Navigation.**

Astronomical Frame. A **frame of reference** used in celestial mechanics. For convenience the sun is fixed in position and does not rotate relative to the fixed stars.

Astronomical Unit (A.U.). The mean distance from the centre of the earth to the centre of the sun: 149,504,000 Km or 92,897,000 miles. Comparisons with other units are:

63,000 A.U. ≡ 1 **light-year**; 3.258 light-years ≡ 1 **parsec.**

Astronomy. The science which studies the phenomena of the heavenly (celestial) bodies: the sun, planets and satellites, comets, etc., of the solar system; other stars of the Milky Way galaxy; other galaxies. The word (Greek, *astron*, star; *ṇemein*, arrange) shows that astronomy began with the mapping of the constellations.

Asymmetric Relation. See **Symmetric Relation.**

Asymmetry. See **Symmetry.**

Asymptote. A line which a curve approaches indefinitely without meeting it at any finite distance. The perpendicular distance from a moving point on the curve to the line approaches zero as the distance from the point to the origin approaches infinity. Technically, it is described as a *tangent at infinity* or a line tangent at an **ideal point**. Hyperbolas are **conic sections** with two asymptotes. The graph of **Boyle's Law** is an example of a rectangular hyperbola with axes of coordinates as asymptotes.

At a Discount. See **Premium.**

At a Premium. See **Premium.**

Atmospheric Pressure. The pressure at a point due to the weight of a column of the atmosphere which contains the point. It is equated to the weight of a column of water about 30 ft high or of mercury 760.1 mm or 29.53 in high. The pressure decreases with elevation above sea-level and decreases with increasing humidity of the atmosphere, water vapour being less dense than dry air. See **Bar Unit of Pressure.**

Attraction. See **Gravitation.**

Auxiliary Circle of Ellipse. The *eccentric circle* of an **ellipse** having the major axis as a diameter. See **Eccentric Angle, Circle of Ellipse.**

Auxiliary Circle of Hyperbola. The *eccentric circle* of a **hyperbola** having the transverse axis as a diameter. See **Eccentric Angle, Circle of Hyperbola.**

Auxiliary Variable. See **Parametric Equations.**

Average. Synonym of mean. See **Arithmetic Means**; **Geometric Means**; **Harmonic Means.**

Average Speed. See **Speed.**

Average Velocity. See **Velocity.**

AVOIRDUPOIS. The standard system of weights in use in Great Britain, parts of the Commonwealth and the U.S.A., for all goods except drugs and precious metals and stones.

1 ounce avoirdupois = 437.5 grains. 1 pound avoirdupois = 1.215 3 pounds troy.

16 drams	= 1 ounce (oz)	20 cwt	= 1 ton
16 oz	= 1 pound (lb)	2,240 lb	= 1 ton (Britain)
28 lb	= 1 quarter (qr) (Britain)	2,000 lb	= 1 ton (U.S.A.)
25 lb	= 1 quarter (qr) (U.S.A.)	14 lb	= 1 stone (st) (Britain)
4 qr	= 1 hundredweight (cwt)	100 lb	= 1 cental

See **Apothecaries' Weight; Troy Weight.**

AXES OF COORDINATES. See **Cartesian Coordinates.**

AXES OF ELLIPSE. In the **ellipse** $x^2/a^2+y^2/b^2=1$, where $a>b$, the diameter of length $2a$ passing through the foci is the *major axis* and the diameter at right angles, of length $2b$, the *minor axis.*

AXES OF ELLIPSOID. In the **ellipsoid** $x^2/a^2+y^2/b^2+z^2/c^2=1$, where $a>b>c$, the diameter of length $2a$ passing through the foci is the *major axis*; the diameter of length $2b$ at right angles is the *mean axis*; the diameter of length $2c$, at right angles to the major and the mean, is the *minor axis.*

AXES OF HYPERBOLA. In the **hyperbola** $x^2/a^2-y^2/b^2=1$, where $a>b$, the diameter of length $2a$, which extended passes through the foci, is the *transverse axis* and the line at right angles to it through the centre, of length $2b$, the *conjugate axis.* The transverse axis bisects the conjugate axis.

AXES OF HYPERBOLOID. Of the three axes of a **hyperboloid,** those which are diameters are called *transverse axes*; those which are not diameters, *conjugate axes.* A hyperboloid of one sheet has two transverse axes and one conjugate; one of two sheets, one transverse axis and two conjugate.

AXIAL SYMMETRY. **Symmetry** with respect to a line.

AXIOM. Synonym of **postulate**. An assumption on which a logical argument is based is an axiom and any proposition accepted as requiring no proof is said to be axiomatic.

AXIS. A line about which a body or a group of bodies may be considered to rotate or revolve. Examples are the centre line of a cone or any diameter of a sphere.

AXIS OF CELESTIAL SPHERE. See **Axis of Earth.**

AXIS OF CIRCLE ON SPHERE. Any plane section of a sphere is a circle with its circumference on the surface of the sphere. The diameter of the sphere passing through the centre of the section is known as the axis of the circle. The ends of the axis are called the poles of the circle.

AXIS OF EARTH. The axis of rotation, coinciding with that diameter terminating in the Earth's geographical North and South Poles. For the purpose of determining directions in space, this axis is extended to points on the **celestial sphere** near the Pole Star and the Southern Cross.

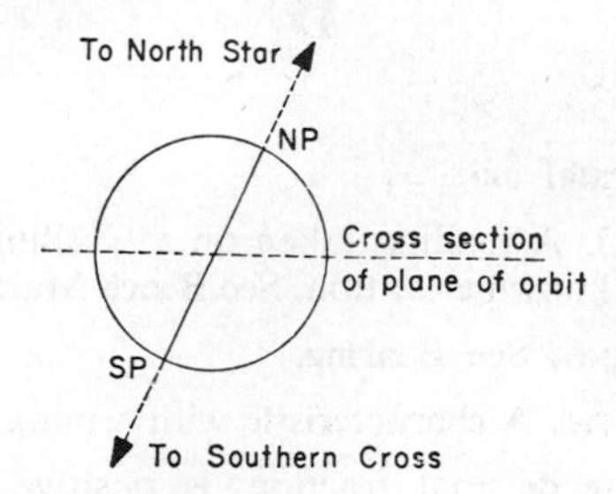

AXIS OF PARABOLA. The line perpendicular to the directrix and passing through the focus. See **Parabola.**

AXIS OF PENCIL. See **Pencil of Planes.**

AXIS OF SYMMETRY. A line about which a figure is symmetrical. A line, that is, which divides a figure into two halves in such a way that the halves fit exactly when the figure is folded along the line. If a plane mirror is placed along the line of **symmetry**, normal to the plane of the figure, then the reflection of one half will appear to coincide with the other half.

AXIS, RADICAL. See **Radical Axis.**

AXONOMETRIC PROJECTION. See **Pictorial Projection.**

AZIMUTH IN ASTRONOMY. If a quadrantal arc on the **celestial sphere** passes from the zenith through a star to a point on the horizon, the azimuth of the star is defined by either the arc of the horizon between this point and the south point or the angle this point and the south point subtend at the centre of the sphere. The azimuth of a star is comparable with the **longitude** of a point on earth.

AZIMUTH IN NAVIGATION. Synonym of whole circle **bearing.**

AZIMUTH IN POLAR COORDINATES. Synonym of polar angle in **polar coordinates in plane.**

AZIMUTH IN SPHERICAL COORDINATES. Synonym of longitude in **spherical coordinates in space.**

AZIMUTHAL ANGLE. The **dihedral angle** between the vertical plane containing the zenith, nadir and south point of the horizon and any other vertical plane through the zenith and the nadir. See **Azimuth in Astronomy.**

B

B.S. **Back sight.**

Btu. **British thermal unit.**

Back Sight (B.S.). A reading taken on a levelling staff at the beginning of **levelling** at a known elevation. See **Bench Mark; Foresight.**

Backward Bearing. See **Bearing.**

Bar Characteristic. A **characteristic** with a minus sign written above it. The **mantissa,** the decimal fraction, is positive. Thus, $\bar{2}.3456$. See **Logarithms.**

Bar Graph. A graphic representation in which a set of frequencies is represented by a set of parallel bars of equal widths and lengths proportional to the frequencies.

Bar Symbol of Aggregation. A line called a *vinculum,* placed above terms which are to be considered together. See **Aggregation.**

Bar Unit of Pressure. The International unit of atmospheric pressure equivalent to that exerted by a column of mercury 760.1 mm or 29.53 in above the free surface in latitude 45° and at zero celsius temperature. It is equal to 10^6 dynes/cm^2 or 14.5 lbf/in^2.

Base Angles of Triangle. The two angles which have the base of a triangle as a common arm.

Base in Logarithms. See **Logarithms.**

Base in Trigonometry. The side of a triangle formally drawn across the page to distinguish it from the hypotenuse and the height, i.e. the side adjacent to the acute angle of reference and entering into four of the six elementary **trigonometric ratios.**

Base of Solid. The plane face on which a solid stands or is supposed to stand. The area of this base and the vertical height enter into the formula for volume.

Base of Triangle. Any side of a triangle to which a perpendicular has been drawn from the opposite **vertex.** The lengths of the base and the perpendicular enter into the formula for area.

Base (Radix) in Number System. Any number can be written in the form $c_nb^n + c_{n-1}b^{n-1} + \ldots + c_2b^2 + c_1b^1 + c_0b^0$, where b is a base of computation and c is some positive integer less than the base. The concept of the number of whole days in a year can be written as $3 \cdot 10^2 + 6 \cdot 10^1 + 5 \cdot 10^0$ or as $1 \cdot 2^8 + 0 \cdot 2^7 + 1 \cdot 2^6 + 1 \cdot 2^5 + 0 \cdot 2^4 + 1 \cdot 2^3 + 1 \cdot 2^2 + 0 \cdot 2^1 + 1 \cdot 2^0$. These are 365 in the **denary number system** and 101101101 in the **binary number system.**

Basis Vectors. Any set of **vector quantities,** two in a plane, three in **Euclidean space**, such that any vector is a linear combination of them. They form a basic framework for all vectors if planar basis vectors are not collinear and spatial basis vectors are not coplanar. The concept can be extended to n-dimensional space.

orthogonal basis vectors. Basis vectors which are mutually perpendicular.

orthonormal basis vectors. Orthogonal basis vectors which are unit vectors (magnitude 1). A common orthonormal basis is provided by three unit vectors **(i, j, k)** along the positive axes of **rectangular coordinates.**

Bearing. The relative direction of a line with respect to the north-south line at any point on the earth's surface. A *true bearing* is the angle between the line and the **meridian of longitude** through the point. A *magnetic bearing* is the angle between the line and the **magnetic meridian** through the point. *Whole circle bearings* (W.C.B.) or *azimuths* employed in *air navigation* are

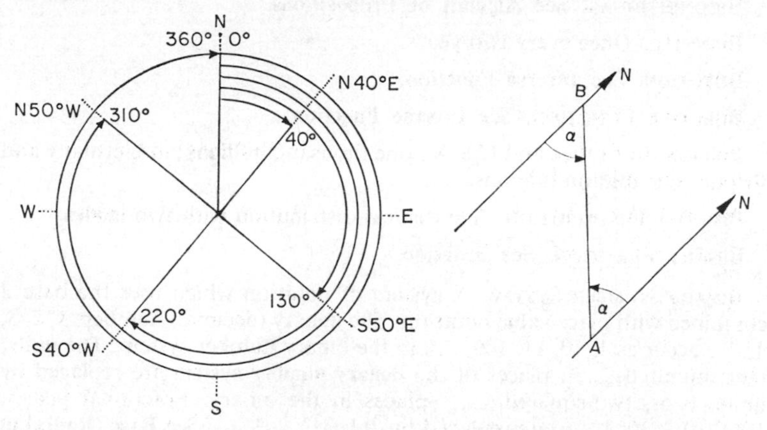

angles measured in degrees (0° to 360°) clockwise from the north direction. *Reduced bearings* (R.B.) or nautical bearings are angles measured clockwise or anti-clockwise from north or south directions. Four pairs of equivalent bearings are: (40°; $N\,40°\,E$), (130°; $S\,50°\,E$), (220°; $S\,40°\,W$), (310°; $N\,50°\,W$). Any line AB has two bearings: if the *forward bearing* of B from A is $N\,\alpha°\,W$, the *backward bearing* of B from A is $S\,\alpha°\,E$.

Bel. See **Decibel.**

Below Par. See **Premium.**

Bench Marks. Marks on a vertical surface to indicate a measured elevation above datum. The symbol used is ⊼ and on official maps it appears in the form BM⊼28.3. All **levelling** begins at a bench mark, which may be official or assumed and ends at another, both marks being official or temporary (T.B.M.), with an assumed value.

BENDING MOMENT, SHEAR IN HORIZONTAL BEAM. At any vertical section of the beam the algebraic sum of the moments of all external forces acting on one side of the section is called the *bending moment* at the section; and the sum of the vertical forces acting on one side is called the *shear*. The fibres of a beam and the molecular structure of a bar set up opposing forces at the section, the *resisting moment* and the *resisting shear*. In the simplest case of a beam of negligible weight carrying a load W at its free end and fixed at the other, the resisting shear at any section of the beam will be a force W, and the bending moment of the weight at a section distant x from it will be Wx and the resisting moment a couple equal to $-Wx$. See **Couple in Mechanics.**

BIANNUAL. Twice a year.

BICIMAL FRACTION. See **Fraction.**

BICONDITIONAL. See **Algebra of Propositions.**

BIENNIAL. Once every two years.

BIJECTION. See **Inverse Function.**

BIJECTIVE FUNCTION. See **Inverse Function.**

BILLION. In France and U.S.A., one thousand **millions**; in Germany and Britain, one million millions.

BIMODAL DISTRIBUTION. A statistical distribution with two **modes.**

BINARY FRACTION. See **Fraction.**

BINARY NUMBER SYSTEM. A system of notation which uses the base 2 combined with place value notation. The denary (decimal) numbers 1, 2, 3, 4, . . . occur as 1, 10, 11, 100, . . . in the binary number system. The units, tens, hundreds, . . . places of the **denary number system** are replaced by units, twos, two-squared, . . . places in the binary. Fractional places, 10^{-1}, 10^{-2}, 10^{-3}, . . . are replaced by 2^{-1}, 2^{-2}, 2^{-3}, . . . See **Base (Radix) in Number System; Fractions.**

BINARY OPERATION. An **operation in set** which combines two elements of the set into a third element of the same set. The sign of an undefined binary operation is $\odot$ or $\circ$ or $*$.

BINOMIAL COEFFICIENTS. The coefficients of the terms which arise from raising a binomial expression to any power. If the expression is in its simplest form, $(1+x)^n$, the first and last coefficients, written in **combination** notation, will be ${}^nC_0$ and ${}^nC_n$, each of which is 1. The coefficient of the typical rth term is written ${}^nC_{r-1}$. If, for simplicity, the coefficients are written $c_0, c_1, c_2, \ldots c_n$, it can be shown that

$$c_0 + c_1 + c_2 + \ \ldots \ + c_{n-1} + c_n = 2^n,$$
$$c_0 - c_1 + c_2 - \ \ldots \ + (-1)^n \cdot c_n = 0.$$

See **Binomial Theorem; Pascal's Triangle.**

Binomial Distribution in Probability. If p is the **probability** of an event happening in a single trial, and q, equal to $1-p$, the probability that it will fail, then the probability of this event happening r times in n trials is ${}^nC_r.p^r.q^{n-r}$, where ${}^nC_r$ is the number of **combinations** of n things taken r at a time. The probability that it will happen *at least* r times in n trials is

$$p^n + {}^nC_1 \cdot p^{n-1}q + {}^nC_2 \cdot p^{n-2}q^2 + \ldots + {}^nC_r \cdot p^r q^{n-r}.$$

As a special case, in five tosses of a coin, one may get 0, 1, 2, 3, 4 or 5 heads. The probability of these results will be 1, 5, 10, 10, 5 and 1 thirty-seconds ($\frac{1}{32}$, $\frac{5}{32}$, etc.) totalling 1, derived from $1(\frac{1}{2})^5$, $5(\frac{1}{2})^4(\frac{1}{2})$, $10(\frac{1}{2})^3(\frac{1}{2})^2$, $10(\frac{1}{2})^2(\frac{1}{2})^3$, $5(\frac{1}{2})(\frac{1}{2})^4$, $1(\frac{1}{2})^5$.

Binomial Equation. An equation of the form $x^n - a = 0$.

Binomial Expansion. An expansion given by the **binomial theorem.**

Binomial Expression. A **multinomial expression** of two terms, e.g. $3x+4y$.

Binomial Series. A **binomial expansion** with an infinite number of terms; the expansion of $(x+y)^n$ where n is not zero or a positive integer. The series is convergent if $|y|<|x|$. See **Convergence, Divergence of Series.**

Binomial Theorem. A formula discovered by Sir Isaac Newton, by means of which a **binomial expression** can be raised to any power on one side of an equation, and on the other, expanded in terms of powers of the two terms and defined coefficients. If n is a positive integer,

$$(a+x)^n = a^n + n \cdot a^{n-1}x + \frac{n(n-1)}{1.2} \cdot a^{n-2}x^2 + \frac{n(n-1)(n-2)}{1.2.3} \cdot a^{n-3}x^3 +$$

$$\ldots + \frac{n(n-1)(n-2)\ldots(n-r+2)}{1.2.3\ldots(r-1)} \cdot a^{n-r+1}x^{r-1} +$$

$$\ldots + n \cdot ax^{n-1} + x^n.$$

If n is fractional or negative, the series on the right is an infinite one which converges to a limiting value only if $|x|$ is less than 1. In its simplest form, this equation may be written in **combination** notation:

$$(1+x)^n = 1 + {}^nC_1x + {}^nC_2x^2 + \ldots + {}^nC_{r-1}x^{r-1} + \ldots + x^n.$$

See **Binomial Coefficients; Pascal's Triangle.**

Bipolar Coordinates. The distances (r_1, r_2) of any point P in a plane from two fixed points A and B respectively. The equation of any curve in bipolar coordinates is of the form $f(x, y)=0$. Thus, the equation of an ellipse is $r_1 + r_2 - 2a = 0$, where A and B are the foci and $2a$ is the length of the major axis.

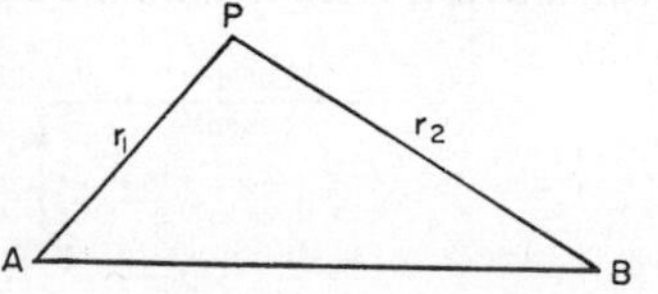

BIQUADRATIC. See **Quartic.**

BIRECTANGULAR POLYHEDRAL ANGLE. A **polyhedral angle** with two right **dihedral angles.**

BIRECTANGULAR SPHERICAL TRIANGLE. A **spherical triangle** with two right angles.

BISECT. To cut into two parts such that the two parts are equal.

BISECTOR OF DIHEDRAL ANGLE. The half-plane which divides a **dihedral angle** into two dihedral angles of equal size. Any one of the set of points forming the half-plane is equidistant from the two faces of the given angle.

BISECTOR OF LINE SEGMENT. The midpoint of the line segment. The **Cartesian coordinates** of the midpoint are the **arithmetic means** of the coordinates of the end points of the segment.

BISECTOR OF PLANE ANGLE. The half-line which divides a plane **angle** into two plane angles of equal size. Any one of the set of points forming the half-line is equidistant from the arms of the given angle. See **External Bisector; Internal Bisector.**

BIT (BINARY DIGIT). A digit in the **binary number system.**

BLOCK GRAPH. A graphical representation in which a set of frequencies is represented by a set of rectangles whose areas are proportional to the frequencies. If the widths are equal, the rectangles can be arranged as a **column graph**: if the heights are equal, as a **bar graph. See Histogram.**

BOOLEAN ALGEBRA. A **ring** with the additional property of having an **identity element**, I, of multiplication such that $x \times I = x$, and $x \times x = x$ for each x. It was introduced by Boole in the middle of the nineteenth century as a mathematical interpretation of the rules of logic. It can be interpreted as an **algebra of propositions** or as an **algebra of sets.**

BOUND OF FUNCTION. A **bound**, upper or lower, of the set of all possible values of the function. The sine and cosine trigonometric functions are bounded by $+1$ (*least upper bound, l.u.b.*) and -1 (*greatest lower bound, g.l.b.*). The exponential function, 2^x, is bounded by 0 (g.l.b.) as $x \to -\infty$, and is unbounded as $x \to \infty$. See **Bound of Set.**

BOUND OF SET. A number that is equal to or greater than every member of the set is an *upper bound*: one that is equal to or less than every member is a *lower bound.* A set which has upper and lower bounds is called a *bounded set*. Thus the infinite set $\{1, \frac{1}{2}, \frac{1}{3}, \frac{1}{4}, \ldots \frac{1}{n}\}$ has 2 as one of its upper

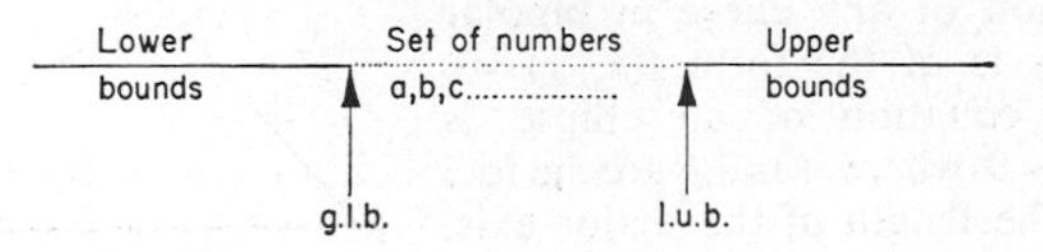

bounds and $-\frac{1}{2}$ as one of its lower bounds. The set of positive integers $\{1, 2, 3, \ldots n\}$ is an *unbounded set* since it has no upper bound.

least upper bound (l.u.b.). The smallest of the upper bounds of a set. It may be the largest member of the set or the smallest number greater than every member of the set, called an **accumulation point.** Thus $\frac{1}{9}$ is the l.u.b. of the set $\{0.1, 0.11, 0.111, \ldots\}$. l.u.b. is sometimes written *sup.*

greatest lower bound (g.l.b.). The largest of the lower bounds of a set. It may be the smallest member of the set or the largest number less than every member of the set, called an accumulation point. Thus 0 is the g.l.b. of the set $\{1, \frac{1}{2}, \frac{1}{4}, \frac{1}{8}, \ldots\}$. g.l.b. is sometimes written *inf.*

See **Algebra of Sets; Bound of Function; Limit of Sequence.**

BOW'S NOTATION. A notation used in naming the vertices of a polygon of forces by assigning letters previously assigned to regions of the plane between the lines of action of the forces. See **Funicular Polygon.**

BOYLE'S LAW. In its simplified version this states that at a given temperature, the product of the volume of a gas and the pressure is constant, expressed $pv = k$.

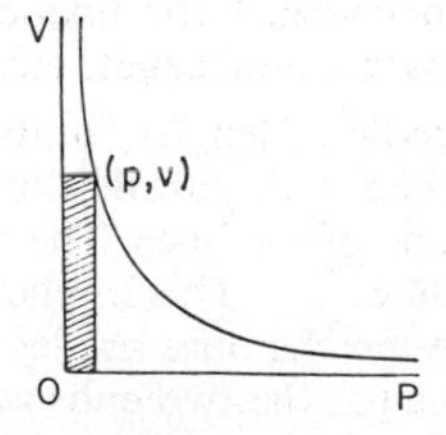

BRACE. See **Aggregation.**

BRACHISTOCHRONE. The problem posed and solved by John Bernoulli in 1696 of finding the path of quickest descent from point A to point B in space. It was one of the early problems of the **calculus of variations.** The path was shown to be part of a **cycloid.**

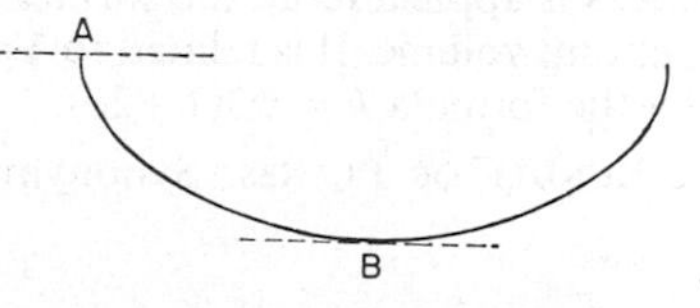

BRACKET. See **Aggregation.**

BRANCH OF CURVE. Any section of a curve distinguishable from another part.

BREADTH (WIDTH). The length of a **cross-section** of a plane or solid figure. The section is usually at right angles to the longest **axis of symmetry** or some central line. See **Dimension of Ordinary Space.**

BRIANCHON'S THEOREM. If a hexagon $ABCDEF$ is circumscribed to a **conic section,** then the diagonals AD, BE, and CF are **concurrent.** See **Circumscribed Polygon.**

BRIGGS'S LOGARITHMS. **Logarithms** with a base 10.

BRITISH THERMAL UNIT (Btu). The amount of heat energy required to raise the temperature of 1 lb of water from 60°F to 61°F (1°F); 100,000 Btu being called a therm. One pound of water loses one Btu if its temperature falls through 1°F. It is equivalent to 252 **calories** or 0.252 *large* calories (written 0.252 Calories), the unit used by dieticians.

BROCARD'S POINTS. In a triangle in which $\alpha_1 = \alpha_2 = \alpha_3$, O is a Brocard point. There are two in every triangle, and $\cot \alpha = \cot A + (a^2/bc) \csc A$. Also $\cot \alpha$ is $\cot B + (b^2/ca) \csc B$ or $\cot C + (c^2/ab) \csc C$.

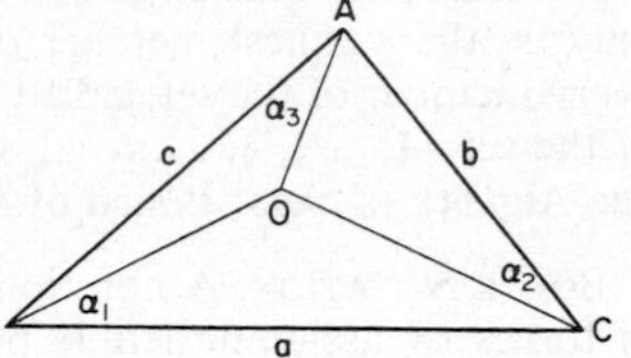

BROKEN LINE. A set of line segments joined end to end but not forming a continuous line.

BROKEN LINE GRAPH. A graph formed by a **broken line** connecting points which represent statistical data.

BROKERAGE. Commission charged for financial contracts, e.g. buying and selling of stocks and shares, mortgages, etc.

BUFFON'S PROBLEM. A needle of length l units falling at random to rest on a horizontal plane marked with parallel lines d units of length apart may or may not lie across one of the lines. The probability that it will is $2l$ to πd. The fraction $\frac{2}{\pi}$ is 0.636 61. . . . This implies that if 1 000 needles each as long as the distance between the lines are let fall at random, about 637 needles will rest across a line (or the two ends may touch two lines).

BULK MODULUS. Synonym of *compression modulus*. For an elastic body, the ratio of the unit **stress** to the unit **strain** associated with the stress. The stress is applied to all the surface of the body, and the strain is the change per unit volume. It is related to **Young's Modulus**, E, and **Poisson's ratio**, σ, by the formula $k = E/3(1-2\sigma)$.

BUNDLE OF PLANES. Synonym of **Sheaf of Planes**.

C

C. (1) The initial letter of Latin *centum*, 100; the Roman symbol for that number. (2) A **tone** emitted by any object vibrating 261 times per second, called *middle C*; if it vibrates 261 times in 2 seconds the tone is *lower C*; if it vibrates 522 times per second, *upper C*. Any positive or negative integral power of 2 times 261 will give a tone C of higher or lower pitch respectively, than *middle C*. See **Pitch in Music**. (3) Symbol for a unit quantity of heat called a **calorie**. (4) A Cantor number. See **Transfinite Number**. (5) Abbreviation to denote **Celsius scale** or centigrade scale. (6) **Field of complex numbers**.

c. International symbol for the speed of all electromagnetic waves *in vacuo*. Especially, the speed of light in interstellar space: 3×10^{10} cm/s or 186,167 miles per second (approx. 670,000,000 m.p.h.). The 3 in the formula is a tolerable value of known value, $2{\cdot}997\,924 \pm 0.000\,001$.

C.G.S. SYSTEM. The system of measurements based on the centimetre, gramme and second as units of length, mass and time respectively. See **Absolute Units; Metric System.**

CALCULATION. The process of obtaining a result from data by using mathematical procedures. Originally the term implied a numerical result, but it is now used as a general term for any procedure which produces a required result, numerical or otherwise, which does not involve the use of measurement or graphical methods. **Computation** is now used to mean numerical calculation.

CALCULUS. INFINITESIMAL CALCULUS. This invention of Newton and Leibniz in the second half of the seventeenth century is one of the greatest achievements in the history of mathematics. It is based on the concepts of **limits**, convergence and **infinitesimals.** *Differential calculus* deals with rates of change of **functions** expressed as the **derivative** of a function and geometrically as the slope of the tangent to a curve. It can be applied to the study of **velocity, acceleration, maximum and minimum values** of functions and **curvature** of curves and surfaces. *Integral calculus* deals with the calculation of lengths of curves, areas of surfaces and volumes of **solids of revolution** by the process of finding the **integral of function.** (Calculus is the Latin word for a pebble: pebbles were one of the earliest aids to counting.) **See Convergence, Divergence.**

CALCULUS OF VARIATIONS. The study which involves the finding of a function $y = f(x)$, for which a given expression containing $f(x)$ attains a greatest or a least value. Historically, the **brachistochrone** gave rise to the calculus of variations. The terms and notation were formulated by Lagrange (1736–1813).

CALENDAR. The partition of **time** into periods of years, months and days based on the natural periods of rotation of the earth about the sun, the moon about the earth and the earth about its axis for which there exist no simple relationships. The calendar year is based on the **tropical year** of 365·2422 days (365 days 5 hr 48 min 46 sec), where the day is the **mean solar day**. To allow for the fractional part of a day, the calendar has 365 days each year and an extra day in each fourth year (leap year) except centurial years, when only those divisible by 400 are leap years. Any calendar is a compromise and this is the best so far devised. The lunar month is widely variable and the relationship between a calendar month and the lunar month has been lost in the arbitrary compromises made in history. The week (sennight from seven-night), 7 days; the fortnight (from fourteen-night), 14 days; the calendar month and the quarter (of a year) are artificial subdivisions of the year having their origins in pastoral pursuits and religious observances.

CALORIE. The unit of quantity of heat in the **c.g.s. system.** The *standard* unit is the amount of heat required to raise the temperature of one gramme of distilled water from 3.5°C to 4.5°C. The *normal* unit is the amount required to do the same from 14.5°C to 15.5°C (58.1°F to 59.9°F). This is equivalent to 0.003 96 . . . **British Thermal Unit.**

One thousand calories is the dietician's Calorie; 1C = 1 000c.

CANCELLATION. The effect of an operation which neutralizes a previous one. The **inverse operations** such as +2 and −2 or +90° and −90° cancel one another. In **fractions,** when the numerator and the denominator contain a common factor, f say, it can be eliminated, since

$$\frac{N_1}{D_1}=\frac{N_2\times f}{D_2\times f}=\frac{N_2}{D_2}\times\frac{f}{f}=\frac{N_2}{D_2}.$$

CANONICAL EQUATIONS. See **Conic Section.**

CANTILEVER. A projecting beam, girder or bracket.

CAP IN SET THEORY. $A \cap B$, read as A cap B, and meaning the intersection of set A and set B. See **Set Theory.**

CAP OF SPHERE. See **Spherical Cap; Spherical Zone.**

CAPACITY. Fundamentally, the concept of internal **volume,** i.e. the amount of space enclosed. Since liquids and materials that pour are usually measured by containers, the volume is given in terms of capacity using different units from those of volume. *Cubic capacity* is a term often used to mean the *internal volume* when expressed in units of volume and not of capacity. The standard units of capacity are the **litre** and the **gallon.** It is also an electrical term.

CAPITAL. The current value of (*a*) buildings, fixtures, furniture, vehicles, tools, etc., (*b*) stock of materials for use and finished articles ready for sale, (*c*) cash in hand or recoverable, (*d*) goodwill: this and any other property that form the total assets of a person or company.

Also the name given to the sum of money loaned at interest.

CARDINAL NUMBER. A whole **number,** denoted $\#A$, used to specify how many elements there are in a set A. See **Countable Set; Inverse Function; One-One Correspondence; Ordinal Number.**

CARDINAL POINTS. The four evenly distributed points round the edge of the mariner's compass: the points N, E, S, W at the ends of two perpendicular diameters, N being in the direction of the North Pole. (Twenty-eight other points, making thirty-two points of the compass name every change of $11\frac{1}{4}°$ from N.)

CARDIOID. An **epicycloid** of one loop and a special case of the **limaçon.** The locus in a plane of a fixed point on the circumference of a circle which rolls without slipping on the outside of an equal fixed circle. The **polar equation** is $r = 2a(1—\cos\theta)$, where r is the radius vector, θ the vectorial angle and the pole O is on the circumference of the fixed circle and the polar axis is a diameter of the fixed circle. The length of all chords through the **cusp** point at O is $4a$, and their midpoints lie on a circle. There are always three tangents with the same gradient.

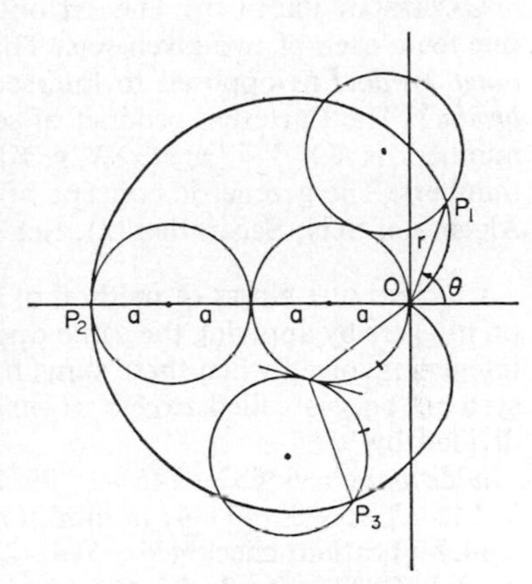

CARTESIAN COORDINATES. These are named after the French philosopher and mathematician, René Descartes (1596–1650), who propounded this concept as a basis for the study of **analytic geometry**. The position of a point in a plane can be determined by two measurements with respect to two intersecting lines in the plane. The position of a point in space can be determined by three measurements with respect to three concurrent lines, not all coplanar, in space. The lines, called **axes of coordinates**, form a **frame of reference** and their point of concurrency is the *origin of coordinates.* The axes are referred to as the *x-axis*, *y-axis* and *z-axis.* The x coordinate of a point in a plane is measured from the point to the y-axis in a direction parallel to the x-axis. The y coordinate is measured from the point to the x-axis in a direction parallel to the y-axis. The two measurements are written as an ordered pair (x, y). The point P' on plane XOY is thus uniquely defined. Measurements parallel to all three axes from a point P in space to the planes made by pairs of axes can be written as an ordered triple (x, y, z). Axes which are mutually perpendicular are *rectangular axes*; those not perpendicular are *oblique axes.*

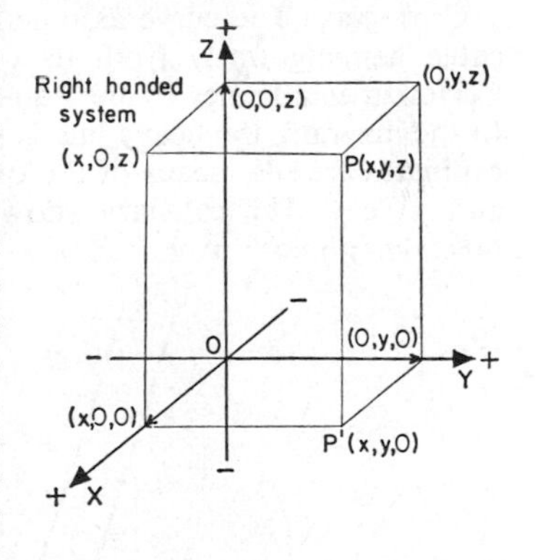

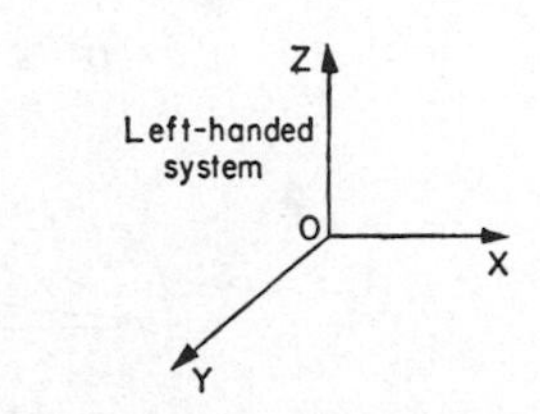

CARTESIAN EQUATION. An equation of a **configuration** when the **variables** are **Cartesian coordinates.**

CARTESIAN PRODUCT. The **set** obtained by selecting pairs of elements, one from each of two given sets. This product is sometimes referred to as *outer product* as opposed to intersection of two sets referred to as *inner product*. The Cartesian product of sets X and Y, each comprising all real numbers, is $X \times Y = \{x, y \mid x \in X, y \in Y\}$ or the set of all ordered pairs of real numbers. The geometric concept of this is all points in the xy-plane. See **Algebra of Sets; Separation (2); Set Theory.**

CASTING OUT NINES. A method of checking the accuracy of an operation on integers by applying the same operation to the sums of the digits of the integers involved when these sums have been written in modulo 9 number system. This is called *excess of nines*, the remainder when a number is divided by 9.

addition: check $653 + 746 = 1\,399$. Sums of digits, 14, 17, 22. In modulo 9, $14 + 17 \equiv 5 + 8 \equiv 13 \equiv 4$; in modulo 9, $22 \equiv 4$.

multiplication: check $479 \times 516 = 247\,164$. Sums of digits, 20, 12, 24. In modulo 9, $20 \times 12 \equiv 2 \times 3 \equiv 6$; in modulo 9, $24 \equiv 6$.

This method of checking can be applied to subtraction and to division involving integral quotients. See **Modulo (Modular) Arithmetic.**

CATENARY. The curve assumed by a uniform, flexible, heavy chain or cable hanging freely from its extremities. It is expressed explicitly in **Cartesian coordinates** by the equations: $y = \frac{1}{2}a(e^{x/a} + e^{-x/a}) = a \cosh (x/a)$. In the diagram, the heavy line is the graph of the catenary when $a = 1$. Its ordinates are the means of the ordinates of the **exponential curves**, $y = e^x$ and $y = e^{-x}$. The catenary shown is thus the graph of the *hyperbolic function*, $y = \cosh x$.

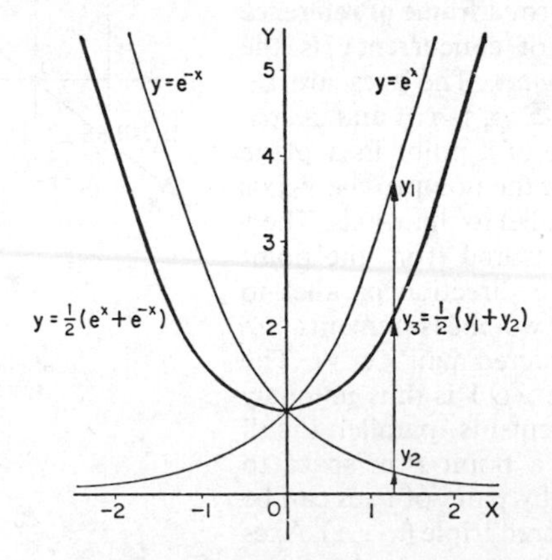

CAUCHY'S TEST. The ordinary **ratio test** for convergence or divergence of an infinite series. The series $\sum_{r=1}^{\infty} u_r$ is convergent or divergent if the **numeri-**

cal value of $\lim_{r \to \infty} (u_r/u_{r-1})$ is respectively less than or greater than unity. See **Convergence, Divergence of Series.**

CAUSTIC. If rays from a ray centre (or radiant point) F, are reflected by a curve C, then the envelope of the reflected rays is the caustic curve of C with respect to F. For example, the caustic of a circle with respect to a point on its circumference is a **cardioid.**

CELESTIAL EQUATOR. See **Spherical Coordinates in Astronomy.**

CELESTIAL HORIZON. The **great circle** in which a plane through an observer's position perpendicular to the **zenith-nadir line** cuts the **celestial sphere.**

CELESTIAL POLES. See **Spherical Coordinates in Astronomy.**

CELESTIAL SPHERE. The dome on which the stars appear to rest; synonymous with sky. A sphere of any useful radius in which the observer occupies the central position used for the purpose of observing the juxtaposition of stars and planets.

CELSIUS SCALE. A scale of temperature named after the Swedish physicist Anders Celsius (1701–1744) who based it on the freezing point of water (0°) and the boiling point of water (100°) as fixed points. The scale is sometimes referred to as the *centigrade scale.* See **Fahrenheit Scale; Kelvin Scale; Temperature Conversion.**

CENT. See **Metric System.**

CENT IN PERCENTAGE. See **Per Cent; Percentage.**

CENT, PER CENT. See **Per Cent.**

CENTAL. A **weight** of 100 pounds.

CENTESIMAL. Associated with 100, as in the centesimal system of measuring an **angle (plane angle)** based on the *grad* (*grade*), 1/100 of a right angle. 1 grad = 100 minutes, 1 minute = 100 seconds.

CENTIGRADE SCALE. A scale of temperature based on the arbitrary allocation of 0°C to the temperature of co-existent ice and water and 100°C to that of co-existent water and steam. A range of five degrees is the equivalent of a range of nine degrees on the **fahrenheit scale.** See **Celsius Scale; Temperature Conversion.**

CENTRAL ANGLE. See **Angle at Centre.**

CENTRAL CONIC. A term usually reserved for the **conic sections, ellipse** and **hyperbola,** which have a **centre of symmetry** in contrast to the **parabola** which has not.

CENTRAL FORCE. A force which acts upon a particle in such a way that the direction of the force is always through a fixed point.

CENTRAL POLYHEDRAL ANGLE. The **polyhedral angle** subtended by a **spherical polygon** at the centre of a sphere.

CENTRAL PROJECTION. See **Geometric Projection; Map Projection; Pictorial Projection.**

CENTRAL QUADRIC SURFACE. An ellipsoid or a hyperboloid having **central symmetry**. The general equation is $Ax^2+By^2+Cz^2=1$.

CENTRAL SYMMETRY. **Symmetry** with respect to a point.

CENTRAL VALUE. See **Class Limits.**

CENTRE. Middle point; chiefly in geometry the **centre of symmetry** of a line, circle, sphere, etc.

CENTRE OF ATTRACTION. The **centre of gravity** of a body.

CENTRE OF BUOYANCY. The **centre of gravity** of that region of liquid displaced by a floating body.

CENTRE OF CURVATURE. See **Curvature.**

CENTRE OF CURVE. The **centre of symmetry** of a curve.

CENTRE OF GRAVITY. That point at which the entire weight of a body may be concentrated, theoretically, to produce the same effect as when distributed. It is the point through which the resultant of the gravitational forces acting on all the particles of the body can be considered to operate irrespective of the orientation of the body. It is the point about which the body has equilibrium. It coincides with the **centre of mass** if the body is situated in a **uniform gravitational field.**

CENTRE OF HOMOTHETY. Synonym of homothetic centre. See **Radially Related Figures.**

CENTRE OF INVERSION. See **Inversion.**

CENTRE OF MASS. That point in a body at which the entire mass may be concentrated, theoretically, to produce the same effect as when distributed.

If (x_1, y_1, z_1), (x_2, y_2, z_2), . . . (x_n, y_n, z_n) are coordinates of masses m_1, m_2, . . . m_n, then the total mass can be considered to be situated, theoretically, at the *centre of mass* $(\bar{x}, y, \bar{z})$, given by:

$$\bar{x}=\frac{\sum_1^n m_r x_r}{\sum_1^n m_r}, \quad \bar{y}=\frac{\sum_1^n m_r y_r}{\sum_1^n m_r}, \quad \bar{z}=\frac{\sum_1^n m_r z_r}{\sum_1^n m_r}.$$

If the body under consideration is placed in a **uniform gravitational field** the centre of mass coincides with the **centre of gravity** and the **centroid.**

CENTRE OF OSCILLATION. If a compound pendulum is supported at P and has its **centroid** at G, O will be the centre of **oscillation** in PG produced if PO is the length of the simple pendulum which swings in the same time as the compound one. The length PO is given by $(l^2+k^2)/l$ where l is PG and k is the **radius of gyration** of the pendulum about G. The time of oscillation is given by $t=2\pi\sqrt{(l^2+k^2)/lg}$. See **Pendulum, Compound; Pendulum, Simple.**

CENTRE OF PRESSURE. The point on a surface submerged in a liquid at which all the pressure may be considered, theoretically, to be concentrated; producing the same effect as when distributed.

CENTRE OF PROJECTION. See **Geometric Projection; Map Projection; Pictorial Projection.**

CENTRE OF REGULAR POLYGON. The common centre of the **inscribed circle of polygon** and the **circumscribed circle of polygon.**

CENTRE OF SHEAF (BUNDLE). See **Sheaf of Planes.**

CENTRE OF SIMILITUDE. The point from which emergent **rays** establish one figure as a central projection of another. See **Radially Related Figures; Similitude.**

CENTRE OF SMALL CIRCLE. See **Small Circle.**

CENTRE OF SYMMETRY. The point about which a figure is symmetrical. See **Symmetry.**

CENTRES OF SIMILITUDE OF TWO CIRCLES. Points on the line of centres which divide it internally and externally in the ratio of the radii. These are the points of intersection of the common tangents and the line of centres. See **Radially Related Figures.**

CENTRIFUGAL, CENTRIPETAL FORCES. When a heavy particle moves in a circular path it must be either attached to a fixed point by an inextensible string, or be constrained to move in a circular path. The tension *in* the string and the reaction *on* the groove are called the *centrifugal force*: the equal and opposite forces exerted on the particle *by* the string and the groove are called the *centripetal forces*. Its measure, in both cases, is $m.v^2/r$, where m is the **mass** of the particle, v is the **velocity** at any instant measured at right angles to the radius, length r.

CENTROID. In a geometric figure, the point having coordinates which are the arithmetic means of the coordinates of all the points making up the figure. If the figure represents a body having uniform density, the centroid coincides with the **centre of mass**; if the body is in a uniform gravitational field the centroid also defines the **centre of gravity**.

CEVA'S THEOREM. In any triangle ABC, AX, BY, CZ are lines, concurrent at P, with X, Y, Z on BC, CA, AB respectively, produced if necessary. The theorem states: $\frac{\overline{AZ}}{\overline{ZB}} \times \frac{\overline{BX}}{\overline{XC}} \times \frac{\overline{CY}}{\overline{YA}} = 1$, where $\overline{AZ}$, $\overline{ZB}$, etc., are **directed line segments**. If, where P is internal, $\overline{AZ}, \overline{ZB}, \overline{BX}$, etc., are considered positive, then $\overline{ZB}$ and $\overline{YA}$ are negative in the example shown when P is external. The converse of this theorem states the condition for three lines through the vertices of a triangle to be concurrent.

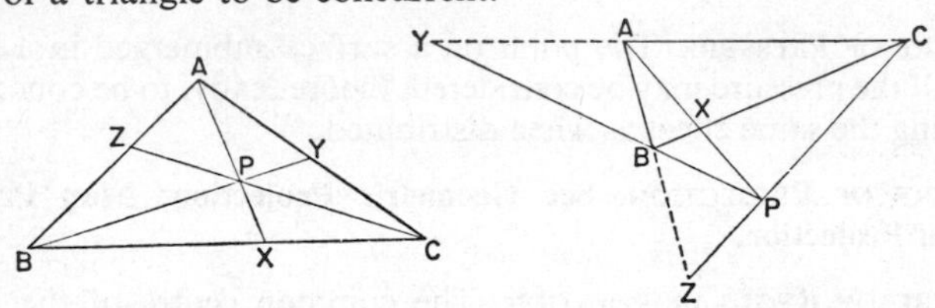

CHAIN. An instrument for measuring distances in **surveys.** The standard chain of 22 yards (66 feet) was introduced by Edmund Gunter in 1620 and its length was convenient in that 10 square chains was equal to 1 acre (4 840 yd²). The metal chain is made of 100 links each measuring 7.92 inches. It is called the Gunter's or surveyor's chain to distinguish it from the engineer's chain of 100 one-foot links.

CHAIN OF SETS. Synonym of **nested sets.**

CHAIN RULE IN DIFFERENTIATION. The general rule for determining the **derivative** of a composite function.

If $y=f_1[f_2\{f_3(x)\}]$, $\frac{dy}{dx} = \frac{df_1[\,]}{df_2\{\}} \frac{df_2\{\}}{df_3(\,)} \frac{df_3(x)}{dx}$;

e.g., if $y=\sin^3 2x=[\sin\{2(x)\}]^3$,

$$\frac{dy}{dx} = 3 \cdot \sin^2 \{\} \cdot \cos \{\} \cdot 2 = 6 \cdot \sin^2 2x \cdot \cos 2x.$$

If the composite function is a function of a function of x, $y=f[u(x)]$, $dy/dx=(dy/du)\cdot(du/dx)$. The chain rule can be extended indefinitely.

CHAIN TRIANGULATION. See **Triangulation.**

CHARACTERISTIC. The integral part of a **logarithm** determined by the **place value** of the digit with the largest place value. Thus, unit numbers have characteristic 0, tens and tenths 1 and -1 respectively, hundreds and hundredths 2 and -2, thousands and thousandths 3 and -3, etc. As the **mantissa,** the decimal fraction, is always positive, the negative sign of the characteristic of a number less than unity is written above the integer and read as 'bar'. Thus: $0.008\,64=10^{\bar{3}.936\,51}=10^{(-3+0.936\,51)}=10^{-2.063\,49}=1/10^{2.063\,49}$. The number $86.4=10^{1.936\,51}$.

CHORD. Derived from the Latin *chorda,* a string, it is associated with an arc. It is the line joining any two points on a curve and is therefore a finite portion of an infinite **secant.**

CHORD OF CONTACT. The line joining the points of contact of a pair of **tangent lines** from a point to a curve.

CIRCLE. The plane figure bounded by a set of points equidistant from a fixed point called the centre. The boundary, called the **circumference**, is sometimes referred to as the circle. In **Cartesian coordinates** the equation of the circumference is $x^2+y^2+2gx+2fy+c=0$ where the centre is at the point $(-g, -f)$ and the radius is $\sqrt{(g^2+f^2-c)}$. If the centre is at the origin and the radius is a, the equation is $x^2+y^2=a^2$. See **Null Circle**; **Parametric Coordinates.**

CIRCLE, IMAGINARY. See **Imaginary Circle.**

CIRCLE OF CLOSEST CONTACT. See **Curvature**; **Osculating Circle.**

CIRCLE OF CURVATURE. See **Curvature.**

CIRCLE OF INVERSION. See **Inversion.**

CIRCULAR FUNCTIONS. Synonym of **trigonometric functions**. The sine, cosine, secant and cosecant functions of the angle $(\theta \pm 2n\pi)$ have values which are repeated for $n=0$, $n=1$, $n=2$, etc. The tangent and cotangent functions have values which are repeated for $n=0$, $n=\frac{1}{2}$, $n=1$, etc. The graphs of the circular functions are obtained by plotting the values of the functions against the values of the angle θ. The sine curve is shown. See **Reduction Formulae in Trigonometry.**

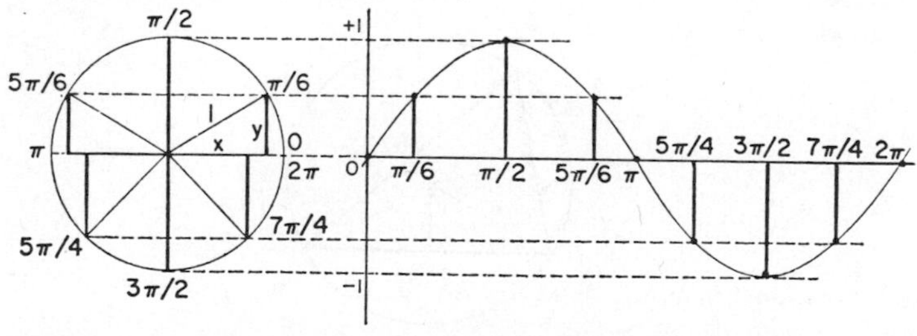

CIRCULAR MEASURE OF ANGLES. See **Radian.**

CIRCULAR MOTION. Motion along the circumference of a circle. See **Uniform Circular Motion.**

CIRCUMCENTRE OF TRIANGLE. The centre of the **circumscribed circle of triangle.**

CIRCUMCIRCLE OF TRIANGLE. The **circumscribed circle of triangle.**

CIRCUMFERENCE OF CIRCLE. The perimeter of a **circle**. Its length cannot be determined rationally in terms of its diameter, but for all circles, the ratio circumference/diameter is always a constant, π. The circumference itself is sometimes loosely referred to as a circle to distinguish it from its interior. See **Pi**, Π, π.

CIRCUMFERENCE OF SPHERE. The circumference of any **great circle** of a **sphere**. See **Circumference of Circle**.

CIRCUMSCRIBED CIRCLE OF POLYGON. The **circle** whose circumference passes through the vertices of a **polygon**. Not all polygons can have circumscribed circles.

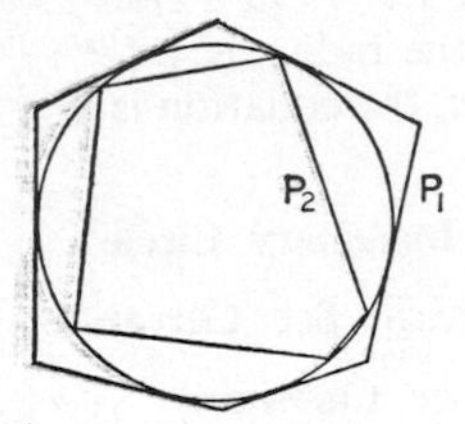

Polygon P_1 circumscribed to circle which is circumscribed to polygon P_2

CIRCUMSCRIBED CIRCLE OF TRIANGLE. The **circle** whose circumference passes through the vertices of a triangle. Its centre is the point of intersection of the perpendicular bisectors of the sides of the triangle. The radius is given by $r = abc/4\triangle = abc/4\sqrt{s(s-a)(s-b)(s-c)}$, where a, b, c are the lengths of the sides and s the semi-perimeter.

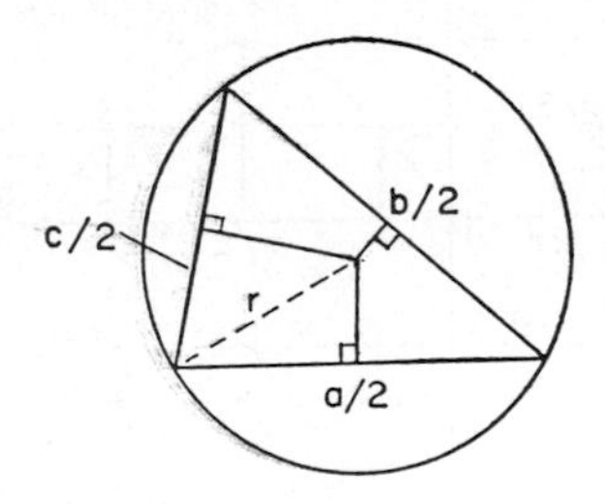

CIRCUMSCRIBED POLYGON. The polygon whose sides are **tangent lines** to some **simple closed curve**.

CIRCUMSCRIBED POLYHEDRON. The **polyhedron** whose faces are tangential to a **closed surface**.

CIRCUMSCRIBED SPHERE OF POLYHEDRON. The **sphere** whose surface passes through the vertices of a **polyhedron**. Not all polyhedra can have circumscribed spheres.

CISSOID OF DIOCLES. Let O be a fixed point on the circumference of a circle of radius a. Let any secant through O cut the circle at Q and the tangent at the extremity of the diameter through O, at R. If P is a point on

OR such that $OP = QR$, the locus of P is called a *cissoid* (from Greek, *kissos*, ivy), named by Diocles about 200 B.C. The **polar equation** is $r = 2a \tan \theta \sin \theta$ and the **Cartesian equation** $y^2(2a - x) = x^3$.

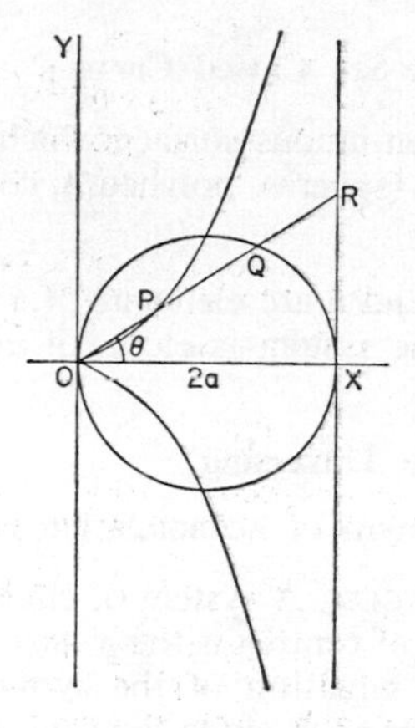

CLASS. Synonym of **set.** A collection of things having some common property.

CLASS INTERVALS. See **Class Limits.**

CLASS LIMITS. All classes of measurements, objects, persons, etc., must be defined between precise limits so that no one member of the class may, inadvertently, be put into more than one class. If the inhabitants of a town are classified for statistical purposes as being in age groups 0-5 years, 5-10 years, etc., the points of demarcation must be clearly understood so that a child who has passed his fifth birthday is included in the *second* group. The upper limits are 5, 10, 15, . . . years; the intervals are five-yearly ones; the central values of the intervals are $2\frac{1}{2}$, $7\frac{1}{2}$, $12\frac{1}{2}$, . . . years.

CLOCKWISE, ANTI-CLOCKWISE (COUNTER-CLOCKWISE). Description of rotation by analogy to the movement of the hands of a clock. In directed rotations, clockwise is negative and anti-clockwise is positive.

CLOSED CURVE. If a plane continuous **curve** is defined by the pair of equations $x = a(t)$, $y = b(t)$, within the range $p \leqslant t \leqslant q$, its end points will be (ap, bq) and (aq, bp). When, simultaneously, $ap = aq$ and $bp = bq$, the end points are coincident and the curve closes. For example, in the curve $x = a \cos \theta$, $y = b \sin \theta$, with $0 \leqslant \theta \leqslant 2\pi$, both end points are $(a, 0)$, and the closed curve is an ellipse. A *closed space curve* needs a third equation for its definition and an analogous reasoning will establish the coincidence of end points. A *simple closed curve* is one which does not intersect itself.

Closed Interval. See **Interval.**

Closed Plane Figure. A figure in a plane whose boundary is a **closed curve.**

Closed Space Curve. See **Closed Curve.**

Closed Surface. A continuous surface which encloses a part of space called the interior. Solids (spheres, polyhedra, cones, etc.) are bounded by closed surfaces.

Closed System. If a and b are elements of a system and $\odot$ denotes a **binary operation,** then the system is closed if $a \odot b$ is also an element of the same system.

Closed Traverse. See **Traversing.**

Cluster Point. Synonym of **accumulation point.**

Coaxal (Coaxial) Circles. A system of circles which have a common **radical axis.** If their line of centres is the x-axis and their radical axis the y-axis, then the general equation of the system is $x^2+y^2+2gx+c=0$, where c is a constant. For each circle the centre is $(-g, 0)$ and radius is $\sqrt{g^2-c}$. When the centre of a circle is at either point $(\sqrt{c}, 0)$ or $(-\sqrt{c}, 0)$ the radius is zero. At these points there are *point circles*, the *limiting points* of the coaxal system. The limiting points of any system of coaxal circles are the points of intersection of all circles which cut any two of the system orthogonally (at right angles).

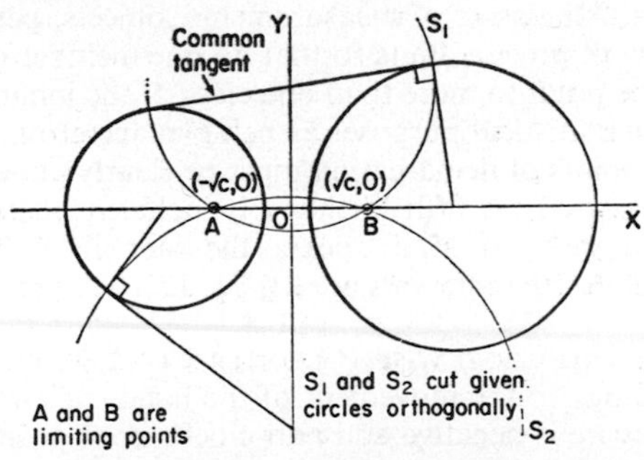

Coaxal (Coaxial) Planes. Synonym of **collinear planes.**

Coding. See **Computer Programming.**

Coefficient. In general, the product of all the factors of an expression but one or more excluded. Thus, in $2hxy$, $2hx$ is the coefficient of y, $2h$ of xy, 2 of hxy. More specifically a coefficient is the product of the constant factors in contrast to the product of the variables. For example, 3 is a *numerical* coefficient in $3x^2$ while b is a literal coefficient in by^2.

Coefficient of Correlation, r. This is the quotient of $\Sigma(xy)$ and $(\Sigma x^2)^{\frac{1}{2}}$ $(\Sigma y^2)^{\frac{1}{2}}$. If σ_x and σ_y are the **standard deviations** of x and y from their respective means of distribution, and n is the number of cases involved in the problem, this reduces to $r = \Sigma(xy)/n \cdot \sigma_x \cdot \sigma_y$, and r is the measure of the **correlation** between the two variables x and y.

Coefficient of Friction, μ. See **Friction.**

Coefficient of Linear Expansion, x_l. If a unit length of a material is raised in temperature 1°C, its length becomes $1 + x_l$. If a length L_0 is raised through t°C, its length becomes $L_0 + L_0(x_l t)$ or $L_0(1 + x_l t)$.

Coefficient of Restitution. See **Laws of Impact in Elasticity.**

Coefficient of Superficial Expansion, x_s. If a unit area of a material is raised in temperature 1°C, its area becomes $1 + x_s$. If an area A_0 is raised through t°C its area becomes $A_0 + A_0(x_s \cdot t)$ or $A_0(1 + x_s t)$. For all practical purposes $x_s = 2x_l$, i.e. $2 \times$ coefficient of linear expansion.

Coefficient of Variability. A term in statistics for the quantity $C = \sigma/\bar{x}$, where σ is the **standard deviation** of a set of x and $\bar{x}$ the **arithmetic mean** of the set.

Coefficient of Variation. A term in statistics for the quantity $V = 100C$, where C is the coefficient of variability.

Coefficient of Volumetric Expansion, x_v. If a unit volume of a material is raised in temperature 1°C, its volume becomes $1 + x_v$. If a volume V_0 is raised through t°C its volume becomes $V_0 + V_0(x_v t)$ or $V_0(1 + x_v t)$. For all practical purposes $x_v = 3x_l$, i.e. $3 \times$ coefficient of linear expansion.

Coefficients, Detached. See **Detached Coefficients.**

Coefficients, Differential. See **Differential Coefficients.**

Coefficients of Binomial Expansion. See **Binomial Coefficients.**

Coefficients of Equation. (1) The **coefficients** of the **variables.** (2) The coefficient of all the terms—variable and constant.

Coefficients, Undetermined. See **Principle of Undetermined Coefficients.**

Cofactor. Synonym of signed minor of a **determinant.**

Coincident. Identical in position.

Collection of Terms. The simplification of expressions by the combination of **similar (like) terms**. The process is governed by the **distributive law**, $ab + ac = a(b + c)$. For example, $3x + 2y + x + 5y = 4x + 7y$.

Collinear Planes. Synonym of *coaxal* (*coaxial*) *planes*. Planes which have a common straight line. The set of planes having a common straight line, called the axis, form a **pencil of planes**.

COLLINEAR POINTS. Points which lie on the same straight line.

COLUMN GRAPH. A graphic representation in which a set of frequencies is represented by a set of columns of equal widths and heights proportional to the frequencies. See **Histogram**.

COLUNAR TRIANGLES. A pair of **spherical triangles** which together form a **lune**.

COMBINATION. If from a given number, say n, of different things a *selection* of r things is to be made, the number of possible selections is represented by ${}^nC_r$, in which C implies the type of selection known as a combination. The value of ${}^nC_r$ is $n!/r!(n-r)!$. Alternative notations are ${}_nC_r$, $C(n, r)$, C^n_r, $\binom{n}{r}$. See **Binomial Theorem**; **Factorial *n***; **Permutation.**

COMBINATORIAL (COMBINATORY) ANALYSIS. The analysis of combinations, examples of which are **permutations, combinations, binomial theorem, magic squares, theory of partitions.**

COMBINED VARIATION. See **Variation.**

COMMENSURABLE. Two numbers or two quantities are commensurable if their ratio is a **rational number.** Two line segments are commensurable if they are **multiples** of a single line segment. The Greeks expressed numbers in terms of line segments, hampered as they were in arithmetic computation by the cumbrous nature of their alphabetic number symbols. The number was rational if its line segment was commensurable with a unit segment. An example of *in*-commensurable line segments is provided by the pair, side and diagonal of a square. If the side is 1, the diagonal is $\sqrt{2}$; if the diagonal is 1, the side is $1/\sqrt{2}$ or $\frac{1}{2}\sqrt{2}$.

COMMON DENOMINATOR. A **common multiple** of the denominators of two or more **fractions**. The least common denominator, L.C.D., is the lowest common multiple of the denominators. Fractions may be added or subtracted when expressed with a common denominator.

COMMON DIFFERENCE. The difference between successive terms of an **arithmetic progression.**

COMMON FRACTION. Synonym of simple or vulgar fraction. A **fraction** of the form N/D or $\frac{N}{D}$, where N and D are integers.

COMMON LOGARITHM. Synonym of Briggs's logarithm. See **Logarithms.**

COMMON MULTIPLE. Any member of an infinite set of **multiples** which contains two or more given integers or polynomials as factors.

COMMON SCALE OF NOTATION. The **system of notation** which has 10 as its base. Synonym of **denary number system.**

COMMON TANGENT. A **tangent line** common to two or more curves. See **Tangents to Two Circles.**

COMMUTATIVE (ABELIAN) Group. A **group** in which the **binary operation** obeys the **commutative law.**

COMMUTATIVE LAW. This law states $a \odot b = b \odot a$. In this equality, a and b are undefined elements of a set and $\odot$ represents an undefined **binary operation.** Examples of operations which satisfy this law are **addition, multiplication, translation** and **rotation** in a plane. The possibility of a system which was non-commutative under multiplication was first considered by Hamilton in the middle of the nineteenth century.

COMMUTATIVE RING. See **Ring.**

COMPARATIVE SCALES. These are designed to show two different systems of units (e.g., f.p.s./c.g.s.) on the same **representative fraction.**

COMPARISON TEST. A test for convergence or divergence of an infinite series. If after some chosen term, the **absolute value** of each term is equal to or less than the corresponding term of some convergent series, the series converges absolutely. If each term is equal to or greater than the corresponding term of some divergent series, the series diverges absolutely. See **Convergence, Divergence of Series; Ratio Tests.**

COMPLEMENT OF SET. See **Set Theory.**

COMPLEMENTARY ANGLES. Any pair the sum of which is a right angle (90° or $\frac{1}{2}\pi$ radians). One angle of the pair is the complement of the other.

COMPLETE QUADRANGLE. A plane figure in **projective geometry** consisting of four points (no three collinear) and six lines they determine. See **Quadrangle.**

COMPLETE QUADRILATERAL. A plane figure in **projective geometry** consisting of four lines (no three concurrent) and six points of intersection. See **Quadrangle; Quadrilateral.**

COMPLETING THE SQUARE. The process of adding a quantity to a given one so that the result is a perfect square. To some square, p^2 is added the rectangle $q(2p+q)$ equal to $(2pq+q^2)$ to give $(p+q)^2$. This process is used in solving a quadratic equation by 'completing the square'. It involves adding the square of half the coefficient of x to the first two terms and subtracting it from the third, thus:

$$\begin{aligned} ax^2+bx \quad &+c=0,\\ x^2+(b/a)x \quad &+c/a=0.\\ x^2+(b/a)x+(b/2a)^2-\{(b/2a)^2-c/a\} &=0,\\ \{x+(b/2a)\}^2-\{(b^2-4ac)/4a^2\} &=0,\\ \{x+(b/2a)\}^2-\{\sqrt{(b^2-4ac)}/2a\}^2 &=0.\end{aligned}$$

This can be factorized as the difference between two squares:

$$\left\{x+\frac{b+\sqrt{b^2-4ac}}{2a}\right\}\left\{x+\frac{b-\sqrt{b^2-4ac}}{2a}\right\}=0.$$

Hence the solution is:

$$x = -\frac{\left(b \pm \sqrt{b^2 - 4ac}\right)}{2a}$$

See **Gnomonic Number; Quadratic Equation Formula.**

COMPLEX. A term used sometimes as a synonym of **set**.

COMPLEX FRACTION. A **fraction** having a fraction or fractions in the numerator and/or the denominator.

COMPLEX NUMBERS. **Numbers** of the form $a+ib$ where a and b are real numbers and $i=\sqrt{-1}$. They can be represented on an **Argand diagram.**

addition, subtraction of complex numbers. $(a+ib)\pm(c+id)=(a\pm c)+i(b\pm d)$.

multiplication of complex numbers. $(a+ib)(c+id)=(ac-bd)+i(bc+da)$.

division of complex numbers. $\dfrac{a+ib}{c+id}=\dfrac{(a+ib)(c-id)}{(c+id)(c-id)}=\dfrac{(ac+bd)+i(bc-da)}{c^2+d^2}$.

See **Conjugate Complex Numbers; Polar Form of Complex Number.**

COMPONENDO. See **Proportion.**

COMPONENT OF VECTOR. A **vector quantity** which, with others, is equivalent to a given vector, representing a force, velocity, acceleration, etc. Thus the vector **v** may be replaced by components **a** and **b**; or by **p, q, r** and **s**. The most useful *pair* is a pair at right angles, **x** and **y**, where $\mathbf{x}=\mathbf{v}\cos\theta$, $\mathbf{y}=\mathbf{v}\sin\theta$.

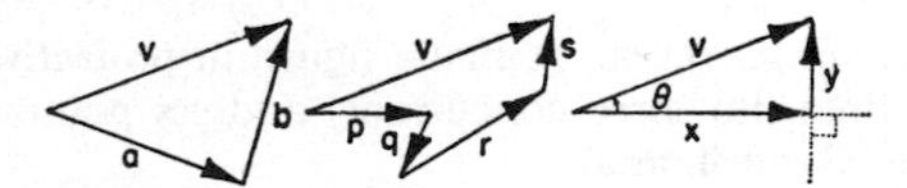

COMPONENT PARTS OF NUMBER. The factors into which a **composite number** or expression may be separated.

COMPOSITE NUMBER. Any **integer** which can be expressed as the product of other integers (its **factors**), as opposed to a **prime number**. All composite **numbers** can be expressed as products of unique sets of prime numbers. This fact was known to the Greeks and was recorded in **Elements of Euclid**. It is sometimes referred to as the fundamental theorem of arithmetic or the unique factorization theorem.

COMPOSITE OF FUNCTIONS. Let f be a **function** which maps set X into set Y, i.e. yfx or $y=f(x)$ and g a function which maps set Y into set Z, i.e. zgy or $z=g(y)$. The composite of function g with function f is a function denoted $g \odot f$ which maps set X into set Z, since for each element $x \in X$ there corresponds an element $z \in Z$ through the intermediary element $y \in Y$. Thus, if yfx $(x \to y=3x)$ and zfy $(y \to z=y-4)$, the composite $g \odot f$ is $x \to z=3x-4$.

COMPOSITE OF RELATIONS. Let R be a **relation** between set X and set Y, i.e. $x\ R\ y$ and S be a relation between set Y and set Z, i.e. $y\ S\ z$. The composite of S with R is a relation $S \odot R$. Thus the composite of $(x>y)$ and $(y>z)$ is $(x>z)$; the composite of (x is parallel to y) and (y is perpendicular to z) is (x is perpendicular to z).

COMPOSITION OF VECTORS. The process of combining (adding and subtracting) **vector quantities** (forces, velocities, etc.). Any pair of vectors acting at a point may be combined graphically by the **parallelogram of vectors** or the **triangle of vectors.** Three or more vectors can be combined by the **polygon of vectors. See Component of Vector; Vector Difference Vector Sum.**

COMPOUND INTEREST. The type of **interest** in which the interest is periodically added to the principal. The formula covering the relationship of amount, principal, rate and the number of times interest has been converted into principal is $A = P(1 + r/100)^n$. Compound interest for any period is the difference between A and P.

COMPOUND INTEREST LAW. The law of change in a function of x such that the rate of change (increment) is proportional to the function itself. If $y = b \cdot e^{ax}$, then $y' = a \cdot b \cdot e^{ax} = a . y$. Many natural phenomona obey this law of change. See **Derivative.**

COMPOUND PENDULUM. See **Pendulum, Compound.**

COMPOUND PRACTICE. See **Practice.**

COMPOUND PROPORTION. When a quantity depends for its value on several **variables** at one and the same time it is said to be in compound **proportion** with the variables. In **simple interest**, for example, the interest is in compound proportion with the principal, the rate and the time. See **Variation.**

COMPOUND TRAVERSE. See **Traversing.**

COMPRESSION. The internal forces brought into play that resist the external forces tending to decrease the length of a body (springs, struts, etc.). **See Hooke's Law; Tension.**

COMPRESSIVE. Appertaining to **compression** as in **stress.**

COMPUTATION. The process of obtaining a numerical result from data by using mathematical procedures which may be arithmetic or analytic and involve addition, multiplication, etc., or the use of slide rule, logarithms, mechanical or electronic **computers.**

COMPUTER. Any instrument or machine designed to perform numerical mathematical operations. Simple **computations** involving the fundamental operations of arithmetic are performed by mechanical devices usually called calculating machines, as opposed to electronic computers which perform the more tedious and the more complex computations when the process of **computer programming** is needed.

analogue computer. One in which the combination of numbers is replaced by the analogous situation of combination of lengths (e.g. slide rule) or voltages.

digital computer. One in which numbers and other data are operated upon as a sequence of binary digits (bits) in the **binary number system**.

COMPUTER PROGRAMMING. The process of planning the sequence of steps to be taken by a **computer** in the solution of a problem. The sequence is often shown in the form of a schematic diagram called a *flow chart*. The information contained in the programmer's instructions and/or flow charts is represented in the form of *coding* which can be interpreted by the computer. See **Mathematical Programming.**

CONCAVE, CONVEX CURVE. A plane curve is concave on one side and convex on the other. It is *concave* on the side nearer to the centre of **curvature**; it is *convex* on the other. Concavity and convexity can be determined by the rotation of the tangent at a point (x, y) on the curve $y=f(x)$. If, as x increases, the rotation is positive (anti-clockwise), the curve is convex on the side nearer to the x-axis and concave on the other. If the rotation is negative (clockwise), the curve is concave on the side nearer to the x-axis, convex on the other.

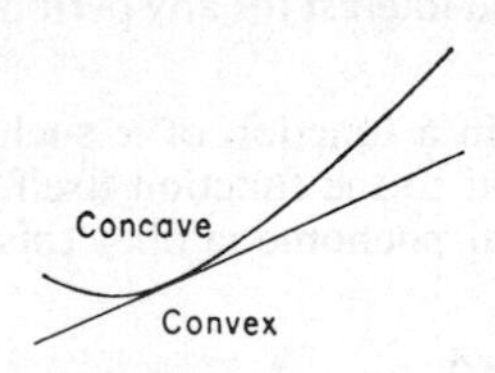

CONCAVE, CONVEX POLYGON. A **polygon** is *convex* if it lies completely on one side of any side extended. This occurs if no interior angle is greater than 180° (π radians). If any interior angle is greater than 180° (a **re-entrant angle**) the polygon is *concave*. All straight lines which cross a convex polygon cut two sides: a straight line may cut the sides of a concave polygon in an even number of points which are not vertices.

CONCAVE, CONVEX POLYHEDRON. A **polyhedron** is *convex* if it lies completely on one side of the plane of any one face. This occurs if **polyhedral angle** (an interior **solid angle**) is less than 2π steradians. A *concave* polyhedron does not have this property.

CONCAVITY, CONVEXITY. The state of a curve, polygon, polyhedron of being **concave, convex.**

CONCENTRIC CIRCLES. **Circles** which have the same centre.

CONCLUSION. See **Hypothesis.**

CONCURRENCY. The property of curves, lines and their segments of being concurrent, that is, of having a common point.

CONCYCLIC POINTS. Points which lie on the circumference of the same circle.

CONCHOID. Let P and P' be points on a line which passes through a fixed point O and intersects a fixed curve C at Q. If $PQ=QP'=b$, a constant, then the locus of P and the locus of P' are the two branches of the *conchoid of C with respect to O.* The conchoid of a circle with respect to a point on its circumference is the **limaçon of Pascal**: if the constant b is the diameter of the circle, the limaçon is a **cardioid.**

CONCHOID of NICOMEDES. The **conchoid** of a straight line L, with respect to a point O not on the line. There are three cases, resulting from b (PQ, QP') being greater than, equal to or less than the perpendicular distance a (OM), from O to the line L. The **polar equation** of the conchoid of Nicomedes is $r = a \sec \theta + b$; the **Cartesian equation** is $b^2x^2 = (x-a)^2(x^2+y^2)$. In all cases, the fixed line $x = a$ is an **asymptote.** The discovery of the conchoid was first attributed to Nichomedes in the second century B.C. In the seventeenth century it was used in analysis: it was employed in duplicating a cube and in **trisection of angle.**

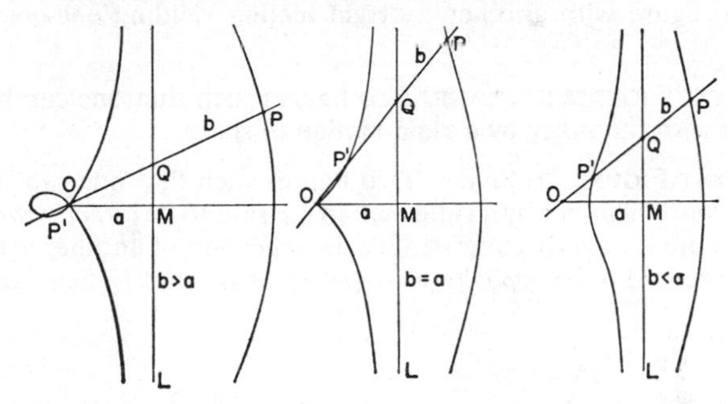

Let angle AOB be the given angle. Let a line QM (Q on OA, M on OB) be perpendicular to OB. Let the conchoid of QM with respect to O be drawn with fixed distance $QP = b = 2\ OQ$. Let QN, the line through Q parallel to OB, cut the conchoid in R. Then angle $ROB = \frac{1}{3}$ angle AOB. (In the diagram $TR = b$.)

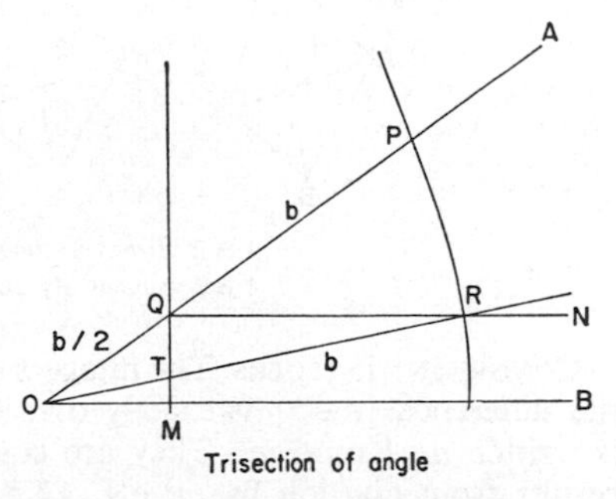

Trisection of angle

CONDITIONAL. See **Algebra of Propositions; Equation; Inequality.**

CONE. The solid bounded by a **conical surface** and a plane end surface. For a **right circular cone**, diameter of base d, height h, slant height (length of an element) s, the mensuration is:

Slant height $s = \sqrt{h^2 + \frac{1}{4}d^2}$; Curved area $\frac{1}{2}\pi ds = \frac{1}{4}\pi d\sqrt{4h^2 + d^2}$; Plane area $\frac{1}{4}\pi d^2$; Total area $\frac{1}{4}\pi d(d+2s) = \frac{1}{4}\pi d(d + \sqrt{d^2 + 4h^2})$; Volume $(\pi/12)d^2h = (\pi/24)d^2\sqrt{4s^2 - d^2}$.

See **Element of Cone.**

CONFIGURATION. An arrangement of points, lines, curves and surfaces which represent diagramatically a geometric concept. See **Geometry**.

CONFOCAL CONICS. **Conic sections** that have the same foci. The equation $x^2/(a^2-k)+y^2/(b^2-k)=1$ represents a family of confocal ellipses when $a^2>b^2>k$ and a family of confocal hyperbolas when $a^2>k>b^2$. The equation $y^2=4a(x-a)$ is the equation of a parabola with focus as origin and its axis as the x-axis, and *latus rectum* $4a$. A family of confocal parabolas is represented by this formula if a is given different values.

CONFOCAL QUADRICS. Quadrics which share the same **principal planes,** each of which plane intersects the surfaces in **confocal conics.**

CONGRUENCE (CONGRUENCY) IN GEOMETRY. The identification of one geometrical figure with another by **rigid motion** yielding **one-one correspondence.**

CONGRUENT FIGURES IN PLANE. Two figures such that one can be made to coincide with the other by a **rigid motion** in space.

CONGRUENT FIGURES IN SPACE. Two figures such that one can be made to coincide with the other by a **rigid motion** are said to be *directly congruent*. If the one figure is directly congruent to the reflection of another in a plane, the figures are said to be *oppositely congruent* or *symmetric*. See **Symmetry.**

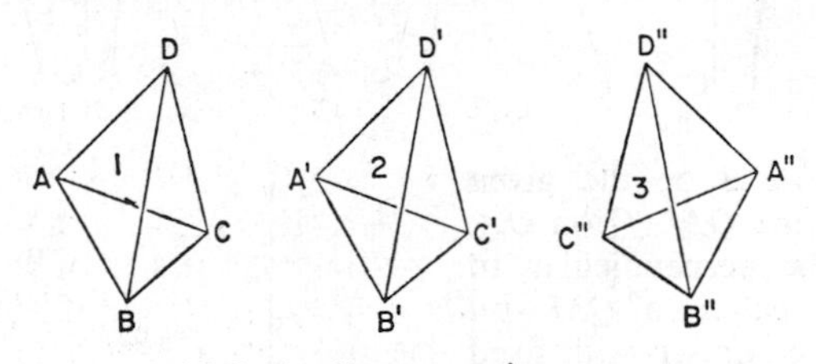

1 & 2 directly congruent
1 & 3 oppositely congruent

CONGRUENT INTEGERS. The integers a and b are congruent (modulo m) if the difference $|a-b|$ is exactly divisible by the number m. The relation is written $a \equiv b(\text{mod } m)$. They are congruent also if the same remainder results from division by m, e.g. $17 \equiv 24(\text{mod } 7)$. See **Modulo (Modular) Arithmetic.**

CONGRUENT TRIANGLES. A special case of **congruent figures.** Two triangles are congruent if there is **one-one correspondence** with respect to (*a*) two sides and the included angle, (*b*) two angles and a side similarly placed with respect to the angles, (*c*) three sides, (*d*) a right angle, the hypotenuse and one other side.

CONGRUENT POLYGONS. **Polygons** which are **congruent figures.** See **Congruence (Congruency) in Geometry.**

CONIC SECTION. A plane section through a **conical surface**. If the cone is right circular the sections are those of **analytic geometry**. Depending upon the inclination of the plane to the axis of the cone and whether or not the plane passes through the apex, seven types of section are distinguished.

In the diagram, the plane is seen edgeways in all cases, and is considered to rotate in an anti-clockwise direction.

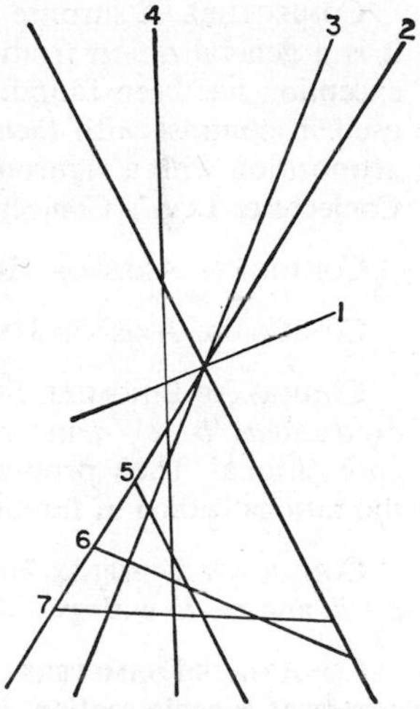

(1) If the plane passes through the apex only, the section is a *point*.

(2) If it passes through the apex and all points on one generating line only, the section is a *straight line*.

(3) If it passes through all points on two generating lines the section is a *pair of intersecting lines* (apex is point of intersection) with maximum possible angle between them, the apex angle of the cone.

(4) If the plane does not pass through the apex, the section may cut both **nappes** of the conical surface to form a **hyperbola**.

(5) When the plane is parallel to a generating line the section is a **parabola**.

(6) If the plane rotates further the section is an **ellipse**.

(7) When the plane is perpendicular to the axis of the cone, the section is a **circle**.

With more rotation of the plane the sequence is repeated in reverse order. If rectangular axes are suitably chosen on the plane, the **Cartesian equations** of all seven are obtained by giving different values to the constants of the general equation of a conic: $ax^2+2hxy+by^2+2gx+2fy+c=0$. If $a=b=h=0$, the *defective equation* $2gx+2fy+c=0$ which results is the equation of a straight line (2). The equations $y^2=4ax$ and $x^2/a^2 \pm y^2/b^2=1$ are *canonical* in that they form the bases for the analytical study of the parabola (5), ellipse (6) and hyperbola (4). If $a=b$, the ellipse becomes circular (7).

See **Focus-Directrix Definition of Conic.**

CONICAL PENDULUM. See **Pendulum, Conical.**

CONICAL PROJECTION. See **Map Projection.**

CONICAL SURFACE. A surface generated by a line (the *generator* or *generatrix*) which passes through a fixed point (the *vertex*) and a fixed closed curve (the *directrix*). If the directrix has a centre, the line through this and the vertex is the axis of the surface. The generator is considered as passing through the vertex and generating twin *nappes* of a complete conical surface. If the directrix is the circumference of a circle and the axis of the surface is perpendicular to the plane of the circle, the surface is a *right-circular conical surface.* Plane sections of such a surface are **conic sections.** A conical surface is both a **developable surface** and a **ruled surface.** See **Element of Conical Surface.**

CONJECTURE. A surmise for which no rigid proof has been found. Since it is a generalization from a set of particular observations for which no exception has been found, the validity is assumed. Conjecture is usually used in contrast with **theorem** when the term theorem is reserved for a proposition with a rigorous proof. See **Four-Colour Problem; Goldbach's Conjecture; Levy's Conjecture.**

CONJUGATE AXES OF HYPERBOLOID. See **Axes of Hyperboloid.**

CONJUGATE AXIS OF HYPERBOLA. See **Axes of Hyperbola.**

CONJUGATE BINOMIAL SURDS. Two binomial **surds** of the form $a\sqrt{b}+c\sqrt{d}$ and $a\sqrt{b}-c\sqrt{d}$, in which a, b, c, d are rational, and $\sqrt{b}$ and $\sqrt{d}$ are not *both* rational. Their product, a^2b-c^2d, is rational. Such pairs are used in the **rationalization of fractions.** See **Rational Number, Quantity.**

CONJUGATE COMPLEX NUMBERS. A pair of **complex numbers** of the form $a+ib$ and $a-ib$ with product a^2+b^2 which is a real number.

CONJUGATE DIAMETERS. A pair of **conjugate lines** that pass through the centre of a **conic section.** If they are given by the equations $y=m_1x$ and $y=m_2x$, where a and b are the semi-major and semi-minor axes of an ellipse or the transverse and conjugate axes of a hyperbola, then $m_1m_2=\pm b^2/a^2$. See **Axes of Ellipse; Axes of Hyperbola.**

CONJUGATE HYPERBOLAS. Two **hyperbolas** for which the transverse and conjugate axes of one are the conjugate and transverse axes respectively of the other. Their standard equations are $x^2/a^2-y^2/b^2=1$ and $x^2/a^2-y^2/b^2=-1$. See **Axes of Hyperbola.**

CONJUGATE IMAGINARY LINES. The **homogeneous equation,** $ax^2+2hxy+by^2=0$, can be written in the form:

$$a[x+(h+\sqrt{h^2-ab})(y/a)][x+(h-\sqrt{h^2-ab})(y/a)]=0.$$

This represents the two straight lines $by=-(h\pm\sqrt{h^2-ab})x$, which pass through the origin. They represent one of the **degenerate conics.** If $h^2>ab$ they represent two generating lines; if $h^2=ab$ they represent two indistinguishable generating lines and the limiting cases of the **ellipse, parabola** and **hyperbola**; if $h^2<ab$ they are a pair of *conjugate imaginary lines.*

CONJUGATE IMAGINARY POINTS. The pair of points in which a circle intersects a straight line which has a perpendicular distance from the centre that is greater than the radius.

CONJUGATE LINES. Two straight lines are conjugate with respect to a **conic section** if each passes through the pole of the other. See **Pole and Polar of Conic.**

CONJUGATE POINTS. Two points are conjugate with respect to a **conic section** if each lies on the polar of the other. See **Pole and Polar of Conic.**

CONJUNCTION. See **Algebra of Propositions.**

CONNECTED SET. If sets P and Q are neither empty sets and each is disjoint from the closure of the other, they are called *separated sets.* A set A, is a connected set if and only if A is not the union of two separated sets. See **Jordan Curve**; **Set Theory.**

CONNECTIVES. See **Algebra of Propositions.**

CONSEQUENT. See **Antecedent, Consequent in Logic**; **Antecedent, Consequent in Ratio and Proportion.**

CONSERVATION OF ENERGY. The general principle in classical mechanics that the total amount of energy in the universe is constant. The principle is understood to apply to energy in all forms, e.g., mechanical energy, heat, light, sound. If a system of particles is in motion under the action of **conservative forces**, the sum of the **kinetic energy** and the **potential energy** of the particles is constant. This is sometimes referred to as the *principle of energy* or the *energy equation.* In modern mechanics the principle has been superseded by one involving the concepts of mass and energy. See **Mass.**

CONSERVATION OF MOMENTUM. The sum of the **vector quantities** $m_1\mathbf{v}_1$ of one body and $m_2\mathbf{v}_2$ of another is unaltered after impact, i.e. $m_1\mathbf{v}_3 + m_2\mathbf{v}_4 = m_1\mathbf{v}_1 + m_2\mathbf{v}_2$. No operation of forces within the system can change the total **momentum.** If the bodies remain in contact after impact, then $(m_1 + m_2)\mathbf{v}_5 = m_1\mathbf{v}_1 + m_2\mathbf{v}_2$, and $\mathbf{v}_5 = (m_1\mathbf{v}_1 + m_2\mathbf{v}_2)/(m_1 + m_2)$. See **Mass.**

CONSERVATIVE FIELD OF FORCE. A field of **force** in which the **work** done in moving a particle from one point to another is independent of the path taken. The earth's gravitational field is a conservative field.

CONSERVATIVE FORCES. If a system of particles moves from some initial position to a standard position under the action of a system of forces, such that the work done depends only on the initial and final positions and not on the manner in which the movement takes place, then the forces are said to form a *conservative field* or system. Examples of such forces are the force due to gravity and the attraction between two particles which is a function of the distance between them. See **Conservation of Energy.**

CONSISTENCY. See **Order Relations.**

CONSISTENT EQUATIONS. **Simultaneous equations** which have a common solution. Their graphs share one or more common points.

CONSTANT. A quantity which permits no variation under specified conditions.

absolute constant. A constant whose value is independent of any conditions; e.g., **numbers.**

arbitrary constant. A constant to which can be assigned different values according to the conditions specified; e.g., **constant of integration, parameter.**

CONSTANT ACCELERATION, SPEED, VELOCITY. See **Uniform (Constant) Acceleration, Speed, Velocity.**

CONSTANT DIFFERENCE SERIES. Synonym of **arithmetic progression.**

CONSTANT OF DILATION. See **Dilation.**

CONSTANT OF GRAVITATION. This is the value of G in the equation due to Sir Isaac Newton: $F = G\ m_1 m_2/r^2$, in which he related the **force** of attraction between two bodies, the product of their **masses** and the square of the distance between their **centres of gravity.** It is thus numerically equal to the force acting between unit masses unit distance apart. The present accepted value is 6.67×10^{-8} in the **c.g.s. system**; and its dimensions in length, mass and time are $\mathbf{L^3 M^{-1} T^{-2}}$.

CONSTANT OF INTEGRATION. The arbitrary **constant** added to any expression arising from **integration.**

CONSTANT OF INVERSION. See **Inversion.**

CONSTANT RATIO SERIES. Synonym of **geometric progression.**

CONSTANT TERM. In any algebraic expression, the term not containing the **variable**; e.g., $y = mx + n$, n being the constant term.

CONSTRUCTION. The drawing of a figure to specification. It is a step in the formal proofs of geometric theorems and riders, often a visual aid to correct procedure.

CONTINUED FRACTION. An integer and a fraction, the denominator of which is also an integer and a fraction, etc. Its general form is

$$a + \cfrac{b}{c + \cfrac{d}{e + \cfrac{f}{g \text{ etc.}}}} \quad \text{for convenience written } a + \frac{b}{c+}\,\frac{d}{e+}\,\frac{f}{g+} \text{ etc.}$$

Such fractions may have a finite number of terms (*terminating*) or an infinite number (*non-terminating*). In certain continued fractions repetitive patterns of digits occur and they are called *recurring* (or *periodic*) continued fractions. The **transcendental numbers e** and π can be expressed in the form of continued fractions. See **Pi, Π, π.**

CONTINUED PRODUCT. A product of more than two factors. It is denoted by $\prod_{r=1}^{r=n} u_r$, which is equivalent to $u_1\ u_2\ u_3 \ldots u_r \ldots u_n$. The **transcendental**

number π may be expressed in the form of a continued product with an infinite number of factors. See **Pi, Π, π.**

CONTINUITY OF FUNCTION. See **Continuous Function.**

CONTINUOUS FUNCTION. If a **function** $f(x)$ tends towards $f(a)$ when x tends towards a from the values $x = a \pm \epsilon$, ϵ being any quantity, however small, then the function $f(x)$ is said to be continuous when $x = a$. A function is continuous in the interval $x = a$ to $x = b$ if it is continuous for all the values x can assume between a and b. A function which is not continuous is *discontinuous*.

CONTINUUM. On a line with an origin, but no end points, every point corresponds to some real **number**; and to every number there corresponds some point on the line. The points on the line are a continuum of one dimension. The continuum of real numbers is thus an **ordered set** as dense as the points on a line. In the same manner we regard time as a continuum, and the combination of these continuums give us our conception of motion.

CONTOURS. Curves drawn on a map through points of the same elevation above some **datum**. They are determined for integral values and at regular intervals, known as the contour or *vertical interval* (V.I.). The corresponding horizontal distance is called the *horizontal equivalent* (H.E.). The angle of slope between successive contours is given by $\tan \alpha = \text{V.I./H.E.}$, the gradient of the incline. See **Angular Levelling.**

CONTRAPOSITIVE. For the conditional 'if p then q', the implication 'if not q, then not p' is the logically equivalent contrapositive statement. See **Algebra of Propositions.**

CONVERGENCE, DIVERGENCE OF SEQUENCE. If (x_n) is an ordered **sequence**, related to a definite number N in such a way that $(x_n - N)$ is a **null sequence**, it is said to converge or to be convergent, and N is called its limiting value. Every sequence not convergent in this sense is said to diverge or to be divergent: it can have no arbitrary constant as a limiting value. If $x_n \to N$ as $n \to \infty$, (or, if $\lim_{n\to\infty} x_n = N$), the sequence is *definitely* (*properly*) *convergent*; if $x_n \to \pm\infty$ as $n \to \infty$ (or, if $\lim_{n\to\infty} x_n = \pm\infty$), the sequence is *definitely* (*properly*) *divergent*. All other sequences are said to be indefinitely divergent or, simply, indefinite. A series formed from a divergent sequence can be summed for only a finite number of terms. If a sequence is not convergent nor definitely divergent it is said to be *oscillating*. If, in the case of an oscillating sequence, there is a constant C for which the numerical values of the partial sum $|S_n| < C$ for all n, then the sequence is said to oscillate finitely; in other cases it oscillates infinitely. See **Comparison Test; Partial Sums of Sequence; Ratio Tests.**

CONVERGENCE, DIVERGENCE OF SERIES. A series, Σa^n, is convergent, definitely divergent, indefinite (indefinitely divergent) according as the sequence of the terms in its partial sums shows the behaviour indicated by these terms. See **Absolute Convergence; Comparison Test; Convergence, Divergence of Sequence; Partial Sums of Sequence; Ratio Tests.**

CONVERSE. For a given statement "if p then q", the converse statement is "if q then p". These are not *logically* equivalent statements. If a set of parallel lines cuts off equal intercepts on one line then the set cuts off equal intercepts on any other line. The converse of this is not true. The converse of **Pythagoras's theorem** is true.

CONVERSION GRAPH. A graph used for the conversion of one scale of measure into another. One scale is marked along the x-axis, the other along the y-axis. A line is drawn through chosen points, each of which represents equivalent values on the scales.

CONVERTENDO. See **Proportion.**

CONVEX. See **Concave, Convex.**

COORDINATE GEOMETRY. Synonym of **analytic geometry.**

COORDINATES OF POINT. Any **ordered set** of numbers which determines the position of a point in space in reference to a **frame of reference.** See **Analytic Geometry; Areal, Cartesian, Cylindrical, Spherical, Trilinear Coordinates.**

COPLANAR LINES. Lines which lie in the same plane.

COPLANAR POINTS. Points which lie in the same plane.

COPUNCTAL PLANES. Three or more planes sharing a common point.

COROLLARY. A statement appended to a **proposition (theorem)** without **proof** because it is immediately deducible from the proof already given.

CORRECT TO n DECIMAL PLACES. A decimal **fraction** expressed with n decimal places where the nth place digit has an assessed value determined by the $(n+1)$th place digit. If this $(n+1)$th place digit is five or more, the nth place digit is increased by one. For example, $\pi = 3.141\ 59\ldots$ is 3.141 6 correct to four places of decimals. See **Approximation; Significant Figures.**

CORRELATION. A term used in statistics for the relation between two **variables**. A concept similar to that of **function**. If the two variables increase or decrease simultaneously the correlation is positive; if one increases while the other decreases, it is negative. See **Coefficient of Correlation.**

CORRELATION COEFFICIENT. See **Coefficient of Correlation.**

CORRESPONDENCE. Synonym of **transformation** or **mapping.** The relationship between the elements in one set of things and the elements in another set or between some elements of one set and other elements of the same set.

When each element of set A is associated with no element or one element of set B, it is a *functional correspondence*. If an element of set A is associated with more than one element of set B, it is a **multi-valued function.** See **Function; One-One Correspondence.**

CORRESPONDING ANGLES. See **Transversal.**

COS. Abr. cosine. See **Trigonometric Ratios.**

COSEC. Abr. cosecant. See **Trigonometric Ratios.**

COSECANT, HYPERBOLIC (csch, cosech). See **Hyperbolic Functions.**

COSECANT RATIO (csc, cosec). See **Trigonometric Ratios.**

COSECH. Abr. hyperbolic cosecant. See **Hyperbolic Functions.**

COSET. The set of all products sa (right coset) or as (left coset), where s is an element of subgroup S of group G, and a is a fixed element of G. See **Group Theory.**

COSH. Abr. hyperbolic cosine. See **Hyperbolic Functions.**

COSINE FORMULAE. In any triangle $AcBaCbA$, $\cos A = (b^2 + c^2 - a^2)/2bc$, and expressions for $\cos B$ and $\cos C$ can be formed by exchange of letters in **cyclic order**. The equation can be transformed into $a^2 = b^2 + c^2 - 2bc \cos A$ with transformed similar expressions for b^2 and c^2. See **Extension of Pythagoras's theorem.**

COSINE, HYPERBOLIC (cosh), $\frac{1}{2}(e^x + e^{-x})$. See **Hyperbolic Functions.**

COSINE RATIO (cos). See **Trigonometric Ratios.**

COT. Abr. cotangent. See **Trigonometric Ratios.**

COTH. Abr. hyperbolic cotangent. See **Hyperbolic Functions.**

COTANGENT, HYPERBOLIC (coth, ctnh). See **Hyperbolic Functions.**

COTANGENT RATIO (cot). See **Trigonometric Ratios.**

COTERMINAL ANGLES. Angles of sizes $(\theta \pm n \cdot 360)$ degrees, where n is any integer. Their arms are the same pair of **initial and terminal lines.**

COULOMB. Unit of quantity of electricity. The quantity of electricity transferred by 1 **ampere** in 1 second.

COUNTABLE SET. Synonym of **denumerable set.**

COUNTABLY INFINITE SET. Any set which can be put into **one-one correspondence** with the set of positive integers. Any such set can be put into a one-one correspondence with any proper subset of itself. This theorem is often known as the *paradox of Galileo*. It is only apparent; thus, the set of positive integers $J^+ = 1, 2, 3, \ldots$ which is countably infinite can be put in one-one correspondence with the set of even integers $2, 4, 6, \ldots$, a proper subset of J^+. See **Countable Set; Set Theory.**

COUNTER-CLOCKWISE. See **Clockwise, Anti-clockwise (Counter-clockwise).**

COUNTING. The process of finding the **cardinal number** of a set by mentally placing the elements of the set into a **one-one correspondence** with the elements of a known **ordered set.** The **natural numbers** are the most widely used set. See **Countable Set.**

COUPLE IN ARITHMETIC. One of many names for two. Others are pair, brace, deuce. Twin and yoke have special connotations.

COUPLE IN MECHANICS. Two equal forces acting in opposite directions and not in the same straight line form a couple if they act upon the same body. The couple causes the body to turn about an axis perpendicular to the plane of the forces. If either of the forces is F, and the arm of the couple is d, the moment of the couple about *any* point in the plane is Fd. See **Moment of Force; Torque.**

CRAMER'S RULE. A rule using **determinant** notation for the solution of simultaneous equations. Given n independent equations involving n unknowns:

$$\begin{array}{l} a_1x_1 + a_2x_2 + a_3x_3 + \ldots + a_nx_n = k_a \\ b_1x_1 + b_2x_2 + b_3x_3 + \ldots + b_nx_n = k_b \\ \ldots\ldots\ldots\ldots\ldots\ldots\ldots\ldots \\ p_1x_1 + p_2x_2 + p_3x_3 + \ldots + p_nx_n = k_p \\ \ldots\ldots\ldots\ldots\ldots\ldots\ldots\ldots \end{array}$$

then, if $D = \begin{vmatrix} a_1\, a_2\, a_3 \ldots a_n \\ b_1\, b_2\, b_3 \ldots b_n \\ \ldots\ldots\ldots \\ p_1\, p_2\, p_3 \ldots p_n \\ \ldots\ldots\ldots \end{vmatrix} \neq 0$, and $D_1 = \begin{vmatrix} k_a\, a_2\, a_3 \ldots a_n \\ k_b\, b_2\, b_3 \ldots b_n \\ \ldots\ldots\ldots \\ k_p\, p_2\, p_3 \ldots p_n \\ \ldots\ldots\ldots \end{vmatrix}$

$D_2 = \begin{vmatrix} a_1\, k_a\, a_3 \ldots a_n \\ b_1\, k_b\, b_3 \ldots b_n \\ \ldots\ldots\ldots \\ p_1\, k_p\, p_3 \ldots p_n \\ \ldots\ldots\ldots \end{vmatrix}$ and $D_3 = \begin{vmatrix} a_1\, a_2\, k_a \ldots a_n \\ b_1\, b_2\, k_b \ldots b_n \\ \ldots\ldots\ldots \\ p_1\, p_2\, k_p \ldots p_n \\ \ldots\ldots\ldots \end{vmatrix}$ etc.

The solution set is: $x_1 = D_1/D$, $x_2 = D_2/D$, $x_3 = D_3/D$, etc. Example:

If $2x + 3y = 18$ and $4x + 5y = 32$, $D = \begin{vmatrix} 2 & 3 \\ 4 & 5 \end{vmatrix} = -2$, $D_1 = \begin{vmatrix} 18 & 3 \\ 32 & 5 \end{vmatrix} = -6$ and $D_2 = \begin{vmatrix} 2 & 18 \\ 4 & 32 \end{vmatrix} = -8$. Hence $x = -6/-2 = 3$; $y = -8/-2 = 4$.

CRITICAL POINT. Any point on the curve of $y = f(x)$ where the **gradient of curve**, dy/dx, is zero or infinite, and the tangent is parallel to the x-axis or the y-axis respectively.

CROSS MULTIPLICATION. The process whereby the proportion, equality of ratios or equality of fractions $a/b=c/d$ is replaced by the equivalent statement $a\ d=c\ b$.

CROSS PRODUCT. Synonym of **vector product.**

CROSS RATIO. Synonym of **anharmonic ratio.**

CROSS-SECTION. The geometric figure made when a solid is cut by a plane, or a plane figure by a straight line, normally perpendicular to an axis. See **Section.**

CROWD ABOUT. A term used to describe the tendency of a sequence of points or values to increase in density as a certain point or value is approached. The sequence, set $N \equiv \{1, \frac{1}{2}, \frac{1}{4}, \frac{1}{8}, \frac{1}{16}, \frac{1}{32}, \ldots\}$, can be represented on a number line as shown. The points crowd about the zero point. The **neighbourhood** of zero can be made smaller by excluding points from set N, the number of points being always finite. Zero is said to be the limit, cluster or **accumulation point** of the set N.

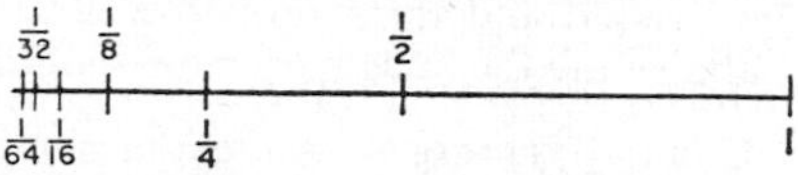

CRUNODE. A **double point** on a curve, at which the tangents are real and distinct. Every **node of curve** is a crunode.

CSC. Abr. cosec, cosecant. See **Trigonometric Ratios.**

CSCH. Abr. hyperbolic cosecant. See **Hyperbolic Functions.**

CTN. Abr. cotangent. See **Trigonometric Ratios.**

CTNH. Abr. hyperbolic cotangent. See **Hyperbolic Functions.**

CUBE IN ALGEBRA. The third power, represented by the index (exponent) 3, as x^3, the cube of x, standing for the continued product $x \cdot x \cdot x$. See **Power of Quantity.**

CUBE IN GEOMETRY. A regular hexahedron; a **regular polyhedron** with 6 congruent square faces, 8 vertices and 12 edges. A cube with edge of unit length (a *unit cube*) is used as a unit of **volume.**

CUBE ROOT. A quantity, the cube of which is a given quantity. The solution of the equation $x^3=n$, written $\sqrt[3]{n}$ or $n^{\frac{1}{3}}$, which is positive or negative as n is positive or negative respectively. See **Evolution.**

CUBIC. Having the properties of a **cube in algebra** or **cube in geometry.**

CUBIC CAPACITY. See **Volume.**

CUBIC EQUATION. An equation of the third degree of the form $a_0x^3 + a_1x^2 + a_2x + a_3 = 0$ where $a_0 \neq 0$. By substituting $(x - a_1/3a_0)$ for x in the above equation it reduces to $x^3 + qx + r = 0$, where $q = (3a_0a_2 - a_1^2)/3a_0^2$ and $r = (4a_1^3 - 9a_0a_1a_2 + 27a_0^2a_3)/27a_0^2$. This simplified form with the square of the unknown missing is used for expressing the three roots of the cubic. The solution was known to Scipio Ferreo in 1505. Having obtained it from Tartaglia, Cardan published it under his own name in his *Ars Magna* in 1545.

If one solution is given by $x=y+z$, $x^3=y^3+z^3+3(y+z)yz$. The simpler equation now becomes: $y^3+z^3+(3yz+q)x+r=0$. If y and z satisfy the equation $3yz+q=0$, they are determinable:

$$y^3+z^3=-r \text{ and } y^3z^3=-q^3/27.$$

Their cubes are the roots of the quadratic equation: $u^2+ru-q^3/27=0$.

Therefore, $x=\left\{-\frac{r}{2}+\sqrt{\frac{r^2}{4}+\frac{q^3}{27}}\right\}^{\frac{1}{3}}+\left\{-\frac{r}{2}-\sqrt{\frac{r^2}{4}+\frac{q^3}{27}}\right\}^{\frac{1}{3}}.$

The second and third roots are given by $\omega y+\omega^2 z$ and $\omega^2 y+\omega z$, where ω and ω^2 are the imaginary cube **roots of unity**.

CUBIC EXPRESSION. An expression of the third degree of the form $a_0x^3+a_1x^2+a_2x+a_3$ in which $a_0\neq 0$.

CUBIC INCH (in^3). The unit of **volume** equal to that of a cube with a one-inch edge (one-inch cube).

CUBIC MEASURE. The measurement of volume based on 1 ft^3 in the **f.p.s. system** and 1 cm^3 in the **c.g.s. system**.

CUBIC NUMBERS. 1, 8, 27, 64, . . . **Numbers** which can be arranged in a cubic (geometric) array. Any cubic number, n^3, is the sum of n consecutive odd numbers beginning with $[n(n-1)+1]$. Thus:

$$8=2^3=3+5,$$
$$27=3^3=7+9+11,$$
$$64=4^3=13+15+17+19.$$

CUBIC PARABOLA. The curve which is the graph of the equation $ay=x^3$ or of the equation $bx=y^3$.

CUBOID. A hexahedron; a **polyhedron** with 6 rectangular faces, 8 vertices and 12 edges; a rectangular **parallelepiped.**

CUP IN SET THEORY. $A\cup B$, read as A cup B, meaning the union of set A and set B. See **Set Theory**.

CUBIT. An ancient measure (Egyptian, Mesopotamian, Biblical) of the length from the elbow to the tip of the middle finger. It varied between the limits of 18 and 22 inches.

CURVATURE. The total curvature of an arc δs is defined as the angle $\delta\psi$ between tangents drawn at its extremities in the same direction with respect to the arc. If the tangents make angles ψ_1, ψ_2 with the positive direction of the x-axis, $\delta\psi=|\psi_1-\psi_2|$. The *mean curvature* along the arc is defined as $\delta\psi/\delta s$. The *curvature at a point* on the arc is the mean curvature of the infinitesimal arc containing the point, and is defined as $d\psi/ds$.

centre of curvature. The centre of the *circle of curvature*: the point of intersection of two consecutive normals to a curve.

circle of curvature. A circle of radius $r = ds/d\psi$, having the same tangent as the curve at a given point, with concavity turned the same way. It is sometimes called the *osculating circle*. Its centre is the *centre of curvature*. The locus of the centres of curvature is called the *evolute* of the curve.

curvature of circle. Since in any circle, $\delta s = r\delta\psi$, the curvature, $d\psi/ds$, at any point is the constant $1/r$. The curvature of a point circle is thus infinite, and the curvature approaches zero as r approaches ∞. The ratio of the curvature of circle A, with radius a to the curvature of circle B, with radius b, is b/a.

curvature of curve. A curve is both the **locus** of a point describing some segmental arc and the **envelope** of a tangential straight line. As the point moves along the arc, the tangent rotates about the point. The curvature of the curve combines the concepts of both movements. If at any point $r \to \infty$ (and $d\psi/ds \to 0$), the rotation of the tangent is arrested and a stationary tangent occurs, defining a point of **inflexion** at the point. If $r \to 0$ (and $d\psi/ds \to \infty$), the motion of the point is arrested and a stationary point occurs, defining a **cusp** at the point.

curvature of straight line. Since $\psi_2 = \psi_1$, $\delta\psi = 0$ at *all* points and $d\psi/ds = 0$. The mean and limiting curvature of a straight line is zero. (Cf. *curvature of circle*.)

curvature of surface. For a sphere of radius R, the curvature is $1/R^2$ at any point on its surface. For a general surface, the curvature at a point P is that of the sphere whose surface approximates most closely to that of the surface at the point P.

radius of curvature. The radius of the *circle of curvature* (osculating circle).

CURVATURE OF EARTH. Outside the polar regions the sea level surface of the earth approximates very nearly to a spherical surface. It departs from the tangential plane at a place:

in the first mile 7.95 inches,
in the second mile 31.83 inches,
in the third mile 71.63 inches,
. . . . . .
in the tenth mile 22 yards.

The variation is directly proportional to the square of the distance. For any survey work involving telescopic sighting this must be taken into account and the triangles used will be **spherical triangles**.

CURVE. A one-dimensional configuration obeying some law. If the law is expressed geometrically, it is a **locus**, or a set of points; if algebraically, the

graph of a function. If the function contains two variables, the curve will lie in a plane and is described as a plane curve; if there are three variables, the function gives a space curve. If the curve has no end-points it is said to be *closed*; and if it does not cross itself, it is *simple-closed*. A space curve which does not lie in a plane is called a *skew* or *twisted* curve. For example, a circle drawn on a sphere is a plane spatial curve; an ellipse projected on a sphere is a skew spatial curve. See **Closed Curve**; **Unicursal Curve**.

CURVE TRACING. The process of plotting points to produce the graph of a **function or relation**. Use is made of the techniques of algebra and calculus with regard to such properties of the function as **curvature**, **limits** and **symmetry**.

CURVILINEAR MOTION. **Motion** along a **curve**. See **Velocity**.

CUSP. Synonym of *spinode*. A **double point** on a curve at which the tangents are real and coincident. It can be (*a*) a single cusp of the first species, (*b*) a single cusp of the second species, (*c*) a double cusp of the first species or point of osculation, (*d*) a double cusp of the second species, (*e*) a double cusp of both species, a point of inflexion and osculation. A cusp separates a pair of arches in the **cycloid**. At the horns of a crescent moon, the points of intersection of a circle and an ellipse with one line segment as diameter and major axis, there are coincident tangents to the circle and the ellipse. The horns are not cusps since *two* curves are involved.

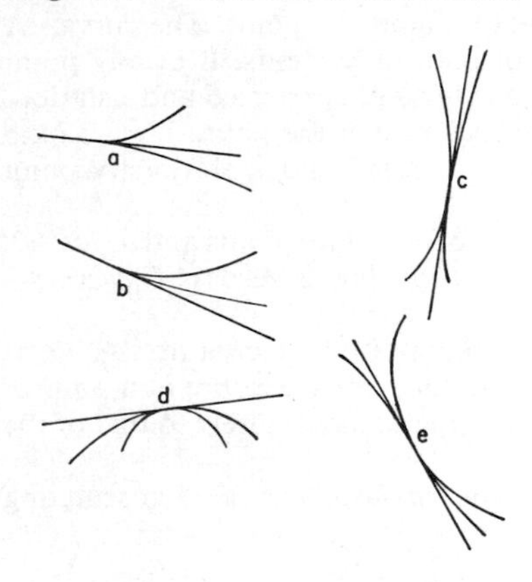

CYBERNETICS. The theory, originated in 1948 by the American Wiener, which studies the mathematical structure of control and communication in such diverse systems as automata, computers and control mechanisms. It is based on the attempts made to find a common structure for naturally occurring systems in the field of biology and those designed in the field of engineering. One important aspect of the study is that of *feedback* in which a system can judge its own effectiveness in achieving a goal and then modify its subsequent actions.

CYCLE. (1) A sequence of events that repeats itself. (2) A revolution as in *cycles per second*. (3) Synonym of **cyclic permutation**.

CYCLIC GROUP. See **Group Theory**.

CYCLIC ORDER. The way in which a **cyclic permutation** is performed. The elements of an **ordered set** are, for convenience, placed round a circle. The advancement of terms is then easily achieved.

CYCLIC PERMUTATION. The advancing of each element of an **ordered set** of elements one position, the last element taking the place of the first. The number of elements in the set being the *degree of cyclic permutation.* When the degree is two, the process is referred to as **transposition**. See **Cyclic Order**.

CYCLIC QUADRILATERAL. A quadrilateral with its vertices on the circumference of the same circle. See **Angle Properties of Circle**; **Ptolemy's Theorem.**

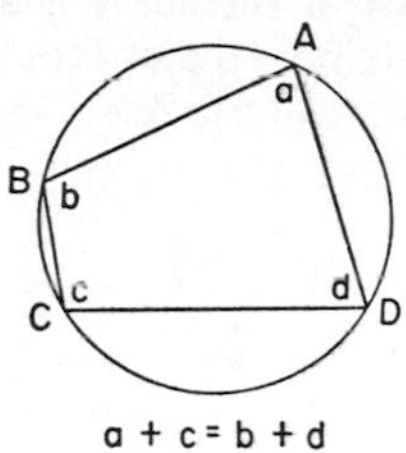

a + c = b + d

CYCLOID. The locus of point P on the circumference of a circle which rolls without slipping along a straight line. If the centre of the circle at any instant is C, the radius a and the angle of rotation between the vertical line through C and CP is θ, the **parametric equations** of the cycloid are: $x=a(\theta-\sin\theta)$, $y=a(1-\cos\theta)$. A **cusp** occurs whenever the point P touches the base line. Between successive cusps, the distance along the base line is $2\pi a$, the circumference of the circle, and the distance along the cycloid is $8a$. The area of the cycloidal arch is $3\pi a^2$, three times the area of the rolling circle. If point P_1 on circle S_1 with radius a performs 1 cycloidal arch between A and B, points $P_2, P_3, P_4, \ldots$ on circles $S_2, S_3, S_4, \ldots$ with radii $a/2, a/3, a/4, \ldots$ will perform 2, 3, 4, . . . arches between A and B. The total cycloidal distances travelled by $P_1, P_2, P_3, P_4, \ldots$ are each equal to $8a$.

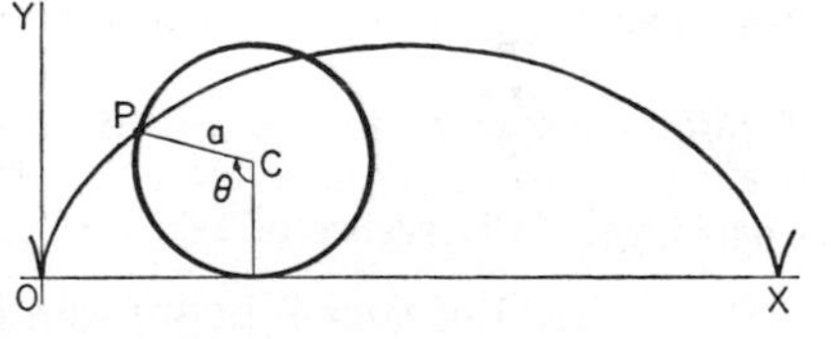

CYCLOSYMMETRIC. See **Symmetric Function.**

CYLINDER. The solid within a **cylindrical surface** and between two plane end surfaces. If it is *right circular*, of length (height) l and diameter d, its mensuration is: curved area, πdl; plane area, $\pi(d/2)^2$; total area, $\frac{1}{2}\pi d(2l+d)$; volume $\pi(\frac{1}{2}d)^2 l$ or $\frac{1}{4}\pi d^2 l$. See **Element of Cylinder.**

CYLINDRICAL COORDINATES. Coordinates of a point, (r, θ, z), which combine **spherical coordinates in space** and **rectangular coordinates**, (r,θ) being the polar coordinates of a point in the xy-plane and z its perpendicular distance from the xy-plane. The coordinates are described as cylindrical since the point describes a **cylindrical surface** when r is constant. The cylindrical and rectangular coordinates have the following relationships: $r=\sqrt{(x^2+y^2)}$, $\theta=\tan^{-1}(x/y)$, $z=z$, $y=r \sin \theta$, $x=r \cos \theta$.

CYLINDRICAL PROJECTION. See **Map Projection.**

CYLINDRICAL SURFACE. One generated by a line (the *generator* or *generatrix*), not in the plane of a closed curve but passing through every point of the curve (the *directrix*), and maintaining one spatial direction. If the curve has a centre, the line through this, parallel to a generating line, is the axis of the surface: if the curve is a circle and the axis is perpendicular to the plane of the circle, the surface is a *right-circular cylindrical surface.* See **Element of Cylindrical Surface.**

D

D. (1) Roman symbol, 500. (2) **Differential operator**: $Dy \equiv \frac{dy}{dx}$ in **derived function.**

D'ALEMBERT'S TEST. The generalized **ratio test** for convergence or divergence of an infinite series. The series $\sum_{r=1}^{\infty} u_r$ is convergent or divergent if, after some term, the **numerical value** of the ratio (u_r/u_{r-1}) is respectively less than or greater than some fixed number less than unity. See **Convergence, Divergence of Series.**

DATA. Plural of *datum*, Latin, with original meaning of 'that which is given'. The information which is accepted as a basis for establishing **propositions, postulates,** deductions, conclusions, **proofs,** etc. The information may be a set of measurements for the purpose of **statistical analysis** or a geometrical configuration together with certain assumptions from which a particular proposition is to be proved. The data may consist of a set of premises on which some logical argument is based. In all cases, the data are accepted as true within the context of the treatment to which they are subjected.

DATUM. **Elevations** of points on the earth's surface are referred to some datum or zero elevation. This may be an assumed horizontal plane called a local datum, or a **spherical surface** such as mean sea level at Liverpool as used in the Ordnance Survey. Sometimes datum refers to a direction of reference as in **inclination.** See **Data.**

DAY. See **Mean Solar Day**; **Sidereal Day**; **Sidereal Year**; **Calendar.**

DECAGON. A **polygon** with ten sides.

DECELERATION. Negative **acceleration.**

DECI-. See **Metric System.**

DECIBEL. Unit of intensity of sound based on a **logarithmic scale** and equal to one-tenth of a bel which is normally a unit too large for practical purposes.

DECIMAL FRACTION. See **Fraction.**

DECIMAL PLACE. See **Denary Number System.**

DECIMAL POINT. See **Denary Number System.**

DECIMAL SYSTEM. (1) Synonym of **denary number system.** (2) Any system of measurement in which all the units can be expressed as powers of ten of any one unit in the system, e.g., the **metric system.**

DECIMALIZATION. The changing of ordinary quantities of measure into a number of some unit and a decimal fraction of this unit to represent the smaller quantities. Thus 5.861 yards for 5 yards 2 feet 7 inches.

DECLINATION IN ASTRONOMY. See **Angle of Declination.**

DECLINATION IN NAVIGATION. See **Magnetic Declination.**

DECLINATION IN SURVEYING. See **Angle of Inclination.**

DECLIVITY. See **Angular Levelling.**

DEDEKIND CUT. A partition of the set of **rational numbers** into two classes by a separator which is not a member of the set. This is a *separation of the second kind* and it permits the extension of the concept of **numbers** to embrace the irrational numbers and produces the concept of the *continuum* of real numbers. See **Separation (1).**

DEDUCTION. A conclusion drawn from **data** by a process of reasoning.

DEFECTIVE EQUATION. One derived from a given one and, in the process, becoming of lower order. At least one **root** cannot be determined. For example, $x^2 - x - 2 = 0$ has two roots, -1 and 2. If it is presented as $x^2 - 4 = x - 2$, then wrongly simplified to become $x + 2 = 0$, this is a defective equation *not* revealing the solution $x = 2$. See **Conic Section; Order of Equation.**

DEFECTIVE NUMBER. See **Perfect Number.**

DEFICIENT NUMBER. See **Perfect Number.**

DEFINITE INTEGRAL. The definite **integral of function,** $f(x)$, over the domain from $x = a$ to $x = b$, is the area enclosed by the curve $y = f(x)$, the x-axis, and the ordinates $x = a$ and $x = b$. Any area above the x-axis is positive, and below, negative. For example the definite integral of x^2 between $x = 2$ and $x = 3$ is written as

$$\int_2^3 x^2 dx = \left[\frac{x^3}{3}\right]_2^3 = [3^3/3 - 2^3/3] = 6\tfrac{1}{3}.$$

The quantity to be integrated, here x^2, is called the *integrand.* In the domain, 3 is called the *upper limit* and 2 the *lower limit*. See **Area under Curve.**

DEFINITE INTEGRATION. The process of finding the **definite integral** of a function, $f(x)$, when *limits of integration* are assigned to the independent **variable** x.

DEFINITION. The description of a new term or idea, using for the purpose only ideas accepted but unproved (**axioms, postulates**) and previously defined terms, **symbols** and **operations.** See **Proof.**

DEGENERATE CONIC. Term used to describe those **conic sections** which are neither hyperbolic, parabolic nor elliptical.

DEGREE IN ALGEBRA. The degree of a term in one **variable** is the **index** (exponent) of the variable; thus x^3 is of the third degree. The number of times a variable is used as a factor to produce a term in one variable. The sum of all the factors making a term in an expression; thus $ab^2c^3d^4$ is a term of the tenth degree. See **Homogeneous Function; Polynomial Expression; Power of Quantity.**

DEGREE IN GEOMETRY. A rotation of one three-hundred-and-sixtieth part of one revolution. See **Angle (Plane Angle).**

DEGREE IN NAVIGATION. One three-hundred-and-sixtieth part of any **meridian of longitude** measures one degree of latitude. One three-hundred-and-sixtieth part of any **parallel of latitude** measures in that latitude one degree of longitude.

DEGREE OF CYCLIC PERMUTATION. See **Cyclic Permutation.**

DEGREE OF DERIVATIVE. See **Derivatives.**

DEGREE OF DIFFERENTIAL EQUATION. See **Order of Differential Equation.**

DEGREE OF HOMOGENEITY. See **Homogeneous Function.**

DEGREE OF POLYNOMIAL. See **Polynomial Expression.**

DEKA-. See **Metric System.**

DE MOIVRE'S THEOREM. For every value of n, positive or negative, integral or fractional, one of the values of $(\cos\theta + i\sin\theta)^n$ is $\cos n\theta + i\sin n\theta$.

DE MORGAN'S LAW. The complement of the union of subsets $A, B, C, \ldots$ of set S is the intersection of the complements of $A, B, C, \ldots$ It can be written in the form $(A\cup B\cup C\cup \ldots)' = A'\cap B'\cap C'\cap \ldots$. See **Algebra of Sets; Set Theory.**

DENARY FRACTION. A common **fraction** using digits in the **denary number system.**

DENARY NUMBER SYSTEM. A **system of notation** using 10 as its base, together with place value assigned to its digits 0, 1, 2, 3, 4, 5, 6, 7, 8, 9. This is sometimes referred to as the common scale of notation or the decimal system, though the latter term should be reserved for the decimal **fraction** part of the denary number system, when place values are used. The integral part of a number is separated from the fractional part by the *decimal point*. Thus 3 247.569 is a conventional representation of

$$3\cdot 10^3+2\cdot 10^2+4\cdot 10^1+7\cdot 10^0+5\cdot 10^{-1}+6\cdot 10^{-2}+9\cdot 10^{-3},$$

place values being thousands, hundreds, tens, units, tenths, hundredths, thousandths respectively. Places after the decimal point are referred to as first place, second place, etc. See **Hindu-Arabic Numerals**; **Base (Radix) in Number System.**

DENIAL. See **Algebra of Propositions.**

DENOMINATE NUMBER. A number of units of measure.

DENOMINATOR. The term D in the common **fraction** N/D.

DENSITY OF SET. See **Order Relations.**

DENSITY OF SUBSTANCE. In a homogeneous body this is the **mass** of unit volume. For water it is one gramme per cubic centimetre (1 000 oz or $62\frac{1}{2}$ lb/ft^3). Mercury which is 13.596 times as heavy has a density of 13.596 g/cm^3 ($849\frac{3}{4}$ lb/ft^3). In the metric system density is numerically the same as **specific gravity** (relative density).

DENUMERABLE SET. Synonym of *countable set.* A set, the elements of which can be put into **one-one correspondence** with the positive integers. For example, the rational numbers can be placed in an array which shows them to be denumerably infinite. See **Inverse Function**; **Set Theory.**

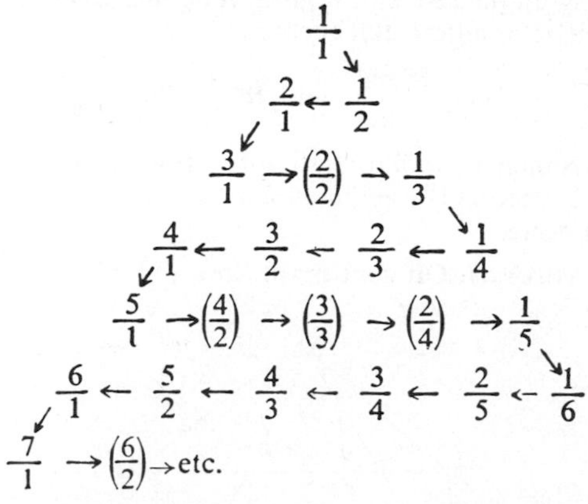

$$\frac{1}{1}\searrow$$
$$\frac{2}{1}\leftarrow\frac{1}{2}$$
$$\swarrow\frac{3}{1}\rightarrow\left(\frac{2}{2}\right)\rightarrow\frac{1}{3}\searrow$$
$$\frac{4}{1}\leftarrow\frac{3}{2}\leftarrow\frac{2}{3}\leftarrow\frac{1}{4}$$
$$\swarrow\frac{5}{1}\rightarrow\left(\frac{4}{2}\right)\rightarrow\left(\frac{3}{3}\right)\rightarrow\left(\frac{2}{4}\right)\rightarrow\frac{1}{5}\searrow$$
$$\frac{6}{1}\leftarrow\frac{5}{2}\leftarrow\frac{4}{3}\leftarrow\frac{3}{4}\leftarrow\frac{2}{5}\leftarrow\frac{1}{6}$$
$$\swarrow\frac{7}{1}\rightarrow\left(\frac{6}{2}\right)\rightarrow\text{etc.}$$

DEPENDENT, INDEPENDENT VARIABLE. When a **function** is defined as a **correspondence** from a set of values of one **variable** to a set of values of another, the first is called the *independent* variable and the second the *dependent*. Thus, for the function $y=x^2$, in general, x is referred to as the independent variable and y as the dependent variable.

DERIVATION. The process of obtaining the derivative of a function.

DERIVATIVE OF PRODUCT. The **derivative** of $y=u\ v$ where u and v are functions of x is $dy/dx=u(dv/dx)+v(du/dx)$.

DERIVATIVE OF QUOTIENT. The **derivative** of $y=u/v$ where u and v are functions of x is $dy/dx=[v(du/dx)\ -u(dv/dx)]/v^2$.

DERIVATIVES. **Differential coefficients** or **derived functions**. The second derivative is the derivative of the first derivative, the third, the derivative of the second, and so on. Thus velocity, the first derivative of distance with respect to time, is ds/dt; and acceleration, dv/dt, is also $d(ds/dt)/dt$, written d^2s/dt^2, the second derivative of distance with respect to time. Theoretically the process is interminable, and the nth derivative of y with respect to x would be written d^ny/dx^n, where n is the *degree* of the derivative.

DERIVED CURVE. A curve whose equation is a **derived equation** obtained by differentiating the equation of a given curve once (first derived curve), twice (second derived curve), etc. Thus for a given curve with equation $y=x^3$, the first and second derived curves have equations $y=3x^2$ and $y=6x$ respectively.

DERIVED EQUATION. (1) In *algebra*, any equation derived from a given equation by use of rules for the modification of equations, e.g., **transposition of terms**, squaring of both sides, etc. (2) In *calculus*, an equation obtained by differentiating a given equation once (first derived equation), twice (second derived equation), etc. See **Derived Curve**.

DERIVED FUNCTION. That function, all terms of which have been found by differentiating all the terms of a given function. If $y=f(x)$, the derived function (**derivative** or **differential coefficient**), is written:

$$D_xy,\ Dy,\ \frac{dy}{dx},\ y',\ f'(x),\ \frac{df(x)}{dx},\ \text{and}\ D^2{}_xy,\ D^2y,\ \frac{d^2y}{dx^2},\ y'',\ f''(x),\ \frac{d^2f(x)}{dx^2},$$

according to whether the differentiation has been performed once, twice, etc. The derived function is used, *inter alia*, as a measure of **rate of change** and **gradient of curve**.

DESARGUES' THEOREM. On **concurrent** lines OA, OB, OC are points A',

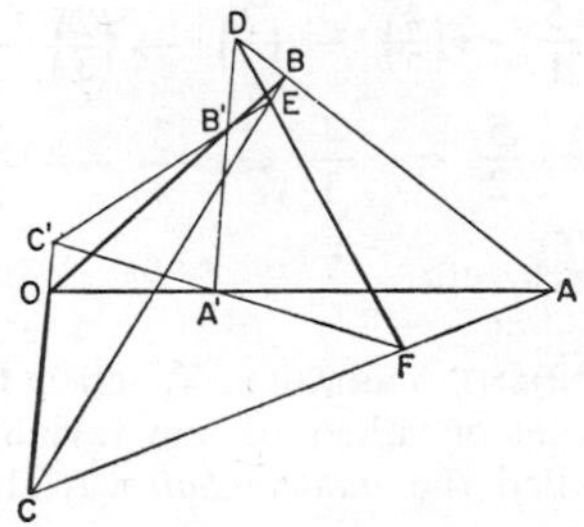

B', C' respectively. If the joins AB and $A'B'$ intersect in D, BC and $B'C'$ at E, CA and $C'A'$ at F, then D, E and F are **collinear points**. This is one of the basic truths used in modern **projective geometry**.

DESCARTES' RULE OF SIGNS. See **Rule of Signs and Roots.**

DESCENDING ORDER. See **Ascending, Descending Order.**

DETACHED COEFFICIENTS. If any algebraical expression is arranged in ascending or descending powers of the unknown quantity and any missing power is included with a zero coefficient, then multiplication and division of expression by expression may be effected expeditiously by writing only the coefficients. The expression $7x^4 - x^2 + 5x + 3$ would be written $7 + 0 - 1 + 5 + 3$. See **Ascending, Descending Order.**

DETERMINANT. A square array of quantities, known as elements, used to represent the sum of specific products of the elements. If there are n^2 elements, n is the number of rows or columns and is called the *order of the determinant*.

second order determinant. $\begin{vmatrix} a_1 & b_1 \\ a_2 & b_2 \end{vmatrix} = a_1b_2 - a_2b_1.$

third order determinant. $\begin{vmatrix} a_1 & b_1 & c_1 \\ a_2 & b_2 & c_2 \\ a_3 & b_3 & c_3 \end{vmatrix} = \begin{matrix} a_1b_2c_3 + a_2b_3c_1 + a_3b_1c_2 \\ - a_3b_2c_1 - a_2b_1c_3 - a_1b_3c_2, \end{matrix}$

$$\text{or } a_1\begin{vmatrix} b_2 & c_2 \\ b_3 & c_3 \end{vmatrix} - a_2\begin{vmatrix} b_1 & c_1 \\ b_3 & c_3 \end{vmatrix} + a_3\begin{vmatrix} b_1 & c_1 \\ b_2 & c_2 \end{vmatrix}$$

(sum of products of elements of first column and their cofactors).

minor of element of determinant. The determinant of next lower order obtained by deleting the row and column containing the given element. If the minor is qualified by a positive or negative sign according to whether the sum of partition order of deleted row and column is even or odd, it is called a *signed minor* or *cofactor*.

Determinants are expanded by the **rule of Sarrus. Cramer's rule** uses this notation for the solution of simultaneous equations. In analytical geometry, the line through the points (x_1, y_1) and (x_2, y_2) can be expressed in the form

$$\begin{vmatrix} x & y & 1 \\ x_1 & y_1 & 1 \\ x_2 & y_2 & 1 \end{vmatrix} = 0.$$

See **Principal, Secondary Diagonals.**

DEVELOPABLE SURFACE. A surface which can be unfolded or rolled out as a plane surface without distortion. See **Development of Solids.**

DEVELOPMENT OF SOLIDS. The unfolding and opening out of a geometric solid so that all faces lie in one plane. The nets from which the five regular solids are formed are examples of development. The cone can be developed but the sphere and the ellipsoid cannot.

DEVIATION IN STATISTICS. Variation from a standard of reference, norm

or the mean in a sequence. The deviation of any element x_r in a set of values $x_1, x_2, x_3, \ldots x_n$, from mean value $\bar{x}$ is $(x_r - \bar{x})$. It follows that $\sum_{r=1}^{n} (x_r - \bar{x}) = 0$. See **Mean in Statistics**; **Quartile Deviation**; **Standard Deviation**.

DIAGONAL OF POLYGON. A line segment joining any two non-adjacent vertices.

DIAGONAL OF POLYHEDRON. A line segment joining any two vertices which do not lie in the same plane.

DIAGONALS, PRINCIPAL AND SECONDARY. See **Principal, Secondary Diagonals**.

DIAGRAM. A general term for any pictorial representation of a geometric **configuration.** Any pictorial representation of a concept in other branches of mathematics, e.g., **number line**; **venn diagram**.

DIAMETER OF CIRCLE. A **chord** that passes through the centre of the **circle**. The centre divides the diameter into two radii.

DIAMETER OF CONIC. A line through the centre terminating on the **conic section.** It is bisected by the centre.

DIAMETER OF SPHERE. A line through the centre ending in **antipodal points** of the **sphere**.

DIFFERENCE. The *remainder* after subtraction. If a difference is zero this becomes a definition of **equality** between the numbers (quantities) having no difference. In the case of **directed numbers**, the numerical value of the difference may be greater than the **numerical values** of the other numbers involved, e.g., $(+5) - (-3) = (+8)$.

DIFFERENCE EQUATION. See **Finite Differences**.

DIFFERENCE IN DIRECTION. See **Angle between Two Straight Lines.**

DIFFERENCE METHOD OF SUMMATION. A method of **summation** of certain series by expressing the general term of the series as the difference between the general terms of two series. In the combination, by addition, of the two series, all terms cancel except the first term of the first series and the last term of the second. For example, the general term of the series $S_n = 1\cdot2 + 2\cdot3 + 3\cdot4 + \ldots + n(n+1)$ can be expressed in the forms $r(r+1)$ or $r(r+1)(r+2)/3 - (r-1)r(r+1)/3$. Hence,

$$\begin{array}{llllll} & 3r(r+1) & = r(r+1)(r+2) & - & (r-1)r(r+1). \\ \text{When } r=1 & 3\cdot1\cdot2 & = 1\cdot2\cdot3 & - & 0\cdot1\cdot2, \\ r=2 & 3\cdot2\cdot3 & = 2\cdot3\cdot4 & - & 1\cdot2\cdot3, \\ r=3 & 3\cdot3\cdot4 & = 3\cdot4\cdot5 & - & 2\cdot3\cdot4, \\ & \cdots\cdots & \cdots\cdots & & \cdots\cdots \\ r=n & 3\cdot n(n+1) & = n(n+1)(n+2) & - & (n-1)n(n+1). \\ \text{Hence} & S_n & = \tfrac{1}{3}n(n+1)(n+2). & & \end{array}$$

DIFFERENCE OF TWO CUBES.

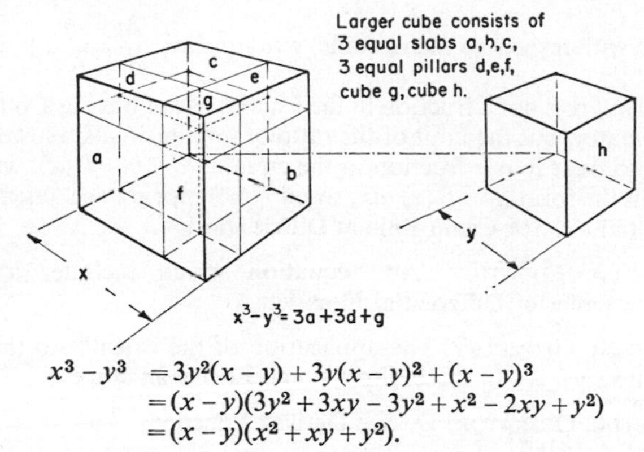

$$\begin{aligned} x^3 - y^3 &= 3y^2(x-y) + 3y(x-y)^2 + (x-y)^3 \\ &= (x-y)(3y^2 + 3xy - 3y^2 + x^2 - 2xy + y^2) \\ &= (x-y)(x^2 + xy + y^2). \end{aligned}$$

DIFFERENCE OF TWO SETS. The set of all elements which are included in set A but not in set B is written $A - B$. The *symmetric difference* of sets A and B is the set of elements which are included in one of the sets but not included in the other, written $A \ominus B$, or $A + B$ when addition of sets A and B is defined as the symmetric difference. See **Algebra of Sets; Set Theory.**

DIFFERENCE of TWO SQUARES. $x^2 - y^2 = (x-y)(x+y)$. See **Completing the Square.**

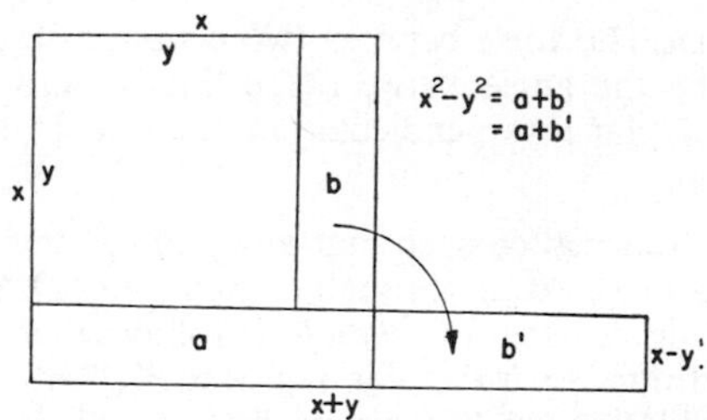

DIFFERENCE OF TWO VECTORS. See **Composition of Vectors.**

DIFFERENCE OPERATORS, Δ, E. See **Finite Differences.**

DIFFERENCES. See **Finite Differences.**

DIFFERENTIAL. Either dy or dx in the expression dy/dx, the **differential coefficient.**

DIFFERENTIAL CALCULUS. See **Calculus.**

DIFFERENTIAL COEFFICIENT. Synonym of **derived function** and **derivative.** For a **function** $y = f(x)$, the ratio of the increment δy to the increment δx is a measure of the rate of change of the function with respect to the independent variable. The limiting value of this ratio as δx approaches zero is called the differential coefficient. It is the instantaneous rate of change of the function with respect to the variable, written $\lim_{\delta x \to 0} \frac{\delta y}{\delta x} = \frac{dy}{dx}$, y' or $f'(x)$. The symbol dy/dx is not a fraction in the usual sense of having a numerator and denominator, but the limit of the ratio of infinitesimals. However, it is convenient to treat it as a fraction in the equality $dy/dx = f'(x)$, which can be written in the form $dy = f'(x)\ dx$, when $f'(x)$ appears as a coefficient of the differential dx. See **Chain Rule in Differentiation.**

DIFFERENTIAL EQUATION. Any equation which includes a **derived function.** See **Order of Differential Equation.**

DIFFERENTIAL GEOMETRY. The application of the **calculus** to the theory of curves and surfaces in three-dimensional **Euclidean space.**

DIFFERENTIAL OPERATOR, D. See **Derived Function.**

DIFFERENTIATION. The determination of the **derivatives** of **functions.** The process is used in finding: instantaneous **velocity; acceleration; gradient of curve; maximum, minimum values.** The **inverse operation** of **integration.** See **Derived Function; Differential Coefficient.**

DIGIT. Derived from the Latin (finger, toe), a word associated with primitive **counting.** Any one of the symbols of the **denary number system.** See **Bit; Hindu-Arabic Notation.**

DIHEDRAL ANGLE. The angle between two planes, called the faces of the angle, measured by the angle between the lines of cross-section of the planes and a third plane perpendicular to the line of intersection. See **Angle of Inclination.**

DILATION. A **transformation** such that any point P and its transform P' can be joined by a line passing through a fixed point O, such that $OP' = k \cdot OP$ where k is the *constant of dilation.* It follows that any line segment PQ is transformed into a parallel line segment $P'Q'$ where $P'Q' = k \cdot PQ$. The principles of dilation are fundamentally those of **similarity.**

DIMENSION OF EQUATION. Synonym of **order of equation.** The highest **dimension of expression** which is a term of the equation. Thus $(x-1)^3 + 2(x+4) = 0$ is third dimension.

DIMENSION OF EXPRESSION. Synonym of degree of expression. The sum of the degrees of the factors which make the expression. Thus, $y+z$, $x(y+z)$, $x^2(y+z)$, $(x-y)(y-z)^3$ are first, second, third and fourth dimension. The definition applies to fractions: $x^3/(y+z) = x^3(y+z)^{-1}$ is an expression of second dimension. See **Degree in Algebra.**

DIMENSION OF ORDINARY SPACE. Coordinates in the system of rectangular **Cartesian coordinates.** In mensuration it denotes length, breadth or width, height or depth in three dimensional **Euclidean space.**

DIMENSION OF PHYSICAL QUANTITY. If **L** is the dimension of length, **M** the dimension of mass and **T** the dimension of time, then in these **fundamental units**, volume is $[L]^3$, speed is $[L][T]^{-1}$, acceleration is $[L][T]^{-2}$, force is $[M][L][T]^{-2}$ and power is $[M][L]^2[T]^{-3}$. See **Homogeneity of Dimensions.**

DIMENSION OF SPACE. The least number of **parameters** needed to represent its points. A curve is one-dimensional; a surface, two-dimensional; ordinary space, three-dimensional: *n*-dimensional space is a set of points having *n* parameters.

DIOPHANTINE ANALYSIS. The determination of integral solutions of certain algebraic equations. They are mainly solved by the employment of **parameters.** See **Indeterminate Equations.**

DIP. (1) The **inclination** of the magnetic needle to the horizon when its pivotal axis is horizontal and at right angles to the isogonal line at that place. (2) The amount of maximum slope of a geological stratum. The *strike* of the stratum is then the horizontal line at right angles to the direction of the dip. Compare the **pitch of roof.** See **Agonic Line; Iso-; Magnetic Declination.**

DIRECT COMMON TANGENTS. See **Tangents to Two Circles.**

DIRECT, INDIRECT, TAXES. See **Taxes.**

DIRECT VARIATION. See **Variation.**

DIRECTED LINE. Any line to which sense of **direction** has been assigned, positive in one direction, negative in the opposite. Axes of coordinates are directed lines.

DIRECTED LINE SEGMENT. Part of a **directed line.** The line segment *AB* is written $\overline{AB}$ or $-\overline{BA}$, as a directed line segment. Directed line segments are used to represent **vector quantities.**

DIRECTED NUMBERS. Numbers prefixed by a positive or a negative sign, to indicate that they can be represented by points on a **directed line** and be computed as **vector quantities.** See **Number Line; Rule of Signs.**

DIRECTION. Through a point *A* pass an infinite number of straight lines, each one defining a direction through *A*. *One only* of these passes through a fixed point *B*, and this line defines the direction $\overline{AB}$ or $\overline{BA}$. Directions are normally related to arbitrarily chosen directions such as geographic and magnetic north, horizontal, vertical, parallel to the polar axis. See **Directed Line Segment; Orientation.**

DIRECTION ANGLES, COSINES. If axes *OX*, *OY*, *OZ* are a frame of reference near a line *l*, and a line parallel to *l* is drawn through *O*, then this will make angles, say α, β, γ with the three axes. These are known as the direction angles of the line *l*, and the direction cosines of the line *l* are $\cos\alpha$, $\cos\beta$, $\cos\gamma$.

DIRECTLY CONGRUENT. See **Congruent Figures in Space.**

DIRECTOR CIRCLE. The locus of the point of intersection of a pair of tangents to a **central conic,** which are at right angles to one another. If the central conic has the general equation $ax^2+2hxy+by^2+2gx+2fy+c=0$, the equation of the director circle is

$$(ab-h^2)(x^2+y^2)-2(hf-bg)x-2(gh-af)y+c(a+b)-(f^2+g^2)=0.$$

DIRECTRIX. See **Conical Surface; Cylindrical Surface; Focus-Directrix Definition of Conic.**

DISCONTINUITY OF FUNCTION. See **Continuous Function.**

DISCONTINUOUS FUNCTION. See **Continuous Function.**

DISCOUNT. (1) Discount on bonds, stocks, shares. See **Premium.** (2) *Cash discount* is a reduction in the selling price in return for prompt payment or payment in cash. (3) *Trade discount* is a reduction in the selling price in consideration for the purchase of large amounts. (4) *True discount* is a reduction of the face value of an agreement by the simple interest on the reduced amount. If S is the face value, n the number of years before settlement is due, and r the rate of interest, then the true discount,

$$D=S-S/(1+nr)=nrS/(1+nr).$$

DISCRIMINANT. For the **quadratic equation** $ax^2+bx+c=0$, the expression b^2-4ac. For the quadratic equation $ax^2+2hxy+by^2+2gx+2fy+c=0$, the expression $abc+2fgh-af^2-bg^2-ch^2$. These expressions determine the nature of the **roots of the equations.**

DISJOINT SETS. See **Overlapping Sets; Set Theory.**

DISJUNCTION. See **Algebra of Propositions.**

DISPERSION. A term in statistics for the variation in or the amount of scatter of a set of data. It is usually measured by the **standard deviation.**

DISPLACEMENT. This may be angular, rectilinear or a mixture of both. See **Rotation; Translation.**

DISTANCE. A measurement of length. Its symbol is s when used for curve segments and the study of motion. In association with time, t, it gives the fundamental **differential equations** of velocity, $v=ds/dt$, and of acceleration, $f=d^2s/dt^2$.

DISTANCE BETWEEN TWO POINTS. If the coordinates of the points are (x_1, y_1) and (x_2, y_2), the distance is $\sqrt{(x_1-x_2)^2+(y_1-y_2)^2}$. If the coordinates are (r_1, θ_1) and (r_2, θ_2), the distance is $\sqrt{r_1{}^2+r_2{}^2-2r_1r_2\cos(\theta_1-\theta_2)}$.

DISTANCE FROM POINT TO LINE. From the point (x', y') to line:
(1) $ax+by+c=0$, distance is $\pm(ax'+by'+c)/\sqrt{a^2+b^2}$;
(2) $x\cos\alpha+y\sin\alpha-p=0$, distance is $\pm(p-x'\cos\alpha-y'\sin\alpha)$.

DISTRIBUTION IN STATISTICS. A term for describing the relative arrangement of a set of elements. It describes the set of values of a variable and the **frequencies** of occurrence of each value. Synonym of *frequency distribution* when it is distinguished from a distribution related to time or place.
normal distribution. One which fits the normal **frequency curve.**
skew distribution. A non-symmetric distribution.
symmetric distribution. One which is symmetric about the median.
See **Frequency Curve, Graph, Polygon; Gauss-Laplace Law.**

DISTRIBUTIVE LAW. In a mathematical system a law which links the operations of addition and multiplication. In general, the law refers to the distribution of multiplication over addition: $a \times (b+c) = (a \times b) + (a \times c)$. In **Boolean algebra** it can also refer to the distribution of addition over multiplication: $a + (b \times c) = (a+b) \times (a+c)$; e.g., in **set theory** $A \cup (B \cap C) = (A \cup B) \cap (A \cup C)$.

DIVERGENCE. See **Convergence, Divergence of Sequence; Convergence, Divergence of Series.**

DIVIDEND. (1) A term in **division.** (2) A term in commerce for the portion of profits returned to shareholders.

DIVIDENDO. See **Proportion.**

DIVISION. A general term for the **binary operation** equivalent to that of breaking up a whole into parts. When used as the reverse operation of **multiplication** the parts are equal in size and the process is expressed in the forms $a \div b = c$, $a/b = c$, where a is the *dividend*, b the *divisor* and c the *quotient*. Fundamentally, the operation is an abstraction from different concrete experiences *Partition* (reduction) corresponds to magnification in multiplication and is the process of sharing a quantity between a number of recipients; for example, the sharing equally of sweets between a given number of children. *Quotition* (grouping) corresponds to successive addition in multiplication and is the process of successive subtraction of equal amounts; for example, the building of equal piles of bricks from a given number. The **ratio** aspect of division is based on the direct comparison of objects and involves the concept of a unit of measure. If the dividend is not a multiple of the divisor a remainder will occur in the process. This is obviated by extending the number concept to embrace **rational numbers** (fractions). When applied to **directed numbers** it is referred to as **algebraic division.** The concept of division is developed still further by its application to the field of **complex numbers**. The term, division, may be applied to fields outside traditional arithmetic. A set of things may be separated in a divisive sense into two sets as a reverse process to intersection (product) in **set theory**. See **Separation**; **Operations in Ordinary Algebra**; **Operator.**

DIVISION OF LINE SEGMENT. If I is a point on the **directed line segment** AB, I divides AB *internally* in the ratio AI/IB. If E is a point on the extension either way of AB, E divides AB *externally* in the ratio AE/EB, where

AE and *EB* are directed line segments. If these ratios are equal, *AB* is said to be divided internally and externally in the same ratio or divided proportionally. See **Harmonic Section**; **Points at Infinity**.

DIVISOR. See **Division**.

DODECAGON. A **polygon** with twelve sides.

DODECAHEDRON. A **polyhedron** with twelve faces. It is one of the **regular solids** when its faces are regular **pentagons**.

DOMAIN OF FUNCTION. See **Function**.

DOMINO. A **polyomino** made of two adjacent squares.

DOT (INNER) PRODUCT. Synonym of **scalar product**.

DOUBLE INTEGRAL. See **Multiple Integral**.

DOUBLE POINT. A second order **multiple point**. A point common to two branches of the same curve, at which there are two tangents. If these are real and distinct they determine a type of double point called a *crunode*; if real and coincident, a type called a *spinode*; if imaginary, an *acnode*. See **Cusp**; **Isolated Point**.

DOUBLE ROOT. See **Repeated Roots**.

DOUBLY RULED SURFACE. See **Ruled Surface**.

DRAG. See **Angle of Attack**.

DUAL. See **Duality**.

DUALITY. A property of **projective geometry**. For each theorem there is a *dual theorem*, in which the roles of point and line (*dual elements*) and point and plane (*space duals*) are interchangeable in a plane and in space respectively. Thus, two points determine a line and two lines determine a point. Operations in which there are dual elements are *dual operations*, e.g., drawing a line through a point and drawing a point on a line. *Dual figures* are obtained by replacing element and operation by dual element and dual operation.

DUODENARY. Associated with twelve as in duodenary **system of notation**.

DUPLICATION OF CUBE. One of the classical problems of geometry, that of finding the dimensions of a cube double the volume of a given one. It involves the solution of the equation $y^3 = 2a^3$. This proved to be impossible by Euclidean methods of construction.

DYNAMICS. The science which studies the effects of forces upon bodies or particles. When the bodies are neither rigid nor elastic the prefix hydro- (Greek, *hydor*, water) is used and hydrodynamics, etc., deal with the effects on fluids and gases. The subject is divided into (1) *statics*, dealing with bodies at rest, i.e. in equilibrium under the action of forces or torques; (2) *kinetics*, dealing with the effects of forces on the motion of bodies; (3) *kinematics*, dealing with abstract motion.

DYNE. A unit of **force** in the **c.g.s. system.** The force which produces an **acceleration** of one centimetre per second in a **mass** of one gramme. Its equivalent in the **f.p.s. system** is 7.233×10^{-5} **poundals.**

E

e. (1) **Eccentricity.** (2) Euler's constant, base of natural **logarithms.** (3) The charge of the electron: 4.774×10^{-10} c.g.s. electrostatic units. (4) **Coefficient of restitution.**

$\mathscr{E}$. Universal set. See **Set Theory.**

E.M.U. **Electromagnetic unit.**

E.S.U. **Electrostatic unit.**

E-CIRCLE. Contraction of **escribed circle.**

ECCENTRIC ANGLE, CIRCLE OF ELLIPSE. The eccentric circles, S_a and S_b, are those circles with radii a and b and centres at the centre of the **ellipse.** The circles have the major and minor axes as diameters; the circle on the

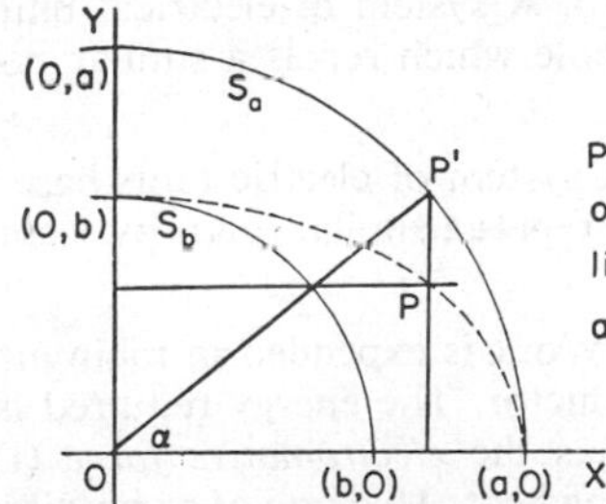

P (a cos α, b sin α) is a member of the set of points on broken line (one quarter shown) called an ellipse

major axis is also called the *auxiliary circle.* If $P\ (x, y)$ lies on an **ellipse** $x^2/a^2 + y^2/b^2 = 1$, the eccentric angle is the angle between OX and OP' where P' is the point on the auxiliary circle with the same abscissa as P. The equation of the ellipse can be transformed into **parametric equations,** $x = a \cos \alpha$, $y = b \sin \alpha$, where α is the parameter. See **Axes of Ellipse.**

ECCENTRIC ANGLE, CIRCLE OF HYPERBOLA. The eccentric circles, S_a and S_b, are those circles with radii a and b and centres at the centre of the **hyperbola.** The circles have the transverse and conjugate axes as diameters; the circle on the transverse axis is also called the *auxiliary circle.* If $P\ (x, y)$ lies on the hyperbola $x^2/a^2 - y^2/b^2 = 1$, the eccentric angle is the angle between OX and OP' where P' is the point of contact of the tangent from

P(a sec α, b tan α) is a member of the set of points on broken line (one of four) called a hyperbola

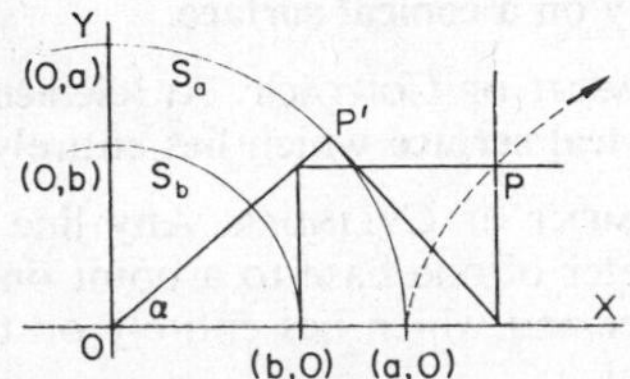

the foot of the ordinate from P to the auxiliary circle S_a. The equation of the hyperbola can be transformed into **parametric equations**, $x = a \sec \alpha$, $y = b \tan \alpha$ where α is the parameter. See **Axes of Hyperbola**.

ECCENTRICITY, e. For any point on a **conic section** the constant ratio (distance to focus)/(distance to directrix). For an **ellipse** $e < 1$; for a **parabola** $e = 1$; for a **hyperbola** $e > 1$. See **Focus-Directrix Definition of Conic.**

ECLIPTIC. A great circle on the **celestial sphere** which is the *apparent* orbit of the sun. It is the great circle that would be seen from the sun marking the earth's orbit on the celestial sphere.

EFFICIENCY. See **Machine.**

ELASTICITY. That property of a body which resists deformation and restores the body to its original shape and size when the deforming forces are no longer operative. The study of elasticity involves the concepts of **stress** and **strain** and their relationship. The ratios of stress to strain are known as **Young's modulus** in longitudinal stress, **rigidity modulus** in shear stress and **bulk modulus** in volume stress.

ELECTROMAGNETIC UNITS (E.M.U.). A system of electrical units based on the concept of a unit magnetic pole which repels a similar pole 1 cm distant with a **force** of 1 **dyne.**

ELECTROSTATIC UNITS (E.S.U.). A system of electric units based on the electrostatic unit of electricity which repels a similar quantity 1 cm distant with a **force** of 1 **dyne.**

ELECTROMOTIVE FORCE (E.M.F.). Work is expended in maintaining the flow of an electric current in a conductor. The energy required is drawn from some external source known as the *electromotive force* (E.M.F.). and measured by the rate of generating heat. This rate of expending energy is a measure of the E.M.F. By definition, the E.M.F. is unity when the current is unity and the rate of working is one **erg** per second. The E.M.F. always acts in one direction only. See **Potential Difference.**

ELEMENT OF ARRAY. A number or letter forming part of an **array. See Determinant; Matrix.**

ELEMENT OF CONE. Any line segment which joins the vertex of the **cone** to a point on the perimeter of the base.

ELEMENT OF CONICAL SURFACE. Any line through the vertex which lies entirely on a **conical surface.**

ELEMENT OF CONTACT. An **element of conical surface** or an **element** of **cylindrical surface** which lies entirely on a **tangent plane.**

ELEMENT OF CYLINDER. Any line segment which joins a point on the perimeter of one base to a point on the perimeter of the other base of a **cylinder** and which lies entirely on the curved surface. All such lines are parallel.

ELEMENT OF CYLINDRICAL SURFACE. Any line which lies entirely on a **cylindrical surface.** All such lines are parallel.

ELEMENT OF DETERMINANT. See **Determinant.**

ELEMENT OF INTEGRATION. The expression following the sign of integration of a **definite integral.**

ELEMENT OF MASS. Given δV, an element of volume of a particular substance, of density ρ, the element of mass is $\rho\ \delta V$.

ELEMENT OF SET. Synonym of member of set. See **Set Theory.**

ELEMENTS. Basic concepts and propositions of a subject.

ELEMENTS OF EUCLID. The contents of thirteen books dealing with **geometry, mensuration, ratio** and **proportion,** written about 300 B.C.

ELEVATION. The vertical distance of a point above some **datum.** The term refers to all such vertical distances but is often used to refer to measurements derived from methods of **levelling.** See **Contours; Plan, Elevation, Section.**

ELIMINATION. The process whereby two or more **simultaneous equations** are combined in such a way that one of the **variables** is removed and the number of equations reduced by one.

ELLIPSE. The **conic section** made by a plane which cuts only one **nappe** of a cone to produce a continuous curve. The set of points which satisfy the equation $x^2/a^2+y^2/b^2=1$, where $b^2=a^2(1-e^2)$, e being the eccentricity of the curve. The foci are the points $(\pm ae, 0)$, and the directrices, the lines $x=\pm a/e$. See **Axes of Ellipse; Focus-Directrix Definition of Conic; Eccentric Angle, Circle of Ellipse; Null Ellipse.**

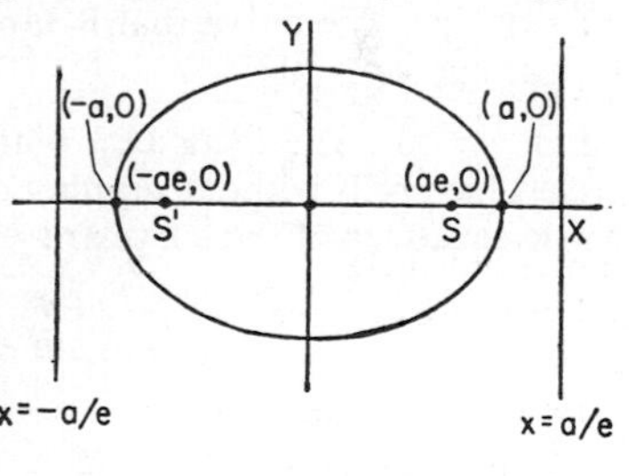

ELLIPSOID. A geometric surface with elliptical cross-sections on planes parallel to its three axes. If $2a$, $2b$ and $2c$ are the lengths of the axes and $a\neq b\neq c$, the equation $x^2/a^2+y^2/b^2+z^2/c^2=1$ represents the surface in **Cartesian coordinates** with axes of coordinates along the axes of the figure. If $b=c$ the *prolate* shape is an ellipsoid of revolution of the **ellipse** $x^2/a^2+y^2/b^2=1$ about its major axis. If $a=b$ the *oblate* shape is an ellipsoid of revolution of ellipse $x^2/a^2+z^2/c^2=1$ about its minor axis. If $a=b=c$ the ellipsoid is a **sphere,** $x^2+y^2+z^2=a^2$. See **Axes of Ellipse; Axes of Ellipsoid.**

ELLIPTIC HYPERBOLOID. See **Hyperboloid.**

ELLIPTIC PARABOLOID. See **Paraboloid.**

ELLIPTIC SPACE. A **non-Euclidean space** based on the postulate that through a point external to a given line there is no line parallel to the given line. The geometry of elliptic spaces was developed by Riemann (1826–1866). One example is spherical geometry, a two-dimensional geometry of the surface of a sphere in which lines are defined as arcs of great circles. There are no parallel lines as any two great circles must intersect. One property of spherical geometry is that the sum of the angles of a **spherical triangle** can be greater than 180°.

ELONGATION. The difference between the celestial longitude of the moon or of a planet and the celestial longitude of the sun. See **Spherical Coordinates in Astronomy.**

EMPIRICAL PROBABILITY. See **Probability.**

EMPTY SET, ϕ. The set with no elements. See **Set Theory.**

ENDOMORPHISM. See **Homomorphism.**

ENERGY. Capacity for doing **work. See Kinetic Energy; Potential Energy.**

ENTIRE FUNCTION. Synonym of **integral function.**

ENTIRE SERIES. A **power series** which converges for all values of the variable.

ENUMERABLE SET. Synonym of **countable set, denumerable set.**

ENUNCIATION. A precise statement.

ENVELOPE. The curve that is tangential to every member of a family of lines or curves.

EPICYCLOID. The locus in a plane of a point on the circumference of a circle which rolls without slipping on the outside of a fixed circle. The **parametric equations** of the curve are

$$x=(a+b)\cos\theta+b\cos[(a+b)\theta/b],$$
$$y=(a+b)\sin\theta-b\sin[(a+b)\theta/b],$$

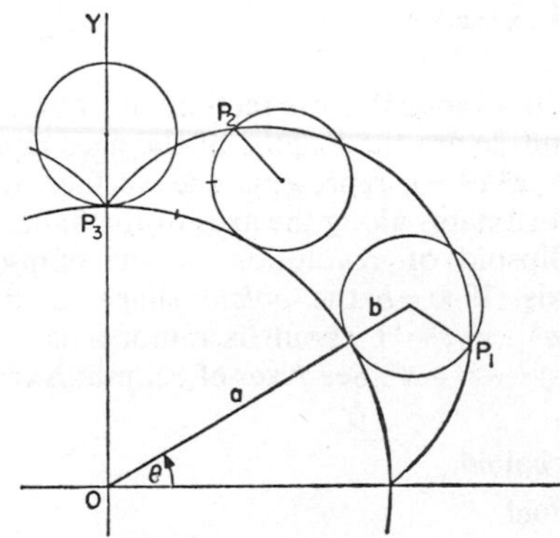

where a is the radius of the fixed circle, b the radius of the rolling circle and θ the angle between the line of centres and the positive direction of the x-axis. There is a **cusp** of the first kind at every point common to the curve and the fixed circle. The number of cusps is given by $n=a/b$. When $n=1$ the curve is a **cardioid.**

Epicycloids were used to explain many of the movements of heavenly bodies by astronomers upholding the Ptolemaic theory. The cutting of gear teeth depends on a knowledge of the *epicyclic curve*. See **Hypocycloid; Pericycloid.**

EPITROCHOID. See **Roulette.**

EQUAL IN ANALYSIS. A description of a **relation** between two quantities which have the property of being alike in some specified way. See **Equivalence Relation.**

EQUAL IN GEOMETRY. The property of two things being alike in some specified way. It may be equal line segments, equal angles, in which case equal is synonymous with **congruent**.

EQUALITY. The property of two things being **equal,** that is that their difference is zero, usually expressed in the form of an **equation.**

EQUALITY IN COMPLEX NUMBERS. The property of two **complex numbers** of having their real parts equal and their imaginary parts equal. Thus, for $a+ib$ and $c+id$, $a=c$ and $b=d$.

EQUATION. A statement of **equality**. If the equality is true only for certain values of the unknown quantities involved, the equation is often called a *conditional* equation and the sign $=$ is used: if the equality is true for all values of the unknown quantities involved, the equation is called an **identity** and the sign $\equiv$ is used.

EQUATIONS OF MOTION. The motion of a rigid body can be considered in two parts: the motion of **translation** of the **centre of gravity** and the motion of **rotation** about the centre of gravity. If (x, y, z) are the coordinates of the centre of gravity with respect to three **rectangular axes, if acceleration** is constant and if **velocity** and displacement are functions of time, then the components of acceleration, velocity and displacement are given by the equations:

acceleration	*velocity*	*distance (displacement)*
$\ddot{x}=a,$	$\dot{x}=at+v_x,$	$x=\frac{1}{2}at^2+v_xt+x_0,$
$\ddot{y}=b,$	$\dot{y}=bt+v_y,$	$y=\frac{1}{2}bt^2+v_yt+y_0,$
$\ddot{z}=c,$	$\dot{z}=ct+v_z,$	$z=\frac{1}{2}ct^2+v_zt+z_0,$

in which x_0, y_0, z_0 are initial displacements and v_x, v_y, v_z initial velocities.

In the case of straight line motion these equations are often expressed in the following formulae, in which f is constant acceleration, v the speed, s the distance from the origin at any time t, and u the initial speed when time and distance are zero:

$$dv/dt=f, \therefore v=u+ft \;;\; ds/dt=v, \therefore s=ut+\tfrac{1}{2}ft^2.$$

These formulae are often combined to produce other formulae:

$$s=\tfrac{1}{2}(u+v)t, \; v^2=u^2+2fs.$$

Components of **angular motion** of the body can be expressed in equations of similar form.

EQUATOR, CELESTIAL. See **Spherical Coordinates.**

EQUATOR OF EARTH. The **great circle** on the surface of the earth equidistant at all points from the North Pole and the South Pole, and dividing the earth's surface into the northern and the southern hemispheres.

EQUATORIAL PLANE. The plane through the centre of a sphere perpendicular to a polar axis.

EQUIANGULAR FIGURES. Figures in which angles in corresponding positions are equal.

EQUIANGULAR POLYGON. A **polygon** having all angles the same size.

EQUIANGULAR SPIRAL. The **spiral** in which the inclination of the radius vector and the tangent at any point is equal.

EQUICONJUGATE DIAMETERS. **Conjugate diameters** of an ellipse that are equal in length. They lie along the diagonals of the rectangle made by the tangents at the ends of the major and minor axes. The diameters have the equations $y = \pm(b/a)x$.

EQUILATERAL POLYGON, TRIANGLE. A polygon, triangle, with all sides of equal length.

EQUILIBRIUM. The state of a body or particle when the **resultant** of all external forces operating on it is zero. There is no **acceleration** of translation or rotation in this state of rest or uniform motion in a straight line. The equilibrium is *stable* or *unstable* if, when slightly displaced, the body tends to return to or move from its original position respectively under the action of the operating forces. If the body remains in the displaced position it is a state of *neutral* equilibrium.

EQUINOXES. The points where the line of intersection of the plane of the earth's equator and the earth's **ecliptic** meet the **celestial sphere**. The sun appears to pass through each point once each year. At these times (vernal equinox, approximately 21/22 March; autumnal equinox, approximately 22/23 September) the earth's axis is perpendicular to its radius vector from the sun, and day and night are equal in length. See **Nodes in Astronomy**.

EQUINUMEROUS, EQUIPOLLENT, EQUIPOTENT SETS. See **One-One Correspondence**; **Inverse Function.**

EQUIVALENCE CLASS. If an **equivalence relation** is defined on a set, the set can be separated into classes such that two elements belong to one class only if they are equivalent. Such classes are called *equivalence classes*. For example, a **vector quantity** can be interpreted as an equivalence class of equal and parallel sensed line segments.

EQUIVALENCE OF PROPERTY. Two properties are equivalent if, logically, they determine the same set. Thus, the set of isosceles triangles is determined by triangles with a pair of equal sides or a pair of equal angles. The symbol $\Leftrightarrow$ is used to denote equivalence.

EQUIVALENCE OF PROPOSITION. See **Algebra of Propositions.**

EQUIVALENCE RELATION. A **relation** between elements of a set which satisfies three conditions: (*a*) *reflexive*, *aRa*; (*b*) *symmetric*, *aRb*, then *bRa*; (*c*) *transitive*, *aRb* and *bRc* then *aRc*, where *a*, *b*, *c* are members of the set and *R* means *in the given relation to*. Various relations are ones of equivalence, e.g., **congruence in geometry**; **equality**. See **Inverse Relation**; **Reflexive Relation**; **Symmetric Relation.**

EQUIVALENT EQUATIONS. Equations which have the same set of solutions. Thus, $x=3y \Leftrightarrow 2x-6y=0$.

EQUIVALENT FIGURES. Synonym of **congruent figures.**

EQUIVALENT SETS. **Sets** which can be placed in **one-one correspondence.**

ERG. Unit of **work** in the **c.g.s. system**; the work done by a **force** of 1 **dyne** acting through a distance of 1 centimetre. 1 erg $=10^{-7}$ **Joule.**

ERROR. For any set of measures of the same thing there is a mean or average which is identified with a theoretical 'true value', a unique value which exists only as a concept. Error is the difference between a particular measure and the mean of a set of measures. In **statistics** it can be the variation due to uncontrollable factors or the sampling error due to the arbitrary nature of the **sample** chosen to represent the **population.** See **Error Curve**; **Relative Error.**

ERROR CURVE. When a variation in measurement due to uncontrollable factors occurs, the frequency distribution curve is referred to as an *error curve* and in general it approximates to the Gaussian bell-shaped curve, the **normal distribution curve.** See **Frequency Curve, Graph, Polygon**; **Gauss-Laplace Law.**

ESCRIBED CIRCLE. A circle drawn externally to a polygon to touch three consecutive sides, the first and third extended, as shown. There are *n* such circles related to an *n*-sided polygon.

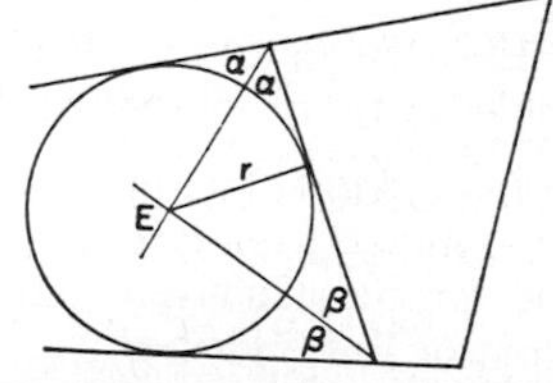

ESCRIBED CIRCLE OF TRIANGLE. The **escribed circle** touching side *a* of triangle *abc* has radius *r*, equal to $\triangle/(s-a)$, where $s=\frac{1}{2}(a+b+c)$ and $\triangle=\sqrt{s(s-a)(s-b)(s-c)}$.

EUCLIDEAN SPACE. A space that is based on the assumptions of Euclid and later modifications. Euclid's *Elements* first appeared about 300 B.C. and were accepted as the basis of all geometrical logic until the parallel postulate of Euclid—later expressed in the form of **Playfair's axiom**—was challenged in the early part of the nineteenth century. In modern mathematics a Euclidean space can have any number of dimensions where the distance between two points is interpreted in the same way as that in two or three dimensional spaces. Euclidean space is sometimes called *parabolic space*. See **Non-Euclidean Space.**

EUCLID'S ALGORITHM. The procedure for finding the **highest common factor** (greatest common divisor) of two numbers. Suppose these are a and b with $a>b$. Divide b into a to give a remainder r_1. Divide r_1 into b to give remainder r_2. Divide r_2 into r_1 to give remainder r_3. Continue until a remainder zero is obtained. The last divisor, i.e. the penultimate remainder, is the h.c.f. (g.c.d.) of a and b. Thus, if a is 187, b 136, the successive remainders are 51, 34, 17, 0. The h.c.f. (g.c.d.) of 187 and 136 is 17.

EUCLID'S TWELFTH AXIOM. See **Playfair's Axiom.**

EULER'S CONSTANT, e. Base of natural **logarithms.**

EULER'S LAW. See **Polyhedron.**

EULER'S THEOREM, EXPONENTIAL FUNCTION. $e^{ix} = \cos x + i \sin x$.

EULER'S THEOREM, HOMOGENEOUS FUNCTIONS. If $u = f(x, y, z, \ldots)$, then $x\ \partial u/\partial x + y\ \partial u/\partial y + z\ \partial u/\partial z + \ldots = nu$, n being number of **partial differentiations.**

EVALUATION. The finding of a numerical value for some expression.

EVEN FUNCTION. A **function** $f(x)$ of a **variable** x which does not change its sign or **absolute value** when the sign of x is changed. Its graph is symmetric with respect to the y-axis of coordinates; e.g. $y = x^2$; $y = \cos x$. See **Odd Function.**

EVEN NUMBER. An **integer** (whole number) which is not odd, and therefore has no remainder when divided by 2. Any number of the form $2n$, where n is an integer (including zero).

EVOLUTE OF CURVE. The locus of the centre of **curvature** of a curve. It is the **envelope** of the normals of the curve.

EVOLUTION. The process of determining the value of x, written $\sqrt[n]{a}$ or $a^{1/n}$, in the given relationship $x^n = a$. It is therefore the process of finding the nth root of a given number a and the **inverse operation** of **involution.** See **Root of Real Number.**

EX-CENTRE. The centre of an **escribed circle.**

EXCESS OF NINES. See **Casting out Nines.**

EXCESSIVE NUMBER. Synonym of abundant number, redundant number. See **Perfect Number.**

EXCOSECANT. See **Exsecant, Excosecant.**

EXPANSION. The process of changing the form of a function into that of a **power series.**

EXPLICIT FUNCTION. In an equation of the form $y = f(x)$, e.g., $y = 3x^2 - 2x + 5$, y is an explicit function of x. The substitution of a value for x determines y *directly*. See **Implicit Function.**

EXPLICIT RELATION. A **relation** between two variables which is expressed as a direct statement. See **Implicit Relation.**

EXPONENT. Synonym for **index.**

EXPONENTIAL CURVE. A mathematical curve important in many branches of science and economics. Its equation is $y = ae^{bx}$. When the constants a and b are unity, the curve $y = e^x$ is the image of the logarithmic curve on the line $y = x$. Otherwise, $y = x$ is a line of symmetry for the curves, $y = e^x$ and $y = \log_e x$ (lon$x$, ln$x$). See **Catenary; Exponential Function.**

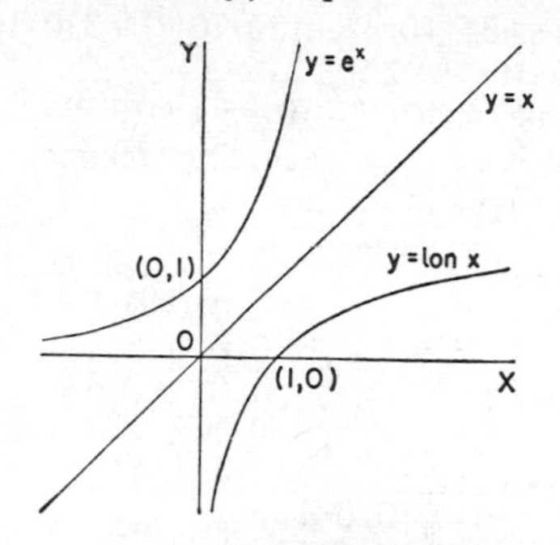

EXPONENTIAL EQUATION. One in which the unknown quantity appears as an exponent (**index**). By taking logarithms of both sides it is resolved into an **algebraic equation.**

EXPONENTIAL FUNCTION. A **transcendental function** with the general form $y = a \cdot b^x$, where a and b are constants and $b \neq 1 \neq 0$. It is sometimes expressed in the form $y = a \cdot e^{cx}$, where a and c are constants and e is the base of natural **logarithms.**

EXPONENTIAL SERIES, THEOREM. The infinite series obtained by the application of **Maclaurin's theorem** to the **exponential function,** e^x, where e is the base of natural **logarithms.**

$$e^x = 1 + x + x^2/2! + x^3/3! + \ldots + x^{r-1}/(r-1)! + \ldots$$

when $x = 1$, $e = 1 + 1 + 1/2! + 1/3! + \ldots$

EXPRESSION. See **Algebraic Expression; Trigonometric Expression.**

EXSECANT, EXCOSECANT. Exsecant θ (exsec θ) = secant $\theta - 1$; excosecant θ (excsc θ) = cosecant $\theta - 1$. See **Versed, Coversed; Trigonometric Ratios.**

EXTENDED COUNT. Synonym of successive addition, one aspect of **multiplication.**

EXTENSION IN ORDER RELATION. See **Order Relations.**

EXTENSION OF PASCAL'S TRIANGLE. Quarter plane A, bounded by line L_1 and line of decimal points D, contains **Pascal's Triangle** arranged such that the last number in each row has units place value, the penultimate number has tens place value, etc. This array can be extended indefinitely, with two main properties: (1) any number is the sum of two numbers in the row above, namely, the one above and the adjacent one to the right; (2) every row is equivalent to an integral power of 11.

Quarter plane B is the reflection of quarter plane A in the line L_2. Property (1) of quarter plane A is retained if numbers in columns and rows are alternatively positive and negative. Property (2) is retained if the numbers are assigned decimal place values. Thus $11^{-3} = .001 - .0003 + .00006 - \ldots$ Both properties are valid when individual place numbers possess more than a single digit. For example:

$$11^{-8} = 1 \cdot 10^{-8} - 8 \cdot 10^{-9} + 36 \cdot 10^{-10} - 120 \cdot 10^{-11} + 330 \cdot 10^{-12} - 792 \cdot 10^{-13} \ldots$$
$$= 2 \cdot 10^{-9} + 240 \cdot 10^{-11} + 2508 \cdot 10^{-13} + \ldots$$
$$= (2+2)10^{-9} + (4+2)10^{-10} + (0+5+1)10^{-11} + \ldots$$

$= 0.00000000466 \ldots$, a recurring decimal fraction containing $2 \cdot 11^7$ (i.e. 39974342) digits in the period.

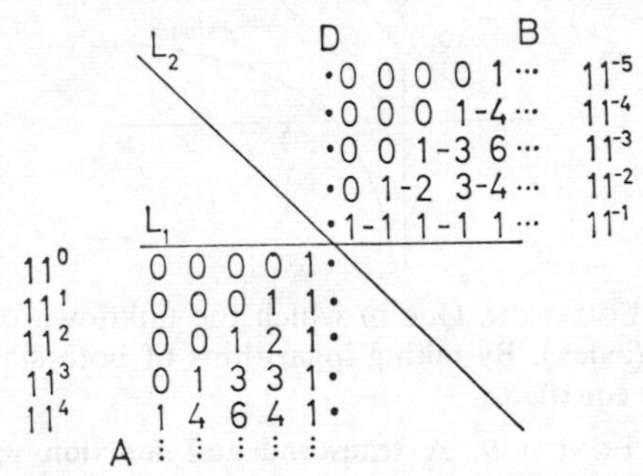

EXTENSION OF PYTHAGORAS'S THEOREM. The generalization from **Pythagoras's theorem** to triangles of any shape. In any triangle the square on any one side is equal to the sum of the squares on the other sides plus or minus twice the rectangle formed by either one of these sides and the projection upon it of the other. Thus, $a^2 = b^2 + c^2 \pm 2cx$, where x is the projection of b on c and the positive or negative sign is used according as a is opposite an obtuse or an acute angle. The theorem can be expressed as the **cosine rule**: $a^2 = b^2 + c^2 - 2bc \cos A$. The size of angle A determines the sign of $\cos A$.

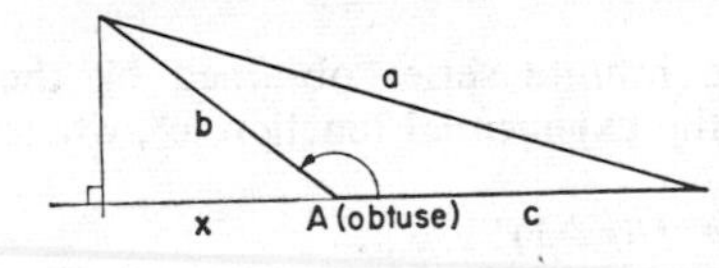

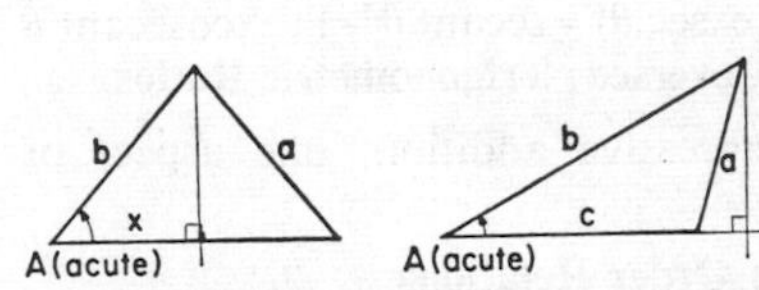

EXTERIOR ANGLE. An angle at a vertex of a **polygon** formed outside the polygon by two adjacent sides, one of which has been extended (produced). The exterior angle and the **interior angle** at any vertex are **supplementary angles.**

EXTERNAL ANGLE. The angle formed outside a **polygon** by any two of its adjacent sides. See **Internal, External Angles.**

EXTERNAL BISECTOR. The bisector of an **exterior angle** of a polygon. It is perpendicular to the **internal bisector**. See **Bisector of Plane Angle.**

EXTERNAL COMMON TANGENTS. See **Tangents to Two Circles.**

EXTERNAL DIVISION. See **Division of Line Segment.**

EXTERNAL VOLUME. See **Volume.**

EXTRAPOLATION. See **Interpolation, Extrapolation.**

EXTRACTION OF ROOT. Synonym of finding the **root of real number.**

EXTREME VALUES. Synonym of **Maximum, Minimum Values.**

F

F.P.S. SYSTEM. The system of measurements based on the foot, pound and second as units of length, mass and time respectively. See **Absolute Units.**

FACTOR. When one number or expression is the product of two or more numbers or expressions, each of the latter is called a factor of the former. Thus 2 and 3 are factors of 6; $x-a$ and $x+a$ are factors of x^2-a^2. To factor, or factorize, a number or expression is to express the number or expression as a product of factors. See **Composite Number**; **Prime Factor.**

FACTORIAL n. ($n!$, $\lfloor n$). The continued product $n(n-1)(n-2)\ldots 3\cdot 2\cdot 1$ is called factorial n. It is the number of **permutations** of n things. If n is large, an approximate value for $n!$ can be obtained by using **Stirling's theorem:** $n!=(n/e)^n\sqrt{2\pi n}$.

FACTORIZATION. The process by which any number or expression is broken into the parts which make it, when the parts are multiplied together. If no such parts, or **factors**, can be found, the number or expression are said to be prime. See **Composite Numbers**; **Prime Factors.**

FACTORIZE. See **Factor.**

FAHRENHEIT SCALE. An obsolescent scale of temperature named after the German instrument maker Daniel Fahrenheit (1686–1736). There are various theories as to the basis of the scale, but it is generally believed that Fahrenheit used the temperature of melting ice and the temperature of a healthy human body as fixed points on his scale following the precedent of the Danish astronomer, Roemer. The fixed points were later freezing point of water (32°) and boiling point of water (212°). See **Celsius Scale**; **Kelvin Scale**; **Temperature Conversion.**

FAMILY OF CURVES. A set of curves which displays a geometrical pattern. The set is represented by an equation containing one or more **parameters,** arbitrary values of which determine members of the set.

FAMILY OF LINES. A set of lines which obeys the same restrictions as a **family of curves.**

FARAD. Unit of electrostatic capacity. A capacity of 1 farad requires 1 **coulomb** of electricity to raise its **potential difference** by 1 **volt**. 1 farad $= 9 \times 10^{11}$ **electrostatic units**. The practical unit, 1 microfarad is 10^{-6} farad.

FERMAT'S NUMBERS. Fermat believed that $2^{2^n}+1$ was a **prime number** for every integral value of n. The first five, ($n=0 \ldots 4$), are 3, 5, 17, 257 and 65,537. A hundred years after Fermat published this, Euler showed that, when $n=5$, the resulting number, 4,294,967,297 is composite, $641 \times 6{,}700{,}417$. Since Fermat's conjecture in 1640, those numbers of the form $2^{2^n}+1$ which have been studied have proved to be **composite numbers.**

FERMAT'S SPIRAL. Synonym of **parabolic spiral**.

FERMAT'S THEOREMS. (1) If a is an integer, and a and p have no common factor, p being prime, then a^{p-1} divided by p leaves a remainder 1. Stated in **modulo arithmetic** $a^{p-1} \equiv 1 \pmod{p}$. (2) The equation in three unknowns, $x^n+y^n=z^n$, can have no solution in positive integers if n is integral and greater than 2. Despite centuries of work on it, this theorem has not been proved.

FEUERBACH'S THEOREM. The **nine points circle** touches the **inscribed circle** and the three **escribed circles** of the triangle associated with it.

FIBONACCI RATIOS. The ratios of any two consecutive numbers from the **Fibonacci sequence**: 1/1, 1/2, 2/3, 3/5, 5/8, 8/13, . . . The limit of this sequence is the **golden section** ratio $(\sqrt{5}+1)/2$. The distribution around the stem of a plant of successive leaves follows a pattern that fits a Fibonacci ratio.

FIBONACCI SEQUENCE. The sequence 1, 1, 2, 3, 5, 8, 13, . . . , in which, if u_n is the general term, $u_n = u_{n-2}+u_{n-1}$. It can be obtained directly from **Pascal's triangle** by summation along the lines indicated. This sequence and others developed similarly have proved to be of great use in interpreting Mendelism in the field of genetics.

```
                1 - - - - - - - - - - - - - - - - - 1
                  - - - - - - - - - - - - - - - - - 1
            1             1 - - - - - - - - - - - - 2
                  - - - - - - - - - - - - - - - - - 3
        1         2             1 - - - - - - - - - 5
                  - - - - - - - - - - - - - - - - - 8
    1         3             3             1 - - - - 13
                  - - - - - - - - - - - - - - - - - 21
1         4             6             4         1 - 34
```

FIELD. See **Field Theory**; **Function**; **Number Field.**

FIELD THEORY. The term *field* was defined in the early part of the twentieth century as any mathematical system consisting of a set of elements subject to two **binary operations** (addition and multiplication) which satisfies the three axioms:

(1) The set is a **commutative group** under the operation addition, with an **identity element** 0 (zero).

(2) The set, with 0 (zero) excluded, is a commutative group under the operation multiplication.

(3) The two binary operations are related by the **distributive law.** See **Ordered Field**; **Subfield.**

FIGURATE NUMBERS. Synonym of **polygonal numbers.**

FIGURE. (1) A symbol used to express an **integer.** (2) A **geometric figure.**

FINITE DIFFERENCES. Let $y_1, y_2, y_3, \ldots y_r, y_{r+1}, \ldots$ be a sequence of values of the function $y=f(x)$, corresponding to the sequence of values of $x=a, a+h, a+2h, \ldots a+(r-1)h, a+rh, \ldots$ The *first order differences* are given by the relation $\triangle y_r = y_{r+1} - y_r$. Thus, the sequence $\triangle$ is: $y_2 - y_1$, $y_3 - y_2, y_4 - y_3, \ldots$ The *second order differences* are obtained from the first in a similar way, and denoted $\triangle^2$; the third from the second, $\triangle^3$; and so on. If for example the function is $y=x^2$, the following related sequences exist:

x	1	2	3	4	5	...
y	1	4	9	16	25	...
$\triangle$	3	5	7	9	11	...
$\triangle^2$	2	2	2	2	2	...

If $Ey_r = y_{r+1}$, then $\triangle y_r = y_{r+1} - y_r = Ey_r - y_r$, and $\triangle = E - 1$.

Hence $$\triangle^2 y_r = (E-1)^2 y_r = (E^2 - 2E + 1)y_r = y_{r+2} - 2y_{r+1} + y_r.$$

And in general:

$$\triangle^n y_r = (E-1)^n y_r = (E^n - nE^{n-1} + \ldots)y_r = y_{n+r} - ny_{n+r-1} + \ldots,$$

a formula for deriving the *nth order differences* of any function. Such formulae are referred to as *difference equations.*

FINITE GROUP. See **Group Theory.**

FINITE, IN GEOMETRY. Bounded, in spatial extent or magnitude, e.g. line segment, closed surface, geometric solid.

FINITE SERIES. A **series** with a finite number of terms.

FINITE SET. A set which cannot be placed in **one-one correspondence** with a part of itself. Theoretically a **denumerable set** with an end number.

FIRST ANGLE, THIRD ANGLE PROJECTIONS. Two conventional ways of showing plans and elevations of objects. Two principal projection planes, vertical (VP) and horizontal (HP) intersect at right angles to form four dihedral angles which are designated first, second, third and fourth, as shown in the diagram. The solid object is considered to be placed in the first or third dihedral angle in order to develop first or third angle projection. The object is projected orthogonally on to the horizontal plane and vertical plane to produce a *plan* and *front elevation* respectively. A *side* (*end*) *elevation* is produced by projecting on to a second vertical plane, referred to as the side vertical plane (SVP) at right angles to the principal one. All three views of the object are shown in one plane by suitable rotations. In the first angle projection the object is in front of the planes of projection; in third angle projection it is behind the planes. In practice a second side elevation is often shown to the left or right of the front elevation; and in third angle projection the plan is sometimes shown below the elevations for

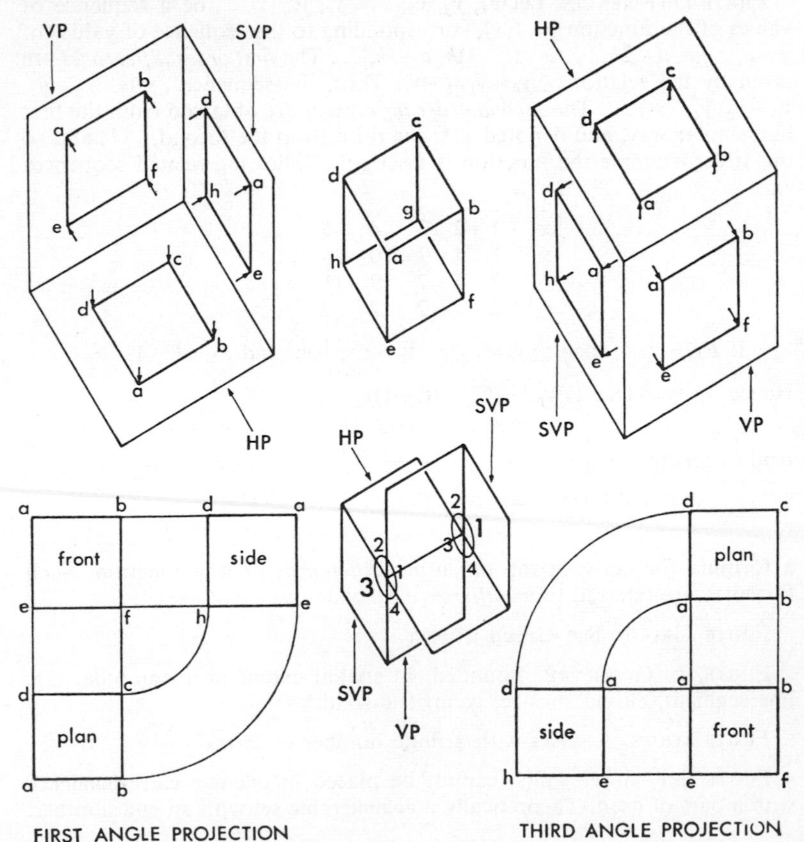

FIRST ANGLE PROJECTION

THIRD ANGLE PROJECTION

convenience. Third angle projection has an advantage over first angle projection in that adjacent faces of the object have their common edges in proximity. See **Plan, Elevation, Section.**

FIRST LAW OF THE MEAN FOR DERIVATIVES. The formula $f(b)-f(a)=(b-a)f'(c)$ in the **mean value theorem in differential calculus.**

FIRST LAW OF THE MEAN FOR INTEGRALS. The formula $\int_a^b f(x)dx=(b-a)f(c)$ in the **mean value theorem in integral calculus.**

FLEXAGON. The term *flexagon* was first used in 1939 by Stone, an English graduate student at Princeton University. He used it to describe paper polygons which were folded from strips of paper and possessed the property that their faces were altered by *flexing*. The flexagons were referred to as *hexaflexagons* because of their hexagonal form. The first was a *trihexaflexagon* folded from a strip of ten equilateral triangles to form three possible faces. The second was a *hexahexaflexagon* folded from a strip of nineteen equilateral triangles to form six possible faces. The theories of **flexure** were developed by Stone and his colleagues, Feynham, Tuckerman and Tukey. They discovered that, by lengthening the chain of equilateral triangles and using zigzag patterns, the number of possible flexagons was apparently unlimited.

FLEXING. See **Flexagon; Flexure.**

FLEXURE. The process of flexing (bending or folding) of a line, curve, surface or solid. See **Flexagon.**

FLOW CHART. See **Computer Programming.**

FLUID OUNCE. **Imperial measure**: 480 minims or one-twentieth of a pint. **Apothecaries' measure**: 480 minims or one-sixteenth of a pint.

If water at 62°F is used for comparison, the following is true:

In U.S.A. 1 fluid ounce $=1.805$ in$^3=455.8$ grains.
In U.K. 1 fluid ounce $=1.732$ in$^3=437.5$ grains.
In U.K. 1 fluid ounce weighs 1 ounce **avoirdupois.**

See **Apothecaries' Fluid Measure.**

FOCUS. See **Focus-Directrix Definition of Conic.**

FOCUS-DIRECTRIX DEFINITION OF CONIC. A conic section is a set of points for which the distances of each from a fixed point called the *focus*, and from a fixed line called the *directrix* are in constant ratio. If P is any point of the set, S the focus, M the foot of the perpendicular from P to the directrix, $PS=e\ PM$, where e is called the **eccentricity**. The value of e determines the conic sections: **ellipse** ($e<1$), **parabola** ($e=1$), **hyperbola** ($e>1$). For each curve, $VS=e\ VD$ where V is the *vertex* of the conic

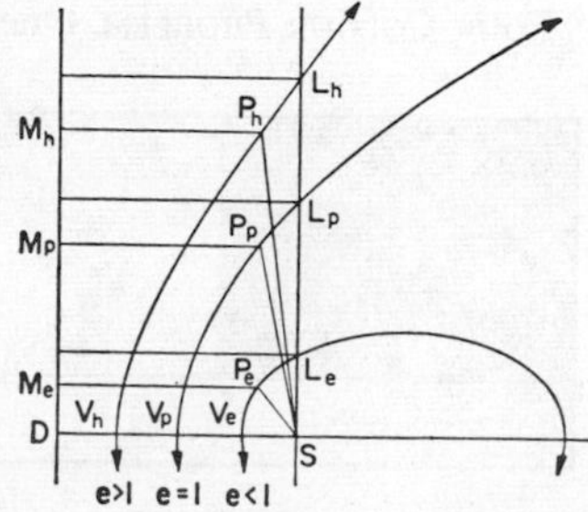

and D is the point of intersection of the axis of the conic and the directrix. For each curve, the **semi-latus rectum** $LS = e\ DS$. See **Conic Section.**

FOOT. (1) Unit of length in the **f.p.s. system**, subdivided into twelve inches, equal to one-third of a **yard**. Probably a standardization of the length of a human foot. (2) The point of intersection of a line (usually a **perpendicular**) with another line or a plane.

FOOT-CUBE. A cube of one foot edge, having **volume** of one cubic foot (1 ft^3) or 1 728 **cubic inches** (1 728 in^3).

FOOT-POUND (ft lbf). The unit of **work** in the **f.p.s. system**; the work done in raising a pound **mass** through one foot against the **force of gravity**.

FOOT-SQUARE. A square of one foot side, having **area** of one square foot (1 ft^2) or 144 **square inches** (144 in^2).

FORCE. The cause of **acceleration or strain** in a body. The unit of force is that force which causes unit acceleration in a body of unit mass. The absolute units of force are the **poundal** in the **f.p.s. system**, the **dyne** in the **c.g.s. system** and the **newton** in the **m.k.s. system.**

FORCE DE CHEVAL. See **Horse-Power.**

FORCE DIAGRAM. Any diagram used to represent the magnitude and direction of forces in a system. See **Bow's Notation**; **Funicular Polygon**; **Parallelogram of Vectors**; **Polygon of Vectors**; **Triangle of Vectors.**

FORCE OF GRAVITY. The **force** that compels all unrestricted matter to fall earthwards, all satellites to fall towards their primaries, all planets towards the sun. It is one aspect of universal **gravitation**. See **Acceleration due to Gravity**; **Constant of Gravitation.**

FORESIGHT (F.S.). A reading taken on a levelling staff when held at a point of unknown elevation. See **Backsight**; **Bench Mark**; **Levelling.**

FORMULA. A concise and comprehensive statement, expressed symbolically, of the **equality** existing between two quantities having a precise **relation**. Examples of formulae: $V=(\pi/6)d^3$; $I=Prn/100$.

FORWARD BEARING. See **Bearing.**

FOUR-COLOUR PROBLEM. One of the **unsolved problems of topology**. Any plane or spherical surface which is divided into regions may have these regions distinguished if they are coloured in four different ways. This has been shown to be true experimentally, but it has never been proved rigorously. The sufficiency of five colours has been proved, but no one has discovered an example where the

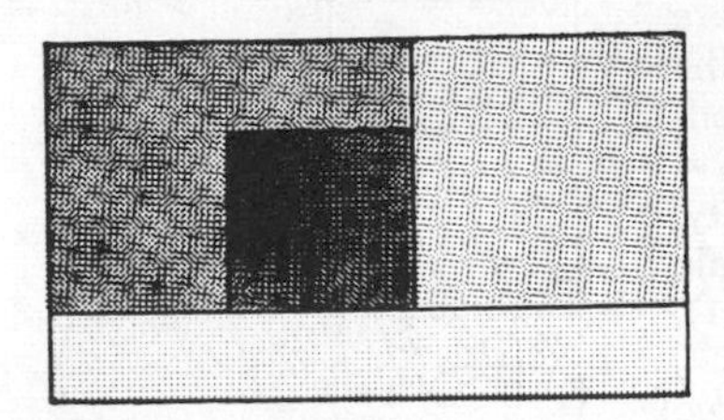

fifth colour is necessary. The problem of determining the number of colours necessary to distinguish regions has been extended to other surfaces: for example, seven are needed on the surface of a **torus**.

FOURIER SERIES. Any **periodic function** of period 2π, can be expressed as an infinite **trigonometric series** of the form:

$$f(x) = \tfrac{1}{2}a_0 + (a_1 \cos x + b_1 \sin x) + (a_2 \cos 2x + b_2 \sin 2x) + \ldots + (a_n \cos nx + b_n \sin nx) + \ldots$$

where coefficients a_0, a_1, b_1, a_2, b_2, etc., are constants which represent the amplitudes of the fundamental and the successive harmonics which together build up the given periodic function. The constants are the integrals

$$a_0 = \frac{1}{\pi}\int_{-\pi}^{\pi} f(x)dx;\ a_n = \frac{1}{\pi}\int_{-\pi}^{\pi} f(x)\cos nx\,dx;\ b_n = \frac{1}{\pi}\int_{-\pi}^{\pi} f(x)\sin nx\,dx.$$

FOURTH PROPORTIONAL. If a, b, c, d are numbers or quantities related by the proportion $a/b = c/d$, then the fourth proportional is d, which is equal to bc/a.

FRACTION. In general, if unity, representing a whole, is divided into D equal parts, each part is expressed as the quotient $1/D$ where D is called the denominator. If N such parts are combined to make a new whole, it is expressed in the form N/D, where N is called the numerator. Alternatively, N/D can be interpreted as the quotient obtained when the quantity N units is divided into D equal parts. The fraction N/D is given different names according to the values of N and D:

bicimal, tercimal, . . . decimal fraction, one in which a **radix fraction** is written in the form $0.abc\ldots$ equivalent to a binary, ternary, . . . denary fraction where D is a power of 2, 3, . . . 10. Thus,

0.101 (base 2) $= 101/1\,000$ (binary) $= 5/8$ (denary),
0.121 (base 3) $= 121/1\,000$ (ternary) $= 16/27$ (denary),
0.371 (base 10) $= 371/1\,000$ (denary).

binary, ternary, . . . denary fraction; one in which N and D are numbers expressed in the **binary, ternary, . . . denary number systems.**

common fraction, simple fraction, vulgar fraction; where N and D are both integers.

complex fraction; where N or D or both contain a common fraction.

proper (improper) fraction; where N is less (more) than D.

rational fraction; where N and D are **real numbers.**

similar fractions or **equivalent fractions**; where the denominators are identical.

unit fraction; where $N = 1$.

See **Continued Fractions; Partial Fractions.**

FRACTIONAL INDEX. **Index** (exponent) that is not an integer. If the fraction is positive, then the denominator represents the **radix** of the root to be extracted and the numerator either the power to which the expression must be raised before the root is extracted or the power to which the extracted root must be raised. For example, $x^{\frac{3}{4}} = \sqrt[4]{x^3} = (\sqrt[4]{x})^3$. If the fraction is negative, the reciprocal of the expression is dealt with similarly, e.g., $x^{-\frac{7}{8}} = 1/\sqrt[8]{x^7} = (1/\sqrt[8]{x})^7$.

FRAME OF REFERENCE. Any set of lines, curves or surfaces to which the position of a point is related. See **Coordinates of Point**.

FREQUENCY. The number of times a value or a phenomenon occurs in some unit of time. When a set of measurements are grouped into categories, the number of measurements in each category.

FREQUENCY CURVE, GRAPH, POLYGON. The *curve* is an idealized form of the *graph* which joins the mid-points of the top sides of the rectangles forming a **histogram**. These joins, the first and last ordinates and the base line form the *frequency polygon*. The bell-shaped curve carrying the names of Gauss and Laplace is the *normal frequency curve*. See **Gauss-Laplace Law**.

FREQUENCY DISTRIBUTION. See **Distribution**.

FREQUENCY OF OSCILLATION. The number of **oscillations** per second in **simple harmonic motion**.

FREQUENCY OF PERIODIC FUNCTION. See **Periodic Function**.

FREQUENCY SURFACE. If on each cell of a **scatter diagram** a prism is erected to represent the frequency that is within the cell, the surface of the solid so formed above the scatter diagram is the frequency surface.

FRICTION. The **force** which opposes relative motion of bodies in contact. In general, when body A, in contact with body B, tends to move relatively to B, the external forces in A are balanced by a normal reaction, N, and a force F, acting in the tangent plane of contact in a direction opposite to that of possible motion. There is a limiting value to the frictional force between any two specified surfaces. This value, called *limiting friction*, is reached when equilibrium is broken and relative movement begins. The ratio of F to N, F/N or $\tan \alpha$, depends only on the nature of the two bodies in contact, and when α reaches its maximum value λ this is called the *angle of friction*, and the corresponding value, $\tan \lambda$, is called the coefficient of friction, symbol μ. The laws and values are determined empirically and the actual nature of the phenomena occurring at the area of contact is believed to depend upon the non-rigidity of bodies.

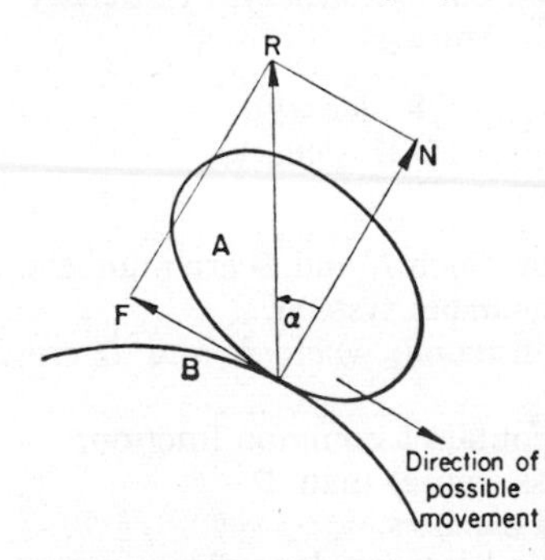

FRUSTUM. See **Truncated Cone, Cylinder, Prism, Pyramid.**

FULCRUM. The fixed point about which a **lever** can move and about which the moments of the force and the weight are calculated. See **Moment of Force.**

FUNCTION. Synonym of **fundamental relation; single valued function.**

function as correspondence. A many-one or **one-one correspondence** between the elements of one set, X (the *domain*), and the elements of another set, Y (the *range*). For example, the correspondence between two sets of **variables,** x the independent variable and y the dependent variable.

function as mapping. The set X can be limited to the domain of the function f, written $y=f(x)$ *is defined on* X. Then the function $y=f(x)$, defined on X, with subset of Y as range, can be defined as a **mapping** of the set X into the set Y (written $X \to Y$), such that to each $x \epsilon X$ there is a unique *image* $f(x) \epsilon Y$.

function as relation. A special case of **relation** in which there is only one y corresponding to each x in the domain. Thus, a function f from set X to set Y is the set $\{(x, y) \mid x \epsilon X, y \epsilon Y, yfx\}$ where yfx specifies that if $y \epsilon Y$ exists for a given $x \epsilon X$, the y is unique.

See **Inverse Function.**

FUNCTION OF SEVERAL VARIABLES. A **function** which has several independent variables. For example, if $z=f(u)$, where u is a number pair (x, y) and x and y are both from the domain of real numbers, the z has a unique value $f(x, y)$. This can be interpreted geometrically as a surface. It is also the **mapping** of points in plane Oxy on to points on the surface.

FUNCTIONAL ANALYSIS. An extension of the study of **functions** of real or complex variables. A generalization which involves the concept of a unique number being associated with a set of functions. The correspondence is referred to as a *functional.* Thus $\int_a^b f(x)dx$ can be interpreted as the functional on the set of all functions which possess an integral between the limits $x=a$ and $x=b$. The value of any functional can only be determined by considering all possible behaviour of the function.

FUNDAMENTAL OPERATIONS. These are: **addition, subtraction, multiplication, division; involution, evolution;** the manipulation of **indices** and **logarithms; differentiation, integration; inversion, transformation, translation, rotation, reflection.** The first four are fundamental to arithmetic; all are processes which lead by means of rules of procedure from **data** (postulates) and accepted (may be unproved) truths to more involved truths.

FUNDAMENTAL UNITS. In the **c.g.s. system** (centimetre-gramme-second), the **m.k.s. system** (metre-kilogramme-second), the **f.p.s. system** (foot-pound-second), the units of length, mass and time are represented by **L, M** and **T**, and most other quantities are expressed in terms of these. See **Dimension of Physical Quantity.**

FUNICULAR POLYGON. Synonym of *link polygon.* An ancillary diagram used for determining the magnitude, direction and position of the resultant of a set of any number of coplanar forces. Consider the forces $\mathbf{F}_1$, $\mathbf{F}_2$, $\mathbf{F}_3$, $\mathbf{F}_4$, represented in *Bow's notation* by **ab, bc, cd, de,** in a polygon of forces *abcdea*, where *ab*, *bc*, etc. are parallel and proportional to $\mathbf{F}_1$, $\mathbf{F}_2$, etc. The resultant **R** is represented by **ae** in magnitude and direction but not position. If any point *o* inside the polygon is joined to the vertices, the force $\mathbf{F}_1$ is seen to be equivalent to **ao** + **ob**. Similarly $\mathbf{F}_2 = \mathbf{bc} = \mathbf{bo} + \mathbf{oc}$, etc. These facts are used in drawing the *funicular polygon.* Any point P_1 is chosen on the line of action of the force $\mathbf{F}_1$ and through P_1 lines are drawn parallel to *ao* and *ob*, *ob* to cut the line of action of $\mathbf{F}_2$ at P_2. Through P_2 is drawn a line parallel to *oc* to cut the line of action of $\mathbf{F}_3$ at P_3. The process is continued until P_5 is found as the point of intersection of lines parallel to *ao* and *oe*. In the funicular polygon the forces $\mathbf{F}_1$, $\mathbf{F}_2$, etc., can be replaced by **ao** + **ob**, **bo** + **oc**, etc. The sum of these forces can thus be arranged as **ao** + (three pairs of equal and opposite forces) + **oe**; and **ao** + **oe**, acting through P_5 is equivalent to the resultant sought. See **Polygon of Vectors.**

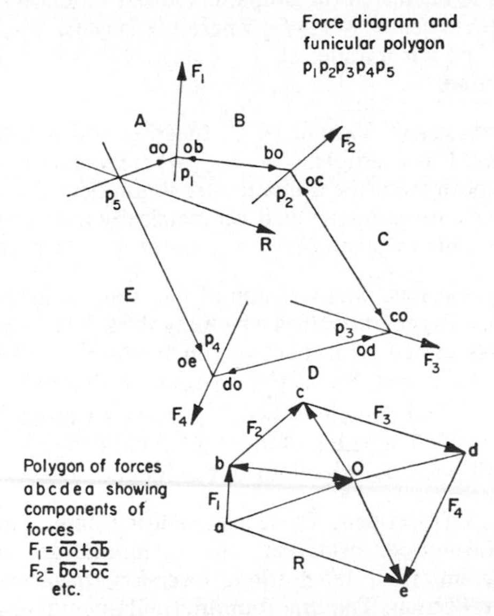

FURLONG. Unit of length; one-eighth of a mile or 220 yards.

FURTHER POLE. See **Pole of Circle on Sphere.**

G

g. (1) **Acceleration due to gravity,** (2) **Gramme (gram) mass.**

gf. **Gramme (gram) force.**

G.C.M. Greatest common measure. See **Highest Common Factor**.

g.l.b. Greatest Lower Bound. See **Bound of Function; Bound of Set.**

G.M.T. **Greenwich mean time.**

G.P. **Geometric progression.**

GALLON. The British standard unit of **capacity**; eight pints. A gallon of pure water weighs ten pounds avoirdupois when the temperature is $16\frac{2}{3}$°C (62°F) and atmospheric pressure 762 mm (30 in). It is equivalent to 4.547 **litres** and has a volume of 4 547 ml or 277.42 in^3.

GAUSS-LAPLACE LAW. The law of frequency of errors. The **Cartesian equation** of the bell-shaped Gaussian curve, the **error curve,** is

$$y = \frac{1}{\sigma\sqrt{2\pi}}\, e^{-(x-m)^2/2\sigma^2}.$$

(σ is the **standard deviation**; m, the abscissa of the mid-ordinate.)

GENERAL EQUATION. An equation which represents a family of curves or lines; e.g., $y = mx + c$ (straight lines), $ax^2 + 2hxy + by^2 + 2gx + 2fy + c = 0$ **(conic sections).**

GENERATING LINE. Synonym of generator, generatrix. The line which, by **translation** or **rotation**, determines a surface. Regular solids are generated by lines which have their movement determined by particular laws. For example, the arc of a semi-circle generates the surface of a sphere in making one revolution about its diameter. See **Conical Surface; Cylindrical Surface.**

GENUS. A **topological property** of a closed surface in space defined as the number of holes in the surface, or the largest number of closed cuts the surface can experience without being separated into two distinct pieces.

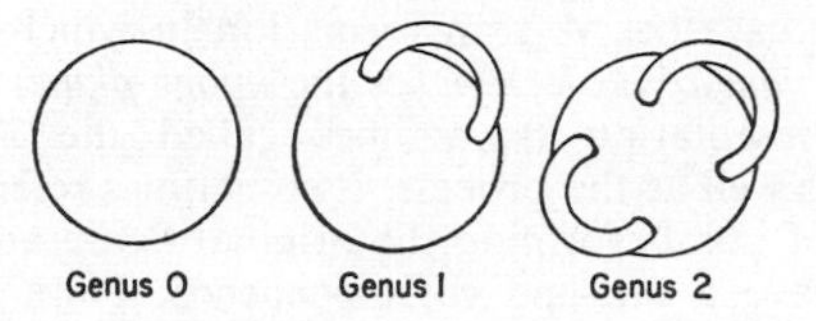

GEODESIC. Pertaining to **geodesy** and its measurements. Also the name of the shortest arc and its length between any two points on a surface. On a sphere, the minor arc of a **great circle.**

GEODESY. The study of the shape, size and weight of the earth.

GEOGRAPHICAL MILE. The length of one minute of arc on the earth's equator. It is equal to 6087.15 feet. See **Nautical Mile**; **Statute Mile**.

GEOGRAPHICAL POLES. The extremities of the earth's axis of rotation. The earth's **meridians of longitude** pass through the north and south geographical poles.

GEOID. The actual shape of the earth—neither **sphere** nor oblate **ellipsoid.**

GEOMETRIC AVERAGE. Synonym of **geometric mean.**

GEOMETRIC FIGURE. A set of lines and a set of points used to elucidate theorems, propositions and problems.

GEOMETRIC MEANS. The terms of a **geometric progression** lying between the first and the last are the geometric means of the first and the last terms. To insert n such means between numbers a and b implies having $n+2$ numbers in **geometric progression**. Hence the common ratio, r, is $(b/a)^{1/(n+1)}$. The n means are therefore $a(b/a)^{1/(n+1)}$, $a(b/a)^{2/(n+1)}$, . . . $a(b/a)^{n/(n+1)}$.

If 4 means are required between 7 and 19 they are: $7 \times (19/7)^{1/5}$, $7 \times (19/7)^{2/5}$, . . . or $7 \times (1.221\ldots)$, $7 \times (1.221\ldots)^2$, . . . or 8.547, 10.430, 12.740, 15.544. When n is 1, the geometric mean of a and b is found to be $a(b/a)^{\frac{1}{2}}$ or $\sqrt{ab}$.

The geometric average of n different quantities, $a_1, a_2, a_3, \ldots a_n$, is the nth root of their continued product: $\sqrt[n]{a_1\ a_2\ a_3 \ldots a_n}$. See **Harmonic, Geometric, Arithmetic Means.**

GEOMETRIC PROGRESSION. A **sequence** of numbers or algebraic expressions each member of which, after the first and second, is obtained from its predecessor by multiplying it by the ratio of term 2 to term 1. It is formally presented as $a, ar, ar^2, \ldots ar^{n-1}$, with n terms, the first being a and the common ratio r. The sum of n terms is given as $S_n = a(1-r^n)/(1-r)$ if $r<1$ and $S_n = a(r^n-1)/(r-1)$ if $r>1$. If $|r|<1$ the geometric **series** can be summed when n increases without bound, for then r^n approaches zero, and $S_\infty = a/(1-r)$.

GEOMETRIC PROJECTION. A transformation in which a **configuration** is mapped on to a *plane of projection* (or the *image-plane*) to produce a two-dimensional representation (sometimes called the *image-figure*). The representation, as well as the process, is sometimes referred to as the *projection.* The set of points forming the original figure and the set forming the projection have a **one-one correspondence**. Pairs of corresponding points can be joined by straight lines called *projection rays.*

central (conical or radial) projection. One in which the rays pass through a fixed point, the *centre of projection*, and form a cone (conical map projection has a different meaning).

parallel (cylindrical) projection. One in which the centre of projection is at an infinite distance, the consequent parallel rays forming a cylinder (this is not to be confused with cylindrical map projection).

oblique projection. A parallel projection in which the rays are not perpendicular to the plane of projection.

orthogonal (orthographic) projection. A parallel projection in which the rays are perpendicular to the plane of projection. Plans and elevations of objects are examples of such projections.

horizontal projection. An orthogonal projection in which a figure is projected on to a horizontal plane as a plan. Vertical distances above or below the plane of projection are marked as *index figures* at corresponding points on the *figured plan.* Elevation drawings are not necessary. Contour maps are such projections.

See **Map Projection**; **Pictorial Projection**; **Plan, Elevation, Section**; **Projective Geometry**.

GEOMETRY. That major branch of mathematics, which, as the Greek word implies, was originally a form of **geodesy**. In Egyptian times the knowledge of the 3:4:5 ratios of the **rope-stretchers' triangle** made possible the building of the pyramids and very accurate surveying. The Greeks made the first formal study; Euclid compiling in his *Elements* all the known synthetic plane and solid geometry, all constructions being performed with a straight-edge and a pair of compasses. Later Greeks added the synthetic geometry of the **conic sections** and a wide variety of spirals and curves. Euclid's twelfth axiom (elucidated in **Playfair's Axiom**) was really a **theorem** for which he could find no satisfactory **proof**. When its validity was questioned in the early nineteenth century *non-Euclidean geometries* were developed and these had repercussions on our concepts of space and on the logic of reasoning. Descartes, in the seventeenth century, initiated the algebraic study of geometry by defining **axes of coordinates**. This departure from practice combined two branches of mathematics into what is now an important field in **analysis**. See **Euclidean Space**; **Non-Euclidean Space**; **Projective Geometry**; **Topology**.

GLIDE-REFLECTION. In **motion-geometry**, the combination of a **reflection** in a line and a **translation** parallel to the line.

GLISSETTE. The locus of a point or **envelope** of a line attached to a curve, which slides along two fixed curves.

GNOMON. (1) The pin on a sundial the sharp edge of which casts the shadow. (2) If $ABCD$ is a parallelogram and $A'B'C'D'$ a smaller similar parallelogram, and A' is placed on A and $A'B'$ along AB, then the uncovered area $B'BCDD'C'B'$ has a shape known as a gnomon. See **Gnomonic Number**.

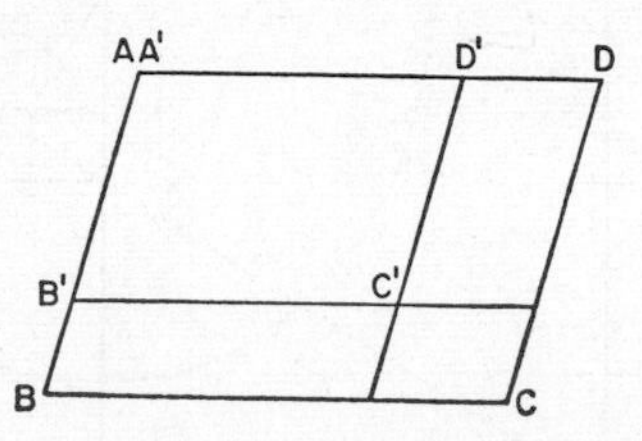

GNOMONIC (CENTRAL) ZENITHAL PROJECTION. See **Map Projection.**

GNOMONIC NUMBER. That odd number which, added to the square of any natural number, produces the square of the next larger natural number. The number $(2n+1)$ in the identity $n^2+(2n+1)=(n+1)^2$, e.g., $3^2+7=4^2$. An odd number in this sense is referred to as a **gnomon** or gnomonic number, because it can be displayed on an L-shape, reminiscent of a shadow stick and its shadow.

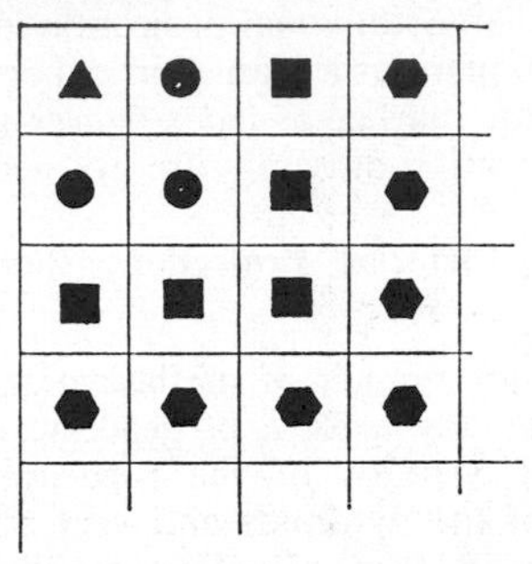

GOLDBACH'S CONJECTURE. Goldbach, a Russian mathematician, postulated in 1742 that every even number greater than 2 is the sum of two **prime numbers**. In 1931 it was proved that every even number is the sum of not more than 300,000 primes! However, Goldbach's conjecture is still assumed to be true as no exception has so far been found. It is now known that every even number is the sum of not more than four primes; every odd number is the sum of not more than three primes. See **Levy's Conjecture**.

GOLDEN RATIO. The ratio $\tau/1$ where τ is the positive root of the equation, $x^2-x-1=0$. Thus the golden ratio is $(\sqrt{5}+1)/2$ or $2/(\sqrt{5}-1)$. Another form is $(1/\tau)/(1/\tau^2)$, or $(\sqrt{5}-1)/(3-\sqrt{5})$. The ratio is the limit of the sequence 1/1, 2/1, 3/2, 5/3, 8/5, 13/8, obtained from the **Fibonacci ratios.** The three **surd** expressions and this limit have the approximate value 1.618/1 or 1/0.618 or less accurately 16/10 or 10/6. It is obtained geometrically by **golden section.** It has aesthetic appeal and is utilized in design, especially in the **golden rectangle** of classic Greek architecture.

GOLDEN RECTANGLE. Any rectangle whose adjacent sides are in the **golden ratio** $(\sqrt{5}+1)/2$. Every golden rectangle consists of a square and another golden rectangle; the lengths of the sides of the square and the longer side of the included golden rectangle being equal to the length of the shorter side of the given golden rectangle. If a sequence of squares and golden

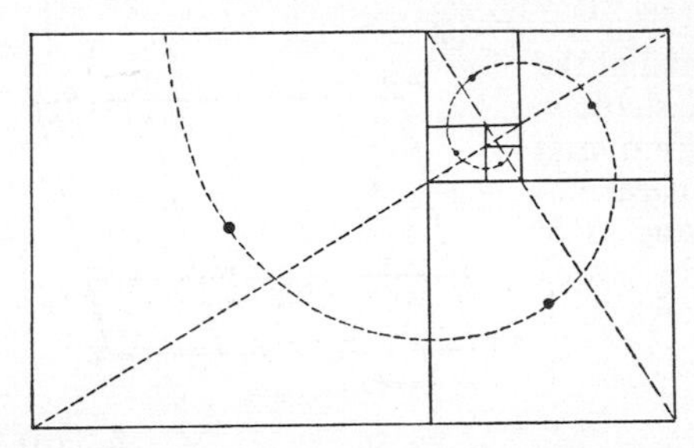

rectangles is drawn inside a given golden rectangle, as shown, the two diagonals drawn intersect at the pole of an equiangular spiral which passes through the centres of the squares. (The diagonals of smaller rectangles are different segments of the two perpendicular ones drawn.)

GOLDEN SECTION. When a line segment AB is divided internally by a point P in **golden ratio,** the line segment is said to be divided in *golden section.* The section is performed geometrically on a right-angled triangle ABC, in which $AB=2$, $BC=1, CA=\sqrt{5}$. In the diagram, P is said to divide AB in mean and extreme ratios, since

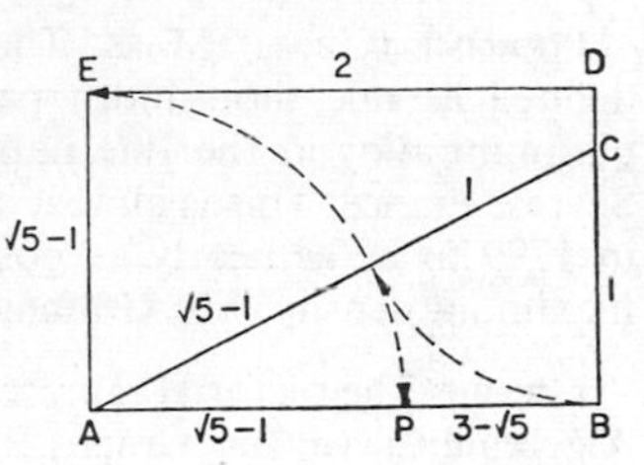

$$\frac{AP}{PB}=\frac{\sqrt{5}-1}{3-\sqrt{5}}=\frac{\sqrt{5}+1}{2}=\frac{2}{\sqrt{5}-1}=\frac{BA}{AP}.$$

The rectangle $ABDE$ is a **golden rectangle.**

GRAD (GRADE). 1/100 of a right angle. See **Centesimal.**

GRADE. (1) Synonym of **grad.** (2) Synonym of **angle of inclination.** (3) Sine of angle of inclination. (4) A **partition of set** of elements for the purpose of classification or quantitative or qualitative comparison.

GRADIENT IN SURVEYING. The ratio vertical interval/horizontal equivalent. See **Contours.**

GRADIENT OF CURVE. The gradient of a curve at a point is the gradient of the tangent (limiting position of the secant) at the point. For the curve $y=f(x)$ it is the value of $f'(x)$, the first **derivative.**

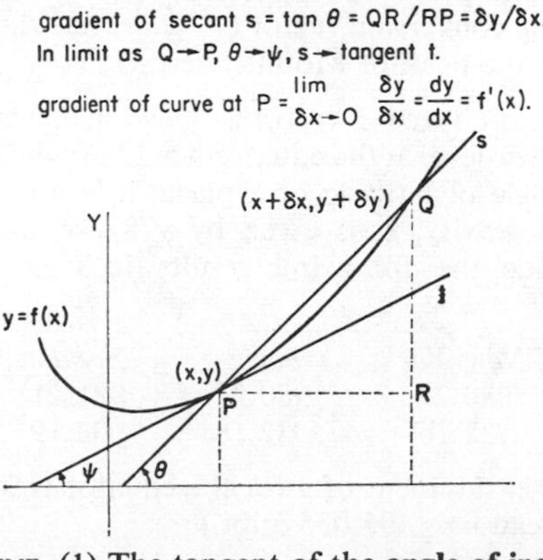

gradient of secant s = $\tan\theta = QR/RP = \delta y/\delta x$.

In limit as $Q \to P$, $\theta \to \psi$, s → tangent t.

gradient of curve at P = $\lim_{\delta x \to 0} \frac{\delta y}{\delta x} = \frac{dy}{dx} = f'(x)$.

GRADIENT OF LINE. (1) The tangent of the **angle of inclination** of the line with the positive direction of the x-axis. (2) The ratio $(y_2-y_1)/(x_2-x_1)$, when (x_1, y_1) and (x_2, y_2) are points on the line.

GRAIN. Unit of weight; the one seven-thousandth part of 1 lb **avoirdupois.** Equivalent to 6.479 892 centigrammes.

GRAMME (GRAM) FORCE, gf. The absolute unit of weight equal to the weight of a gramme (gram) mass under standard gravity, 980.665 cm/s².

GRAMME (GRAM) MASS. The unit of **mass** in the **metric system.** It is defined as one thousandth part of the standard Kilogramme mass of platinum alloy at the International Bureau of Weights and Measures at Sèvres, France. This is almost a duplicate of the original mass constructed in 1799 to be as nearly as possible the mass of 1 dm³ of pure water at maximum density. See **Gramme (Gram) Force.**

GRAPH. The pictorial representative of a relation between variables. See **Algebraic Curve; Bar Graph; Block Graph; Column Graph; Coordinates of Point; Frequency Curve, Graph, Polygon; Histogram; Trigonometric Curve.**

GRAVITATION. The law of attraction formulated by Newton as $F=k\ m_1m_2/d^2$, where F is the **force** of attraction between two bodies of **masses** m_1 and m_2 when the distance between their **centroids** is d. Cavendish evaluated the constant k as 6.67×10^{-8}; hence the force of attraction between **gramme masses** one centimetre apart is 6.67×10^{-8} **dynes.** See **Force of Gravity.**

GRAVITATIONAL LEVELLING. **Levelling** which depends upon the use in some way of the spirit level for determining the horizontal direction at any point. See **Elevation.**

GRAVITY VARIATION WITH ALTITUDE. If g_a is the value of g, the **acceleration due to gravity,** at some altitude a, and g_s is the value of g at sea-level in the same latitude, then, r being the radius of the earth, $g_a=g_s(1-2a/r)$. The conversion factor $(1-2a/r)$ has the value of 0.997 25 at 29,000 feet (near the peak of Mount Everest).

GRAVITY VARIATION WITH LATITUDE. Newton surmised that a **mass** weighing 194 lbf at sea-level at the equator would weigh 195 lbf at the North Pole. If θ is the angle of latitude of a place, it is now assumed that the **acceleration due to gravity,** g, is given by $978.039\ (1+0.005\,295\ \sin^2\theta)$. This formula yields the following results in c.g.s. and f.p.s. units respectively:

The Equator	*New York*	*Paris*	*London*	*The N. Pole*
978.04	980.27	980.97	981.21	983.22
32.09	32.16	32.18	32.19	32.26

(The most recent measurement of g for the equator is 977.99 (32.08). For Newton's 195 lbf read *now* 195.035 3 lbf.)

GREAT CIRCLE. The line of intersection with the surface of a sphere made by a plane passing through the centre. On the Earth, the equator, all **meridians of longitude** and the full 'circles' made by extending all routes of shortest length **(geodesics)** are referred to as great circles, though the earth is not a perfect sphere.

GREAT CIRCLE NAVIGATION. The movement across the surface of the earth, or in the air, in the direction of a great circle. This navigation uses the shortest route from place to place.

GREATEST LOWER BOUND (g.l.b.). See **Bound of Function; Bound of Set.**

GREENWICH MEAN TIME (G.M.T.). The sun transits at the same time for all points on a meridian, hence every separate point on an E–W line has a different noon-time, which fixes the **local time**. The earth is now divided into zones each with a standard time based on the local time of some meridian crossing the zone. Theoretically, each hour covers a **lune** extending across 15° of longitude. The edges of the zones are very irregular. Noon at Greenwich (London) is taken as noon in the British Isles, Portugal, Morocco and much of West Africa. Noon in central Europe is at 1 p.m., G.M.T. (13.00 hr), and this zone covers most of the Sahara Desert and stretches southwards into the Congo. Any one place can have two local times at different times of the year for civil reasons. See **International Date Line.**

GREGORY'S SERIES. One of the **series** which, summed to infinity, yields a value of π. It was first published by Leibniz.

$$\pi = 4[1 - \tfrac{1}{3} + \tfrac{1}{5} - \tfrac{1}{7} + \tfrac{1}{9} - \dots + (-1)^{n-1}/(2n-1) \pm \dots]$$

GROSS PROFIT, LOSS. See **Profit, Loss.**

GROUND LINE. The intersection of a horizontal plane such as a floor with a vertical plane such as a wall, when a front or side elevation or a plan of some object is made in orthogonal projection. Also the intersection of picture plane and the ground when a perspective projection is being made. See **Geometric Projection; Pictorial Projection.**

GROUND SPEED. The speed of an aircraft with respect to the ground. It is computed from information based on radio, radar or visual observation; or it is estimated from readings of air speed and wind speed. See **Air Navigation (Avigation).**

GROUP DIFFERENCES. Differences between groups which have been treated by statistical methods.

GROUP IN STATISTICS. Any collection of data which is considered together for the purpose of statistical analysis. See **Statistics.**

GROUPING. Synonym of quotition, one aspect of **division.**

GROUP OF SYMMETRIES. The collection of **rigid motions** which transform a geometric figure into itself. Such motions can be **rotations** or **reflections** (flips). A square has eight rigid motions: R_0, R_1, R_2, R_3, corresponding to rotations of 0°, 90°, 180°, 270°, and F_1, F_2, F_3, F_4, corresponding to reflections in the axes of symmetry l_1, l_2, l_3, l_4. Any such collection of symmetries is subject to **binary operation** to form the mathematical system known as a group. See **Group Theory; Symmetry.**

GROUP THEORY. The term group was introduced by Galois in the early nineteenth century and later systematized by Cayley to denote any mathematical system consisting of a set of elements subject to one **binary operation** satisfying these axioms: (1) it is a **closed system**; (2) it obeys the **associative law**; (3) there is an **identity element**, e, such that $e \odot x = x \odot e = x$ for any element x in the group ($\odot$ represents the binary operation); (4) each element, x, has an inverse, y, such that $x \odot y = y \odot x = e$.

If the group obeys the **commutative law** it is a **commutative (Abelian) group**. A *subgroup* is any subset of a group which is itself a group. The concept of group is of fundamental importance in modern mathematics.

alternating group. A group of all even **permutations** of n letters. Its *order* is $n!/2$.

cyclic group. A group of elements which are all powers of one element. Such a group is Abelian.

finite group. A group with a finite number of elements.

infinite group. A group with an infinite number of elements.

order of group. The number of elements in a finite group.

symmetric group. A group of all **permutations** of n letters. Its *order* is $n!$

See **Coset; Quotient Group.**

GUDERMANIAN (*Gdx*). The **function** u of the **variable** x given by $\sin u = \tanh x$, $\cos u = \operatorname{sech} x$ and $\tan u = \sinh x$.

GULDINUS'S THEOREMS. Two theorems, published by Guldinus between 1635 and 1642, which had appeared in Pappus's *Treatise on Mechanics* in about A.D. 300. See **Pappus's Theorems, 1 and 2.**

H

h.p. **Horse-Power.**

H.C.F. **Highest common factor.**

HP. Horizontal plane. See **First Angle, Third Angle Projections.**

HALF-LINE. That part of a **straight line** on either side of a point on that line. A ray.

HALF-PLANE. That part of a plane on either side of a **straight line** in that plane.

HARMONIC CONJUGATES. See **Harmonic Section.**

HARMONIC, GEOMETRIC, ARITHMETIC MEANS. The **harmonic mean, geometric mean** and **arithmetic mean** of two numbers may be illustrated geometrically. Let AP and AQ be measured along one line in the same direction to represent two given numbers, p and q. On PQ as diameter draw a circle having centre O. Draw tangents AG, AG', and chord GG' crossing $APOQ$ in H. If AH is of length h, AG is g and AO is a, then h, g and a are the H.M., G.M. and A.M. of p and q.

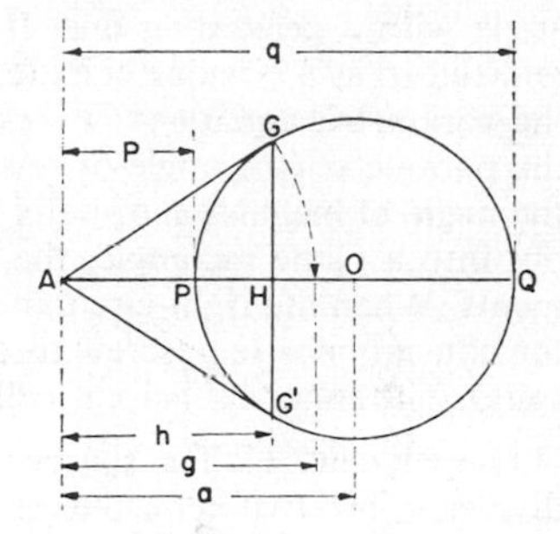

HARMONIC MEANS. The terms of a **harmonic progression** lying between the first and the last are called the harmonic means of the first and the last terms. To insert n means between numbers a and b implies having $n+2$ numbers in H.P. The common difference of the reciprocals of the terms is $(1/b - 1/a)/(n+1)$, and the n means are the reciprocals of $a+d, a+2d, \ldots a+nd$, d being the common difference. If *four* means are needed between 7 and 19, $(1/19 - 1/7)/5$ being $-12/665$, they are the reciprocals of 83/665, 71/665, 59/665 and 47/665 or 8.012 . . . , 9.366 . . . , 11.271 . . . and 14.148. . . When $n=1$, the harmonic **average** of a and b is the reciprocal of $1/a + \frac{1}{2}(1/b - 1/a)$ or $2ab/(a+b)$. **See Harmonic, Geometric, Arithmetic Means.**

HARMONIC MOTION. See **Simple Harmonic Motion.**

HARMONIC OSCILLATION. The **oscillation** of a body which is disturbed slightly from a position of stable **equilibrium.**

HARMONIC PROGRESSION. A series of numbers in which every term is the **reciprocal** of the corresponding term of a series in **arithmetic progression.**

HARMONIC SECTION. If AB is a line segment, it will be divided in harmonic section at I, internally, and at E, externally, if $AI/IB = AE/EB$. The points I and E are called the *harmonic conjugates* of A and B. When I is the midpoint of AB, E is not a finite point; in **projective geometry** E is interpreted as the **ideal point** of the line AB. See **Anharmonic Ratio.**

HAVERSINE. See **Versed, Conversed.**

HECTO-. See **Metric System.**

HEIGHT. Synonym of **altitude.** See **Dimension of Ordinary Space.**

HEIGHT FINDING. The determination of **elevation** which employs the techniques of **levelling.**

HELIX. A cylindrical, conical, or spherical helix is a space curve lying on a cylinder, cone or sphere respectively, and maintaining a constant angle with a **generating line**. If the cylinder is right-circular, the helix is referred to as a circular helix (e.g., bolt thread) and can be represented by the **parametric equations**: $x = r\cos\theta$, $y = r\sin\theta$, $z = (r\theta)\cot\alpha$, where θ is the parameter (the angle of revolution), r the radius of the cylinder and α the **angle of inclination** of helix to generator. When the cylinder is opened out into a plane rectangle, the helix appears as a set of parallel line segments. When the right-circular cone is opened out into a sector of a circle, the conical helix (e.g. screw thread) appears as segments of an **equiangular spiral**. The spherical helix is called a **loxodromic spiral**.

HEMISPHERE. Half a **sphere**. Any plane through the centre of sphere divides it into two hemispheres.

HEPTAGON. A **polygon** with seven sides.

HERONIC TRIPLES. Any three integers which express the lengths of the sides of a triangle which has an integral area. Any triangle in which an altitude is drawn is seen as the sum or the difference between two right-angled triangles. If *both* of these are **Pythagorean triangles** the sum and the difference will be triangles of integral area. Consider, for example, the triangles (3, 4, 5), (5, 12, 13). If the first is made nine times as big by multiplying its lengths by 3, it may be written (9, 12, 15), and now $\triangle_1$ (9, 12, 15) has area 54; $\triangle_2$ (5, 12, 13) has area 30. If they are drawn with one line of 12, they combine to give an area of 84 with sides 13, 14, 15; or they differ by an area of 24 with sides 4, 13, 15. These, (13, 14, 15) and (4, 13, 15), and six others found similarly are eight Heronic triples developed from two **Pythagorean triples.**

HERON'S (HERO'S) FORMULA. A formula connecting the lengths of the sides of a triangle with its area. If a, b, c, s are the lengths of the three sides and the semi-perimeter,

$$\triangle = \sqrt{s(s-a)(s-b)(s-c)}.$$

The formula is used in dealing with the radii of the **inscribed circle of triangle,** the **escribed circle of triangle** and in the trigonometrical ratios of the half angles of a triangle. If r, r_1, r_2, r_3 are the radii of the inscribed and the three escribed circles of a triangle, then the area of the triangle is given by $\triangle = \sqrt{r\, r_1\, r_2\, r_3}$. It is useful, when applying the formula, to know the identity between products and sums:

$$(a+b+c)(b+c-a)(c+a-b)(a+b-c)$$
$$= 2b^2c^2 + 2c^2a^2 + 2a^2b^2 - a^4 - b^4 - c^4.$$

HEXAFLEXAGON. See **Flexagon.**

HEXAHEXAFLEXAGON. See **Flexagon.**

HEXAFOIL. A **multifoil** bounded by six congruent arcs of a circle.

Hexagon. A **polygon with six sides.**

Hexagonal Numbers. See **Polygonal Numbers.**

Hexahedron. A **polyhedron** with six faces. It is one of the **regular solids** when its faces are regular tetragons (squares).

Hexomino. A **polyomino** made of six adjacent squares.

Highest Common Factor (H.C.F.). The largest number that will divide without remainder every one of a set of numbers is the H.C.F. of the set. Sometimes called the greatest common measure, G.C.M., when applied to weights and measures.

Hindu-Arabic Numerals. The numerals 1 2 3 4 5 6 7 8 and 9 together with 0, were invented by the early Hindus. A **denary number system** using **zero** and **place value** was known about the seventh and eighth centuries A.D. Leonardo Fibonacci of Pisa was responsible for the introduction of this system into Europe in the early thirteenth century and the Arabs were mainly associated with its popularization, so much so that the numerals are often referred to as Arabic numerals. There was tremendous opposition to the introduction of the system with its immense computational possibilities as Roman numerals and the **abacus** were so well established. The extension of the system to decimal notation is attributed to Stevin (1548–1620). The numerals had various forms until they were standardized by printing; but variations still exist, e.g., I and 1, 4 and 4, 7 and 7. The decimal point can appear as 2.3 (American), 2·3 (British), and 2˙3 and 2,3 in various countries.

Histogram. A graphic representation in the form of a **block graph** or a **column graph** in which the widths (or *intervals*) are proportional to the widths (or **class limits**) of the variable and the areas are proportional to the frequencies within the intervals. If the intervals are equal the heights of

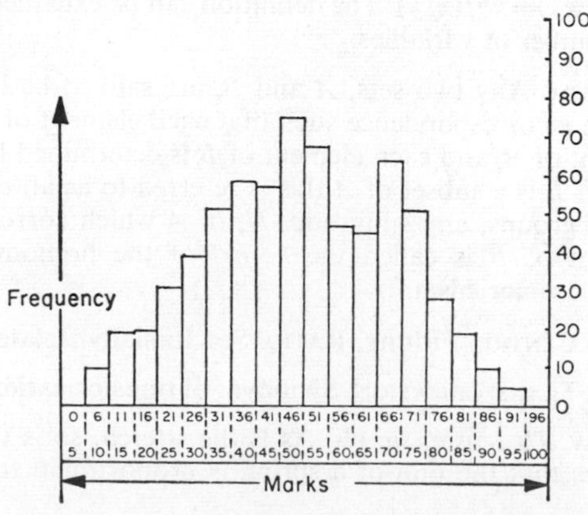

the columns or the lengths of the blocks are proportional to the frequencies. The diagram shows a histogram of the frequency distribution of students/ marks (720 students; maximum possible mark 100). See **Frequency Curve, Graph, Polygon.**

HODOGRAPH. If **directed line segments** representing the **velocities** of a point are drawn from a fixed point for varying values of time, t, the set of points at the other ends of the line segments form the hodograph of the motion. The tangent to the hodograph at any point gives the direction of the acceleration vector at the point. See **Triangle of Vectors.**

HOMALOGRAPHIC (EQUAL AREA) PROJECTION. See **Map Projection.**

HOMEOMORPHISM. See **Topological Transformation (Homeomorphism).**

HOMOGENEITY OF DIMENSIONS. All separate additive terms in a physical equation must have the same dimensions. For example, time of a complete oscillation of a pendulum is given by $t = 2\pi\sqrt{[l/g]}$. Here, dimensions of the left-hand side $= [T]$ and dimensions of the right-hand side $= [L]^{\frac{1}{2}}\{[L]^{\frac{1}{2}}([T]^2)^{-\frac{1}{2}}\}^{-1} = [T]$, also. See **Dimension of Physical Quantity.**

HOMOGENEOUS COORDINATES. The **Cartesian coordinates,** x and y, of a point in a plane can be expressed as $x = X/Z$ and $y = Y/Z$, where X, Y, Z are numbers described as the homogeneous coordinates of the point. Any equation in Cartesian coordinates becomes **homogeneous** by these substitutions. See **Homogeneous Expression.**

HOMOGENEOUS EQUATION. An equation which can be written in the form of a **homogeneous function** equated to zero.

HOMOGENEOUS EXPRESSION. A series of terms in which the sum of the **indices** of the **variables** in each term is the same. In the **general equation** of the **conic sections** the first three terms $ax^2 + 2hxy + by^2$ are homogeneous.

HOMOGENEOUS FUNCTION. The functions $f(x, y)$ is homogeneous of degree n if $f(kx, ky) = k^n f(x, y)$. The definition can be extended to functions of any finite number of variables.

HOMOMORPHISM. Any two sets, A and B, are said to be homomorphic when they have a correspondence such that each element of A determines *only one* element of B, and each element of B is determined by *at least one* element of A. If B is a subset of A this is referred to as an *endomorphism.* If A and B are groups, any subgroup, N, of A which corresponds to the identity element of B is called the *kernel* of the homomorphism. See **Group Theory**; **Isomorphism.**

HOMOTHETIC CENTRE, FIGURE, RATIO. See **Radially Related Figures.**

HOMOTHETIC TRANSFORMATION. Synonym of **transformation of similitude.**

HOOKE'S LAW. *Ut tensio, sic vis.* As is the stretch, so is the force. This succinctly states that the pull of a spring is proportional to its extension beyond its normal length. This enables forces or masses to be compared by measurements of extension. $T=\lambda\frac{L-L_0}{L_0}$, where L_0 is the length of the spring, L the length under tension and λ the modulus of **elasticity**. See **Young's Modulus.**

HORIZON. From the Greek, *horos*, a limit; the line at ground level, limiting one's vision. Ideally, the line is a circle of radius approximately $1{\cdot}23\sqrt{h}$ miles when the centre of observation is h feet above ground level.

HORIZON, CELESTIAL. See **Celestial Horizon.**

HORIZONTAL. Referring to or having properties like the **horizon**. At right-angles to the direction of the plumb-line (a vertical line).

HORIZONTAL EQUIVALENT. See **Contours.**

HORIZONTAL LINE, PLANE. The horizontal line is any line lying on a horizontal plane. This is a plane tangential to the surface of the earth or parallel to a tangential plane. It can be tested for accuracy by a spirit level.

HORIZONTAL PROJECTION. See **Geometric Projection.**

HORSE-POWER. A rate of doing **work** : 550 foot-pounds per second. A mechanical horse with a tractive power of 25 lbf and moving at 15 m.p.h. would be exerting 1 h.p. The metric h.p., *force de cheval*, is equal to 1.014 h.p.

HOUR. One twenty-fourth of a **mean solar day**; 3 600 seconds. See **Second of Time-Interval.**

HYDRODYNAMICS. The **dynamics** of fluids as compared with the dynamics of particles and rigid bodies. Hence it is the study of motion in bodies such as water, oils and gases.

HYDROSTATICS. The **statics** of fluids: the study of the conditions of equilibrium in bodies such as gases, oils and water. See **Dynamics.**

HYPERBOLA. The **conic section** made by a plane which cuts both **nappes** of a cone. The set of points which satisfy the equation $x^2/a^2-y^2/b^2=1$,

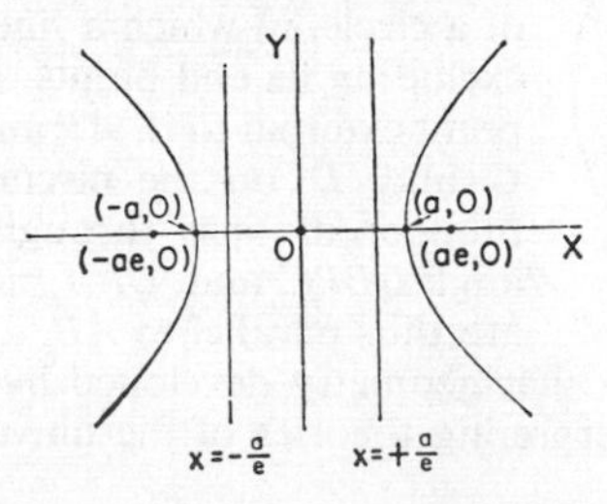

where $b^2 = a^2(e^2 - 1)$, e being the **eccentricity** of the curve. The *foci* are the points $(\pm ae, O)$, and the *directrices*, the lines $x = \pm a/e$. See **Axes of Hyperbola; Focus-Directrix Definition of Conic; Eccentric Angle, Circle of Hyperbola.**

HYPERBOLIC EXPANSIONS. The expressions on the right-hand side of the equal signs in

$$\sinh x = x + x^3/3! + x^5/5! + \dots$$

and

$$\cosh x = 1 + x^2/2! + x^4/4! + \dots$$

derived from the identities $\sinh x \equiv \frac{1}{2}(e^x - e^{-x})$ and $\cosh x \equiv \frac{1}{2}(e^x + e^{-x})$ by substituting the expansions

$$e^x = 1 + x + x^2/2! + x^3/3! + x^4/4! \dots$$

See **Hyperbolic Functions.**

HYPERBOLIC FUNCTIONS. Functions analogous to the **circular functions,** containing combinations of e^x and e^{-x}. Geometrically they are related to the **hyperbola**. The two basic formulae are:

$$\text{hyperbolic sine } x \ (\sinh x) = \tfrac{1}{2}(e^x - e^{-x}),$$
$$\text{hyperbolic cosine } x \ (\cosh x) = \tfrac{1}{2}(e^x + e^{-x}).$$

From these are derived: hyperbolic tangent x ($\tanh x = \sinh x/\cosh x$); hyperbolic cotangent x ($\coth x$ or $\text{ctnh}\, x = 1/\tanh x$ or $\cosh x/\sinh x$); hyperbolic secant x ($\text{sech}\, x = 1/\cosh x$); and hyperbolic cosecant x ($\text{csch}\, x$ or $\text{cosech}\, x = 1/\sinh x$).

The hyperbolic functions share the same pattern as the circular functions with the exception that wherever there is a product or implied product of sines, there is a sign change. Thus, $\cos^2 x - \sin^2 x = \cos 2x$, but, $\cosh^2 x + \sinh^2 x = \cosh 2x$.

HYPERBOLIC LOGARITHMS. **Logarithms** to the base e.

HYPERBOLIC PARABOLOID. See **Paraboloid.**

HYPERBOLIC SPACE. A **non-Euclidean space** based on the postulate that through a point external to a given line there are several lines parallel to the given line. The geometry of hyperbolic space was developed by Bolyai (1802–1860) and Lobachevsky (1793–1856). One example is that of the two-dimensional geometry of the interior of a circle, in which a line is defined as a chord excluding its end points. If AB is a line and P a point external to it, AP and BP determine points C and D on the circumference. An infinite number of lines through P, lying within the angles BPC and DPA, never intersect AB and are thus parallel to AB.

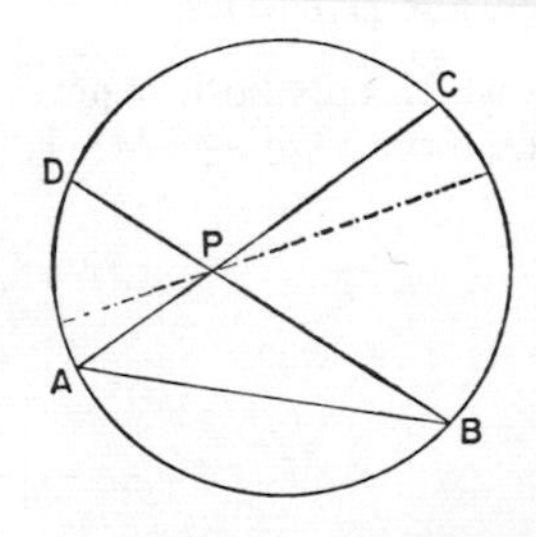

Another example is that geometry developed by Poincaré (1852–1912) which is useful in interpreting theories of the universe in astronomy and physics. A line is defined as an arc of a Euclidean circle cutting the circumference at right angles. One property of this geometry is that the sum of the angles of a triangle is less than 180°.

HYPERBOLIC SPIRAL. Synonym of reciprocal spiral. The **spiral** whose **polar equation** is $r\theta = a$.

HYPERBOLOID. A geometric surface of which certain sections are **hyperbolas**. The surface represented by $x^2/a^2 + y^2/b^2 - z^2/c^2 = 1$ is an *elliptical hyperboloid of one sheet*. It has one surface and its sections parallel to the coordinate planes are one set of **ellipses** and two sets of hyperbolas. If $a = b$, the elliptical set becomes a set of circles and the hyperboloid is equivalent to a hyperboloid of revolution obtained by revolving a hyperbola around its conjugate axis. The surface represented by $x^2/a^2 - y^2/b^2 - z^2/c^2 = 1$ is an *elliptical hyperboloid of two sheets*. It has two surfaces and its sections parallel to the coordinate planes are one set of ellipses and two sets of hyperbolas. If $b = c$, the ellipses become circles and the hyperboloid is equivalent to a hyperboloid of revolution obtained by revolving a hyperbola around its transverse axis. See **Axes of Hyperbola; Axes of Hyperboloid.**

HYPOCYCLOID. The locus in a plane of a point, P, on the circumference of a circle which rolls without slipping on the inside of a fixed circle. The **parametric equations** of the curve are

$$x = (a - b) \cos \theta + b \cos [(a - b)\theta/b],$$
$$y = (a - b) \sin \theta - b \sin [(a - b)\theta/b],$$

where a is the radius of the fixed circle, b the radius of the rolling circle and θ the angle between the line of centres and the positive direction of the x-axis. There is a **cusp** of the first kind at every point common to the curve and the fixed circle. The number of cusps is given by $n = a/b$. When $n = 4$ the curve is an **astroid** with **Cartesian equation** $x^{\frac{2}{3}} + y^{\frac{2}{3}} = a^{\frac{2}{3}}$. See **Epicycloid; Pericycloid.**

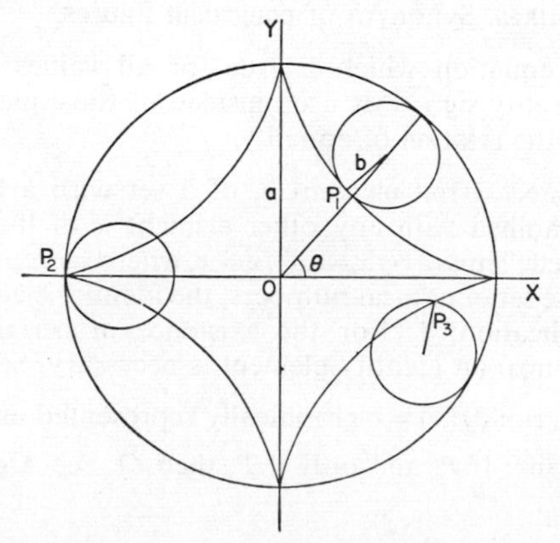

HYPOTENUSE. The longest side of a right-angled triangle. The *five* side of the **rope stretchers' triangle**. The diagonal of any rectangle is the common hypotenuse of two congruent triangles. See **Pythagoras's Theorem**.

HYPOTHESIS. In general, an assumption or premise from which a theory is developed. In the conditional proposition, if P, then Q, P is the hypothesis. Q is the conclusion. See **Algebra of Propositions.**

HYPOTROCHOID. See **Roulette.**

HYPSOMETRICAL LEVELLING. **Levelling** based on variations in **atmospheric pressure.**

I

I, J. The points $(1, i, 0)$ and $(1, -i, 0)$ in **homogeneous coordinates** on the **ideal line** of a plane through which all circles in the plane pass, and in which, therefore, all circles intersect. All conic sections which pass through I and J are circles.

i, j. Alternative symbols for $\sqrt{-1}$. See **Complex Numbers.**

ICOSAHEDRON. A **polyhedron** with twenty faces. It is one of the **regular solids** when its faces are regular trigons (equiangular, equilateral triangles).

IDEAL LINE. Synonym of line at infinity. The line represented by the equation $z=0$, in **homogeneous coordinates.**

IDEAL POINTS. Synonym of **points at infinity**. A term used to permit certain concepts to be universally true. Thus, a **pencil of parallel lines** is a special case of **pencil of lines** when an ideal point is the point of intersection.

IDEMPOTENT LAW. See **Algebra of Sets.**

IDENTICAL FIGURES. Synonym of **congruent figures.**

IDENTITY. An equation which is true for all values of the variables involved. The identity sign $\equiv$ is used instead of the equality sign $=$. The term, in logic, is the **relation** of equality.

IDENTITY ELEMENT. That element e, of a set with a **binary operation,** which, when combined with any other element x of the set, leaves that element unchanged. Thus, $e \odot x = x \odot e = x$, where $\odot$ represents the binary operation. For the set of natural numbers, the identity element for addition is 0; for multiplication, 1. For the existence of certain mathematical systems (e.g., groups) an identity element is necessary. See **Group Theory.**

IDENTITY FUNCTION. $f(x)=x$ graphically represented as the line $y=x$.

iff. P iff Q means: If P, and only if P, then Q. See **Algebra of Propositions.**

IMAGE. The **reflection** of a point in a point, line or plane. In **mapping,** if set X is mapped into set Y and $x \in X$ is associated with $y \in Y$, y is the image of x.

IMAGINARY AXIS. The y-axis in the **Argand diagram.**

IMAGINARY CIRCLE. The equation $(x-a)^2+(y-b)^2-c^2=0$ represents the circumference of a real circle, centre at (a, b) and of radius c. The equation $(x-a)^2+(y-b)^2+c^2=0$, represents the circumference of an imaginary circle, centre at (a, b) and radius $\sqrt{-c^2}$ or ic.

IMAGINARY NUMBER. The **complex number,** $a+ib$, when $b \neq 0$. When $a=0$ the number is purely imaginary; when $b=0$, the number is real.

IMAGINARY ROOT OF EQUATION. A root which is an **imaginary number.** In general, the roots of the **quadratic equation,** $ax^2+bx+c=0$, are $\alpha=(-b+\sqrt{b^2-4ac})/2a$ and $\beta=(-b-\sqrt{b^2-4ac})/2a$. When $b^2-4ac<0$, that is the discriminant is negative and has no real square root, α and β are a pair of conjugate **complex numbers** of the form $p+iq$ and $p-iq$, where i is $\sqrt{-1}$ and a, b, c are real. If $p=0$, α and β are purely imaginary.

IMPACT. The sudden application of a **force** for a very short time as in the case of collision of bodies. See **Laws of Impact.**

IMPERIAL MEASURES. Units prescribed in the United Kingdom.

IMPLICATION. (1) A conclusion deduced from a given assumption. (2) A conditional statement of the kind: 'If a, then b'. See **Algebra of Propositions.**

IMPLICIT DIFFERENTIATION. The evaluation of dy/dx when $f(x, y)$ is an **implicit function.** If, for example, $x^3+3x^2y-4xy^2-6y^3=0$, differentiating with respect to x gives:

$$3x^2+6xy+3x^2\ dy/dx-4y^2-4x\cdot 2y\ dy/dx-18y^2\ dy/dx=0.$$

This may now be written:

$$\frac{dy}{dx}=-\frac{3x^2+6xy-4y^2}{3x^2-8xy-18y^2}.$$

IMPLICIT FUNCTION. If an equation of the form $f(x, y)=0$, cannot be solved for y by substitution of a value of x, without solving a resulting equation in y, then y is said to be an implicit function of x. The variable y may be a **multi-valued function** of x. For example, in the equation $y^2-3x+y-2=0$, y is a two-valued implicit function of x; for $x=5$, $y=-\frac{1}{2}\pm\frac{1}{2}\sqrt{69}$.

IMPLICIT RELATION. The relation between two variables expressed indirectly by their relations to a third variable called a **parameter.** For example, from the **parametric equations** $y=at+b$, $x=ct+d$, y can be found in terms of x by eliminating the parameter t, yielding the **explicit relation** $y=(a/c)x+b-ad/c$.

IMPROPER FRACTION. A common **fraction,** N/D, in which N is greater than D. Such a fraction can be converted into a mixed number containing an integer and a proper fraction. Thus $22/7 = 3\frac{1}{7}$.

IMPULSE (IMPULSIVE FORCE). A product of **force** and time. It is measured by $\int_0^t P\,dt$, which is Pt if P is constant. If the force is regarded as a product of **mass** and **acceleration**, or $m\,dv/dt$, then:

$$\int_0^t P\,dt = \int_0^t m\ (dv/dt)\ dt = \int_u^v m\ dv = mv - mu$$

The impulse is thus measured by the change of **momentum** it produces during its operation.

INCENTRE. The centre of an **inscribed circle of polygon.** The centre of an **inscribed circle of triangle** or the common point of intersection of the bisectors of the angles of a triangle.

INCH CUBE. A cube of one inch edge having a **volume** of one **cubic inch** (1 in^3).

INCH SQUARE. A square of one inch side having an **area** of one **square inch** (1 in^2).

INCIRCLE. See **Inscribed Circle of Polygon; Inscribed Circle of Triangle.**

INCLINATION. In general, the **dihedral angle** between two planes. The difference in direction of a line and some **datum** direction. See **Angle of Inclination.**

INCLINATION OF LINE IN PLANE. The angle known as the **angle of inclination,** measured in an anti-clockwise direction, between the line and a **datum** line in the plane. In analytical geometry, the datum line is the positive x-axis.

INCLINATION OF LINE TO PLANE. The angle between the line and its **orthogonal projection** in the plane.

INCLINATION OF TWO LINES. Any one of the four angles formed at the point of intersection of two lines in a plane. **Skew lines** do not have such angles.

INCLINATION OF TWO PLANES. The smaller **dihedral angle** between the planes.

INCLINED PLANE. A plane that is not tangential to the surface of the earth, i.e. is not **horizontal.**

INCOMMENSURABLE. See **Commensurable.**

INCREMENT. A small positive or negative change in the value of the independent variable of a **function,** usually written δx (or Δx). The name is also given to the consequent change in the value of the function, δy (or Δy). Hence $\delta y = f(x + \delta x) - f(x)$. See **Differential Coefficient.**

INDEFINITE INTEGRATION. See **Integration of Function.**

INDEPENDENT VARIABLE. See **Dependent, Independent Variable.**

INDETERMINATE EQUATION. An equation with more than one variable, e.g., $ax+by+c=0$, for which there is an infinite set of values which satisfy the equation. If the coefficients and the set of values are integral the equation is a *Diophantine equation.*

INDEX, INDICES. Positive or negative integers or fractions placed above and to the right of quantities to express the power to which the quantity is to be raised or lowered. See **Power of Quantity; Theory of Indices.**

INDIRECT COMMON TANGENTS. See **Tangents to Two Circles.**

INDIRECT VARIATION. See **Variation.**

INDUCTION. A method used in proving that a proposition $P(n)$ is true for all values of n. The steps involved are: (1) $P(1)$ is shown to be true; (2) it is proved that if $P(k)$ is true, so also is $P(k+1)$. These two steps establish the truth of $P(1), P(2), P(3)\ldots$

INEQUALITY. A statement that one quantity is greater than or less than another, written $a>b$ or $a<b$ respectively. Inequalities are unaltered when the same number is added to or subtracted from both sides and when each side is multiplied by or divided by the same positive quantity. Multiplication or division of both sides by a negative quantity changes the sense of inequality.

conditional inequality. One that is *not* true for all values of the variable, e.g., $3x<5$.

unconditional (absolute) inequality. One that is true for *all* values of the variable, e.g., $x^2+1>0$.

sense of inequality. Two or more inequalities have the same sense if $a>b$ and $c>d$ or $b<a$ and $d<c$. They have opposite sense if $p>q$ and $r<s$.

INERTIA. One of the few properties common to all kinds of matter and shown in the resistance of any body to forces tending to change its state of equilibrium. A **force**, proportional to the **mass** of the body and the amount of **acceleration** required, has to be applied to overcome inertia. One speculation concerning the nature of inertia is that it may be an innate property of the environment and be related to the electromagnetic field in the neighbourhood of the body.

inf. Abr. greatest lower bound. See **Bound of Set.**

INFERENCE IN LOGIC. A deduction from a premise which involves only one step in any process of reasoning. Any logical **proof** in mathematics is a succession of inferences. There are various rules of inference. *Substitution* permits the identity $(2a)^2-(3b)^2\equiv(2a-3b)(2a+3b)$ to be inferred from the identity $x^2-y^2\equiv(x-y)(x+y)$. The *rule of detachment* states that the premises 'If P then Q' and 'P' permit the inference 'Q'. Thus, 'Lions are predators' and 'This is a lion', hence 'This is a predator'.

INFERENCE IN STATISTICS. The process of drawing conclusions about a **population**, based on random **samples** in such a way that the **probability** of a correct inference can be determined according to various hypotheses concerning the population under study. See **Significance Tests.**

INFINITE. Not finite; limitless; unbounded. See **Infinity.**

INFINITE FUNCTION. A function $f(x)$ is said to approach plus or minus infinity as x approaches a, if, for any number N, there is a neighbourhood of a for all points of which the **absolute value** of the function is greater than or less than N respectively. It is written:

$$\lim_{x \to a} f(x) = +\infty \quad \text{or} \quad \lim_{x \to a} f(x) = -\infty.$$

INFINITE GROUP. See **Group Theory.**

INFINITE SERIES. A series with an **infinite** number of terms.

INFINITE SET. A set with an **infinite** number of elements which can be put into **one-one correspondence** with part of itself. For example,

$$\begin{array}{ccccccccc} 1 & 2 & 3 & 4 & 5 & 6 & 7 & 8 & 9\ldots \\ 2 & 4 & 6 & 8 & 10 & 12 & 14 & 16 & 18\ldots \end{array}$$

See **Countably Infinite Set.**

INFINITESIMAL ANALYSIS. Synonym of **infinitesimal calculus.**

INFINITESIMAL CALCULUS. An alternative name for **calculus,** which is based on the study of **infinitesimals.**

INFINITESIMALS. **Variables** which approach zero as a **limit.**

INFINITY. The state of being infinite, unlimited, denoted by the symbol ∞. The concept can involve the concepts of numerical or spatial boundlessness. In geometry, the term sometimes has the quality of definiteness when it is convenient to refer to infinity as a place at an infinite distance, as in **points at infinity.** See **Infinite Set.**

INFLEXION. A point on a curve where the tangent crosses the curve is a point of inflexion. At such a point the convexity of the curve changes to concavity with respect to any line not through the point of inflexion, and the tangent changes its sense of rotation. For the curve $y = f(x)$, at a point of inflexion dy/dx may be positive, zero or negative, but d^2y/dx^2 is always zero. This fact distinguishes such a point from a turning point where the value of y may be a maximum or minimum. At a point of inflexion the gradient (given by dy/dx) has a maximum or minimum value. See **Curvature; Concave, Convex Curve; Maximum, Minimum Values.**

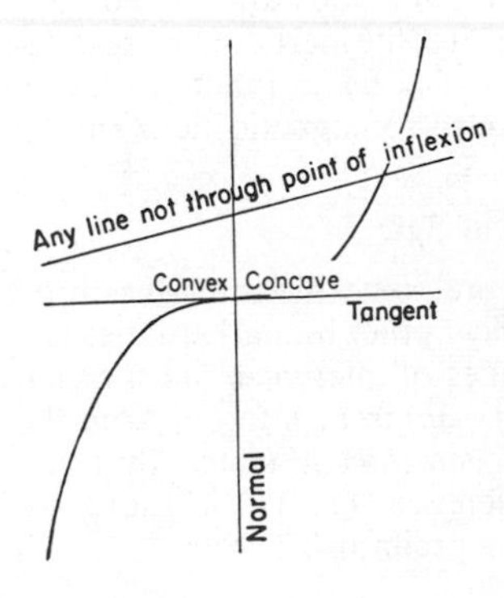

INITIAL AND TERMINAL LINES. If a radius vector OP rotates from the position OP_i to the position OP_t, then OP_i is called the initial line and OP_t the terminal line forming the arms of the angle P_iOP_t. See **Polar Coordinates in Plane; Spherical Coordinates in Space.**

INITIAL CONDITIONS. The **vector quantities** that apply to a body at the commencement of some study of its movements or of the forces that act upon it. Generally the zero suffix after the symbol indicates the initial condition: $x_0, y_0, z_0, r_0, t_0, v_0, \theta_0, \phi_0$, etc.

INNER PRODUCT OF SETS. See **Algebra of Sets; Cartesian Products.**

INNER PRODUCT OF VECTORS. The product $l_1l_2 \cos \theta$, where l_1 and l_2 are lengths of two **vector quantities** and θ the angle between them. Two vectors are perpendicular when the inner product is zero.

INSCRIBED CIRCLE OF POLYGON. The circle which touches the sides of a **polygon** internally. Not all polygons can have inscribed circles.

INSCRIBED CIRCLE OF TRIANGLE. The circle that touches the sides of the triangle internally. The centre of the circle is the point of **concurrency** of the **internal bisectors** of the angles and its radius is $\triangle/s$ where $\triangle$ is the area of the triangle and s the semi-perimeter. See **Bisector of Angle.**

INSCRIBED POLYGON. The **polygon** whose sides are chords of some **simple closed curve.**

INSCRIBED TRIANGLE OF CIRCLE. A triangle whose vertices lie on the circumference of a circle.

INSTANTANEOUS CENTRE OF ROTATION. The point about which every particle of a body is moving in a circular arc during any instant of time. If the actual paths of two particles are known, then **normals** to their paths meet in the instantaneous centre of the body. Two examples are shown.

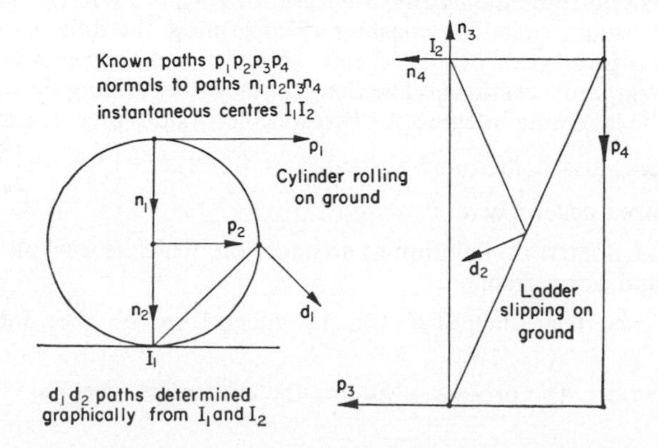

INSTANTANEOUS RATE OF CHANGE. See **Rate of Change of Function.**

INSTANTANEOUS SPEED. See **Speed.**

INSTANTANEOUS VELOCITY. See **Velocity.**

INTEGER. Any one of the numbers 1, 2, 3, 4, . . . (positive integers) and $-1,\ -2,\ -3,\ -4,$ . . . (negative integers). 0 (zero), neither positive nor negative, is usually included to complete the pattern.

INTEGRAL CALCULUS. See **Calculus.**

INTEGRAL CURVES. A family of curves whose equations are the solutions of a given **differential equation**, e.g., the family of circles $x^2+y^2=c$, which is the solution of the equation $x+y\ dy/dx=0$.

INTEGRAL DOMAIN. A set of elements subject to two **binary operations** ('addition' and 'multiplication') which satisfies these axioms: (1) it is a commutative **ring**; (2) there is an **identity element**, e, such that $e\ \ x=x$, for any element x; (3) when a product $x\ \ y=0$ (0 being the identity element for addition) then x and/or $y=0$.

INTEGRAL EXPRESSION. An **algebraic expression** in which there are no **variables** in the denominator when all **indices** are made positive.

INTEGRAL FACTOR. Any **factor** which is a whole number (integer).

INTEGRAL FUNCTION. Synonym of *entire function.* One which when expanded by the application of **Maclaurin's theorem** is valid for any assignable value of the **variable.**

INTEGRAL INDEX (EXPONENT). An **index** which is a whole number (integer) and not a fraction.

INTEGRAL NUMBER. Synonym of **integer.**

INTEGRAL OF FUNCTION. Given any **function** of a **variable,** $f(x)$, and another function, $g(x)$, obtained from it by the processes of **differentiation,** then $f(x)$ is said to be the *indefinite integral* of $g(x)$, written $\int g(x)dx=f(x)$. If c is a constant, called the **constant of integration,** the differentiation of $(x)+c$ also gives $g(x)$, hence $\int g(x)dx=f(x)+c$ is the *general form.* The *definite integral* of a **continuous function** is obtained by finding the difference between the indefinite integrals for two specified values of x. If a and b are these values, it is written and evaluated as $\int_a^b g(x)dx=\Big[f(x)+c\Big]_a^b=f(b)-$ $f(a)$. See **Area under Curve.**

INTEGRAL SOLUTION. Solution of an equation, which is a whole number (integer) and not a fraction.

INTEGRAND. The function $f(x)$ in the integral $\int f(x)dx$. See **Integral of Function.**

INTEGRATION. The process of finding the **integral of function.**

INTEGRATION BY PARTS. The process of integrating a product by use of the formula for differentiating a product: $d(u\ v) = u\ dv + v\ du$ is rearranged as $u\ dv = d(u\ v) - v\ du$. Integration of both sides produces the formula for integration by parts:

$$\int u\ dv = u\ v - \int v\ du.$$

See **Integral of Function.**

INTEGRATION BY SUBSTITUTION. The process of integrating a function by changing the **integration variable.** See **Integral of Function.**

INTEGRATION VARIABLE. The variable x in the integration $\int f(x)dx$. See **Integral of Function.**

INTERCEPT. Part of a curve, line, surface, plane or solid cut off by any other curve, line, surface, plane or solid.

INTERCEPT EQUATION OF LINE. If a straight line crosses the coordinate axes in points $(a, 0)$ and $(0, b)$, and $P\ (x, y)$ is any point on the line, then for all points on the line, $x/a + y/b = 1$.

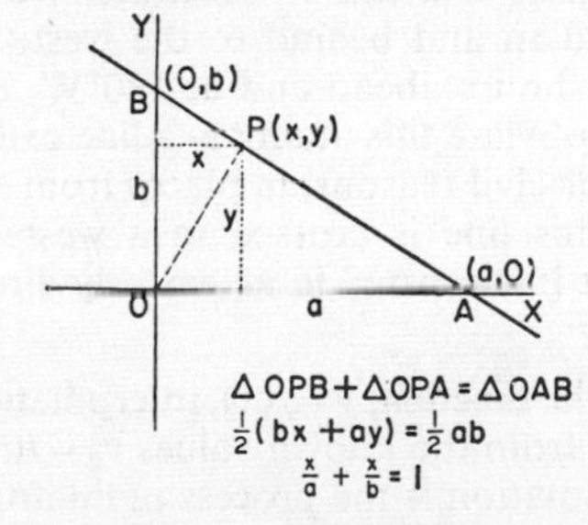

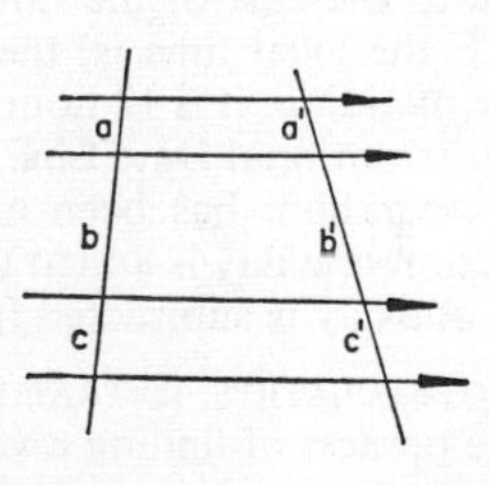

INTERCEPT THEOREM. If a set of parallel lines is crossed by two **transversals**, the ratios of the intercepts on one transversal are equal to the ratios of corresponding intercepts on the other. Thus

$$a/b = a'/b',\ (a+b)/c = (a'+b')/c',$$

etc.

Also, the ratio of the intercepts on the transversals between one pair of parallel lines is equal to the ratio of the intercepts between any other pair of parallel lines in the set. Thus

$$a/a' = b/b' = c/c' = (a+b)/(a'+b'),$$

etc.

INTEREST. The sum of money paid for the use of a loan, the *principal.* In **simple interest** it is based on the principal alone; in **compound interest** on the principal and accumulated interest. The *rate* of interest is expressed as a **percentage** for an agreed period of time. At the end of each period the sum of the principal and the interest is called the *amount.*

INTERIOR ANGLE. An angle at a vertex of a **polygon** formed inside the polygon by two adjacent sides. The interior angle and either of the **exterior angles** at a vertex are **supplementary angles.**

INTERIOR ANGLES. A description of two angles that occur on the same side of a **transversal** and between a pair of lines crossed by the transversal. When the pair of lines are such that the sum of the angles is 180°, the lines are **parallel.**

INTERNAL BISECTOR. The bisector of an **interior angle** of a **polygon.** It is perpendicular to the **external bisector.** See **Bisector of Plane Angle.**

INTERNAL COMMON TANGENTS. See **Tangents to Two Circles.**

INTERNAL DIVISION. See **Division of Line Segment.**

INTERNAL, EXTERNAL ANGLES. The pair of angles at a vertex of a **polygon,** whose sum is 360°.

INTERNAL VOLUME. See **Volume.**

INTERNATIONAL DATE LINE. The **local time** is ahead of **Greenwich Mean Time** to the east of the Greenwich meridian and behind to the west. At 180°E the local time is, theoretically, 12 hours ahead and at 180°W, the same meridian, it is 12 hours behind. To obviate this anomaly a line called the International Date Line, departing, for civil reasons, in places from the 180° meridian, has been adopted. As this line is crossed in a westerly direction one day is added to the date; as it is crossed in an easterly direction one day is subtracted from the date.

INTERPOLATION, EXTRAPOLATION. For the **function,** $y=f(x)$, interpolation is the process of finding a value $y_i=f(x_i)$ from the known values $y_a=f(x_a)$ and $y_b=f(x_b)$, where $x_a<x_i<x_b$. Extrapolation is the process of finding a value $y_e=f(x_e)$ from y_a and y_b, where $x_e<x_a\nleqslant x_b$ or $x_a<x_b<x_e$. Linear interpolation and extrapolation is the common form of the process by which these estimates are obtained. This assumes that $y=kx$ for $x_{e1}<x_a<x_i<x_b<x_{e2}$. But there are systematic methods in use by which a polynomial function of any degree is fitted to calculated values of y in the neighbourhood of the required interpolated or extrapolated position. In all cases the estimated values reflect the nature of the function being used to find them. The methods of **finite differences** for interpolation and extrapolation depends upon the knowledge of the values of y corresponding to a sequence of values of x in the **neighbourhood** of x_i and x_e.

INTER-QUARTILE RANGE. The whole range of **dispersion** is halved by the median and quartered by the lower quartile point, the median and the upper quartile point. The inter-quartile range is the portion of the whole range of dispersion lying between the lower and upper quartile points. See **Median in Statistics.**

INTERSECTION OF TWO CURVES. Those parts which the curves have in common. The intersection may be a single point, a finite number of points or parts of curves. Intersection of curves is expressed algebraically as two or more simultaneous equations.

INTERSECTION OF TWO SURFACES. The parts which are common to the two surfaces. The intersection may consist of isolated points, a curve or parts of the surfaces.

INTERSECTION (PRODUCT) OF SETS. See **Set Theory.**

INTERVAL. If x represents a **set** of numbers or points, and a and b represent end values, then set x is a *closed interval* within the inequality $a \leqslant x \leqslant b$, an *open interval* within the inequality $a < x < b$, closed and open within the inequality $a \leqslant x < b$ and open and closed within the inequality $a < x \leqslant b$. Thus an interval $[a, b]$ is the set $\{x | x \text{ a real number}, a \leqslant x \leqslant b\}$ for given real numbers a and b $(a \leqslant b)$.

INTRANSITIVE RELATION. See **Transitive Relation.**

INTRINSIC EQUATIONS. The equations which determine the position of a curve in space without reference to particular coordinate axes. Such equations are often in terms of the radius of *curvature* and the arc length.

INTRINSIC GEOMETRY. The geometry of surfaces which is independent of the surrounding space. If a surface is deformed without extension or compression, the property that the distance between two points is unaltered is an example of the intrinsic geometry of the surface.

INTRINSIC PROPERTIES OF CURVE. Properties of a **curve** which remain unaltered with any change in the **system of coordinates.**

INTRINSIC PROPERTIES OF SURFACE. Properties of a **surface** which remain unaltered with any change in the **system of coordinates.**

INVARIANT PROPERTY. That property of a **function,** an **equation** or a geometric figure which remains unaltered under a particular transformation. Thus, in the general equation of the conic sections,

$$ax^2 + 2hxy + by^2 + 2gx + 2fy + c = 0,$$

the expressions $(a + b)$, $(h^2 - 4ab)$ and the **discriminant** are unaltered under translation and/or rotation and are called the invariants. See **Rigid Motion.**

INVERSE CURVES. See **Inversion.**

INVERSE ELEMENT. In any set of elements, the inverse of a is a' such that $a \odot a' = e$, where e is the **identity element** and $\odot$ an undefined **binary operation.** Thus, in the set of numbers $\{0, \pm 1, \pm 2, \ldots \pm n\}$ under addition, $-n$ is the inverse of $+n$, i.e. $(+n) + (-n) = 0$, the identity element. See **Set Theory.**

INVERSE FUNCTION. If two sets X and Y are related by the **function** f, denoted yfx or $y = f(x)$ there is an inverse (reciprocal) function f^{-1}, denoted

$xf^{-1}y$ or $x = f^{-1}(y)$ if and only if the sets X and Y have a **one-one correspondence.** Any function which permits such an inverse function is called a *bijective function* or a *bijection* and the sets X and Y are called *equipotent sets* (written $X \sim Y$). This is equivalent to stating that the sets X and Y have the same **cardinal number,** denoted $\#X$ or $\#Y$. Thus $X \sim Y \Leftrightarrow \#X = \#Y$. Every set equipotent to the set of all integers $N = \{0,1,2,\ldots\}$ is a **denumerable set** (or **countable set**). The cardinal number $\#N$ is a *denumerable infinity.*

If a function is seen as a mapping, an inverse function may be defined in the following way. If, for two sets X and Y, function f, defined on X maps X on to Y, i.e., $f: X \to Y, x \to f(x)$, the function $y = f(x)$ has an inverse function denoted f^{-1}, the reciprocal of f. Thus $f^{-1}(y)$ represents the set of all elements of X which have y as an image. Hence the function $y = f(x)$ has as its inverse $x \epsilon f^{-1}(y)$.

INVERSE HYPERBOLIC FUNCTION. If $y = \sinh x$, then the inverse statement that x is the number with hyperbolic sine y is written $x =$ inverse $\sinh y$, or arc $\sinh y$ or more usually $\sinh^{-1}y$. Hence, $\sinh^{-1}y$ is the **inverse function** of $\sinh x$. A similar notation applies to the other hyperbolic functions. The -1 in the notation is *not* an **index** and $\sinh^{-1}y$ must not be confused with $(\sinh y)^{-1}$.

INVERSE OF MATRIX. See **Matrix.**

INVERSE OF NUMBER. The reciprocal of the number. If n is the number, the inverse is $1/n$ or n^{-1}.

INVERSE OF SET. See **Set Theory.**

INVERSE OPERATION. The reverse process which cancels the result of a **binary operation.** Addition and subtraction are inverse operations, as also are multiplication and division, involution and evolution: e.g., $(a+b) - b = a$; $(b \times)a(\div b) = a$; $\sqrt[n]{a^n} = a$. See **Cancellation; Operator.**

INVERSE POINTS. See **Inversion.**

INVERSE PROPORTION. Synonym of **inverse variation.**

INVERSE RATIO. Synonym of reciprocal *ratio.* The ratio of the reciprocals of two numbers. Thus, the inverse ratio of a/b is $(1/a)/(1/b) = b/a$.

INVERSE RELATION. If two sets X and Y are related by the **relation** R, denoted xRy, there is an inverse (reciprocal) relation R^{-1}, denoted $yR^{-1}x$. In the diagram representing the graph of the relation R seen as a mapping, the inverse relation is represented by reversing the directions of the arrows. As **equivalence relations,** $xRy \Leftrightarrow yR^{-1}x$. For example, $x < y \Leftrightarrow y > x$. See **Symmetric Relation.**

INVERSE SQUARE LAW. In general, the intensity of energy (heat, light, sound, magnetism or electricity) received from a source of energy varies inversely as the square of the distance, d, from the source. The variation is of the form $I=k/d^2$ or $I \propto 1/d^2$. The law can be illustrated by the projection of a unit square, at a distance l, then at distance $2l$, $3l$, etc. The same amount of energy is received over a larger area, so the intensity is reduced.

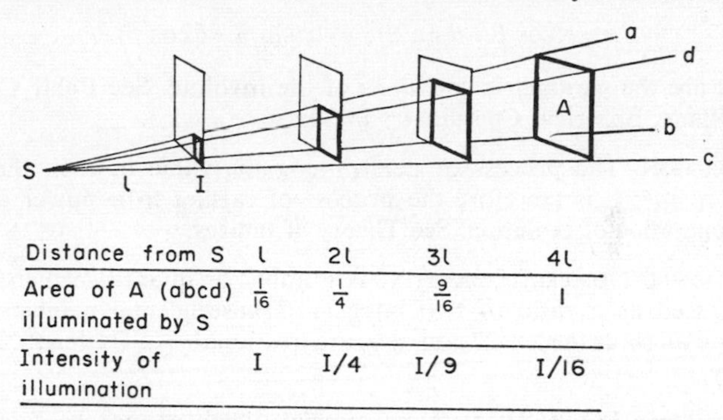

Distance from S	l	$2l$	$3l$	$4l$
Area of A (abcd) illuminated by S	$\frac{1}{16}$	$\frac{1}{4}$	$\frac{9}{16}$	1
Intensity of illumination	I	I/4	I/9	I/16

INVERSE TRIGONOMETRIC FUNCTIONS. If $y=\sin x$, then the inverse statement x is the angle whose sine is y is written $x=$inverse sin y or arc sin y or more usually, $\sin^{-1}y$. Hence $\sin^{-1}y$ is the **inverse function** of $\sin x$. A similar notation applies to the other trigonometric functions. The -1 in the notation is not an **index** and $\sin^{-1}y$ must not be confused with $(\sin y)^{-1}$.

INVERSE VARIATION. A **variation** of the form $y=k/x$ or $xy=k$ in contrast to direct variation $y=kx$.

INVERSION. If OP is a radius vector and P' a point on it such that $OP \cdot OP'=r^2$, P' is the inversion of P with respect to a circle with centre O and radius r. P and P' are *inverse points* and their loci *inverse curves*. The circle is the *circle of inversion* and the square of its radius the *constant of inversion*. If P lies on the circumference of a circle which passes through the centre of inversion, the inversion of P will lie on a straight line, L. The concept of inversion can be extended to space when a sphere is substituted for the circle. In space, the inverse points associated with all points on a sphere passing through the centre of inversion lie in a plane. See **Polar Coordinates in Plane; Spherical Coordinates in Space.**

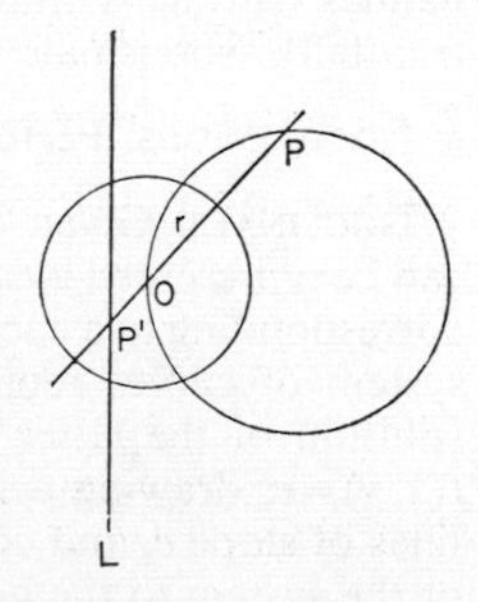

INVOLUTE. The curve traced by any point on a flexible, inextensible thread, kept taut while being unwound from or wound on to another curve. If a is the involute of b, b is the involute of a.

INVOLUTE OF CIRCLE. If the centre of the circle is the origin of coordinates and unwinding of the **involute** begins on the x-axis at the end of radius, r, the length of the tangent from a point on the involute to the circle is the length of the unwound arc. If the radius vector of a point on the involute makes an angle θ with the positive direction of the x-axis, the **Cartesian coordinates** of the point will be

$$x = r(\cos\theta + \theta\sin\theta);\ y = r(\sin\theta - \theta\cos\theta).$$

These are the **parametric equations** of the involute. See **Polar Coordinates in Plane; Spherical Coordinates in Space.**

INVOLUTION. The process of determining the value of x in the given relation $x = a^n$. It is therefore the process of raising to a power and the **inverse operation of evolution.** See **Theory of Indices.**

IRRATIONAL NUMBER, QUANTITY. A number or quantity which cannot be expressed as a ratio of two integers. **Transcendental numbers** like e and π and all **surds** (e.g. $\sqrt{27}$ and $\sqrt[3]{9}$) are irrational. See **Rational Number, Quantity.**

IRREDUCIBLE POLYNOMIAL. A **polynomial** which cannot be factorized. Thus, in the field of rational numbers, quadratics $x^2 - 2$ and $x^2 + 2$ are irreducible. In the field of **complex numbers** both can be factorized: thus, $x^2 - 2 = (x - \sqrt{2})(x + \sqrt{2})$, $x^2 + 2 = (x - i\sqrt{2})(x + i\sqrt{2})$.

ISALLOBAR. A line on a weather map through places where the change of pressure (pressure tendency) with time is constant. See **Iso-.**

ISO-. (Greek, *isos*, equal.) When lines are drawn on maps through places experiencing equal intensity or frequency of natural phenomena, they are referred to as isobars (atmospheric pressure), isobaths (depth of sea-bed), **isoclinals** (dip of magnetic needle), isogons (magnetic declination), isohalines (surface salinity of oceans), isohels (hours of sunshine), isohyets (rainfall), isoseismals (earthquake shock), isotherms (temperature).

ISOCHRONOUS. Performed in equal periods of time.

ISOCLINALS. Given a differential equation, $dy/dx = f(x, y)$, the solution can be represented by a series of curves. The curves of the system, $c = f(x, y)$, corresponding to a succession of values of c, are called the isoclinals of the system of curves representing the solution of $dy/dx = f(x, y)$. Hence the solution of the latter may be approximated to graphically by sketching $f(x, y) = c$, drawing across each member of this a series of short parallel lines of slope c, and connecting up these short parallels from one member of the system to the next.

ISOLATED POINT. Synonym of *acnode*. A point contiguous to which there is no other point of the set of points under consideration. The origin, (0, 0) for example, can be an isolated point since it satisfies the equation $x^3 = x^2 + y^2$, although the graph of this equation does not pass through the origin. Such a point is at least a **double point** since a quadratic is the lowest degree **homogeneous equation** which satisfies these conditions.

ISOMETRIC AXES. Three lines with a common point, separated by angles of 120°. In diagrams, at the lowest point, the vertical one is drawn in reverse and the other two make angles of 30° with the ground line, drawn horizontally. Such axes are used in **pictorial projection.**

ISOMETRIC PROJECTION. See **Pictorial Projection.**

ISOMETRY. A one-one transformation in space which leaves the distance between any two points unchanged. The size and shape of any object are unaltered. **Translation** and **rotation** are direct isometries as they do not alter the **orientation**: **reflection** is an opposite isometry as the orientation is reversed. See **Transformation in Geometry**.

ISOMORPHIC. Having the property of **isomorphism.**

ISOMORPHISM. Any two mathematical systems which have a **one-one correspondence** are said to be isomorphic. For example, the **algebra of propositions** and the algebra of **set theory** are isomorphic because both are **Boolean algebra**. The binary, ternary, quaternary, . . . denary number systems are isomorphic in that they have a one-to-one correspondence and all represent the real number system. See **Homomorphism.**

ISOSCELES. Having at least two sides of equal length.

ISOSCELES SPHERICAL TRIANGLE. A **spherical triangle** with two equal sides or two equal angles.

ISOSCELES TRAPEZIUM. A **trapezium** with its non-parallel sides equal.

ISOSCELES TRIANGLE. A triangle with two equal sides. The third side is called the base; and the two equal angles adjacent to it, the base angles.

J

J. (1) **Integral domain** of **integers,** (2) **Joule.**

J+. Set of positive **integers** (natural numbers).

J, j. See **I, J; i, j.**

JET FORCE. The **force** exerted by a jet of water or small particles striking a flat surface, measured by calculating the change of **momentum** per second. If m is the **mass,** v the **velocity,** and t the **time,** then $F=mv/t$ poundals or $(m/g)(v/t)$ pounds weight.

JOIN. A straight line segment lying between two given terminal points.

JOINTS. The theoretical points where struts and ties in structures are joined by devices which permit movement under differing loads, wind pressures, etc. See **Strut, Tie.**

JORDAN CURVE. In the field of **topology**, a simple closed curve in a plane is called a Jordan curve. It divides the plane into an interior region and an exterior region, each region being a *connected set*. The Jordan curve theorem, due to C. Jordan (1838–1922), states that: any point A in the interior can be joined to any other point A' in the interior and any point B in the exterior to any other point B' in the exterior by arcs which do not cross the curve; and any arc joining an interior point to an exterior point crosses the curve. Jordan's proof of his theorem contained errors, but it was eventually established by Veblen in 1905.

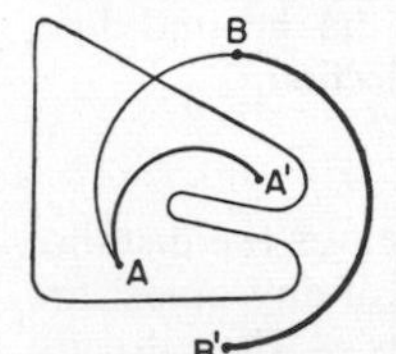

JOULE. Unit of **work** in the **m.k.s. system**; the work done in one second by a current of one **ampere** flowing through a resistance of one **ohm.** 1 joule $=10^7$ **ergs.** Work done by 1 **newton** in moving 1 metre.

K

KELVIN SCALE. Synonym of *absolute scale of temperature.* Named after Baron Kelvin (1824–1907), physicist, second son of British mathematician James Thomson. He proposed in 1848 his scale of temperature, independent of the properties of any particular substance. The *absolute zero*, 0° Kelvin, equal to $-273.18°$ Celsius or centigrade and $-459.72°$ Fahrenheit, is based on the concept of heat as the **kinetic energy** of molecular motion and not arbitrarily chosen as is the zero associated with the freezing point of water. The degree intervals are equal to those on the Celsius or centigrade scale. See **Celsius Scale; Fahrenheit Scale.**

KEPLER'S LAWS. (1) All the planets move in elliptical orbits, the sun being at one focal point. (2) The radius vector sweeps out equal areas in equal times. (3) The square of the period of revolution of a planet is proportional to the cube of the mean distance of the planet from the sun. The first and second laws were published in 1609, the third in 1619.

KERNEL. See **Homomorphism.**

KILO-. See **Metric System.**

KILOWATT. 1 000 **watts**, equivalent to about 1.34 **horse-power.**

KILOWATT-HOUR. An energy production or consumption of 1 000 **watts** for 1 hour, equivalent to $3\,600 \times 10^3$ **joules.**

KINEMATICS. The study of the various modes of motion in **dynamics.**

KINETIC ENERGY. If the **velocity** of a body of **mass** m is changed from v_0 to v_1 by the action of some **force**, then $\frac{1}{2}m(v_1^2 - v_0^2)$ is the measure of its change of kinetic energy, the **energy** associated with its *movement.* See **Conservation of Energy; Potential Energy.**

KINETICS. The study of motion in **dynamics** in relation to the forces which cause or arrest it.

KITE. A **quadrilateral** with two pairs of equal adjacent sides, and consequently one pair of equal angles. The diagonals intersect at right angles.

KLEIN BOTTLE. The bottle made by taking a tapering tube, with the usual meanings of outside, inside and rim edges, bending it, and penetrating the broad end outside surface by the narrow end, then sealing the ends together. A cross-section is shown. The bottle has a single surface and the properties of a three-dimensional **Möbius band (strip)**.

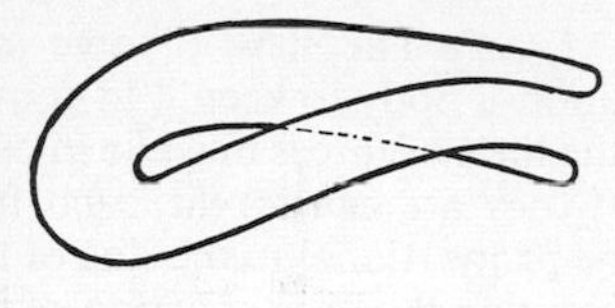

KNOT. A speed of one **nautical mile** per hour; equivalent to $1.\dot{1}\dot{5}$ **statute miles** per hour.

KÖNIGSBERG BRIDGES PROBLEM. The problem was: to cross all of the seven bridges at Königsberg once only in a single journey. One of the first problems arising in **topology**, this was analysed by Euler (1707–1783). He replaced the geographical map by the schematic diagram called a topological graph with points for mutually exclusive land areas and line segments for bridges. He defined a point on the graph as odd (even) if an odd (even) number of segments terminate at the point. He demonstrated three general propositions about such graphs: (1) If *all* the points are even, the journey is possible and ends at the starting point; (2) if *one* or *two* points are odd, the journey is possible but ends at a point different from the starting point; (3) if *more than two* points are odd, the journey is impossible. The third proposition applies to the problem of the bridges at Königsberg and determines the impossibility of making the journey.

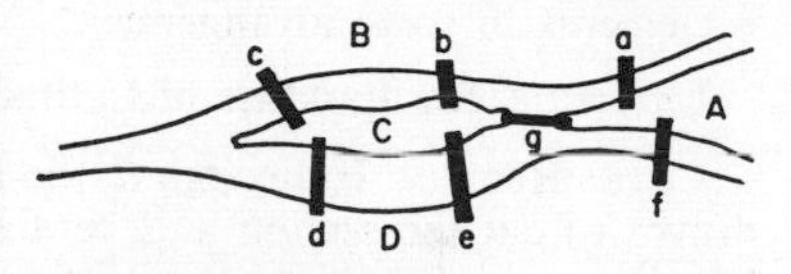

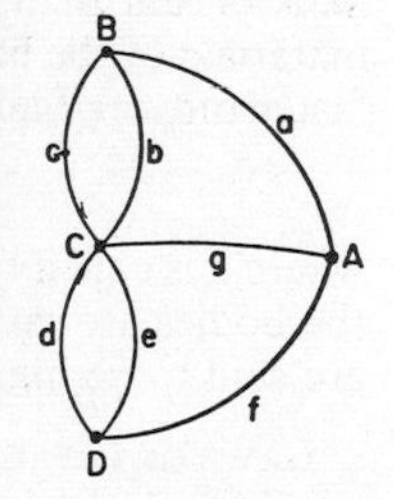

L

lbf. **Pound force.**

L.C.D. Least (lowest) common denominator. See **Common Denominator.**

L.C.M. **Least (lowest) common multiple.**

l.u.b. Least upper bound. See **Bound of Function**; **Bound of Set.**

LAGRANGE'S THEOREM. The order of a subgroup, S, of a finite group, G, is a divisor of the order of G. See **Group Theory.**

LAMINA. A solid with two plane parallel faces separated by a distance negligible in comparison with the linear dimensions of the faces. A device used in the study of **centroids** of areas.

LAMI'S THEOREM. If three forces act upon a body to keep it in equilibrium, the lines of forces must lie in one plane. If they are **concurrent**, each force will be proportional to the sine of the angle between the lines of action of the other two.

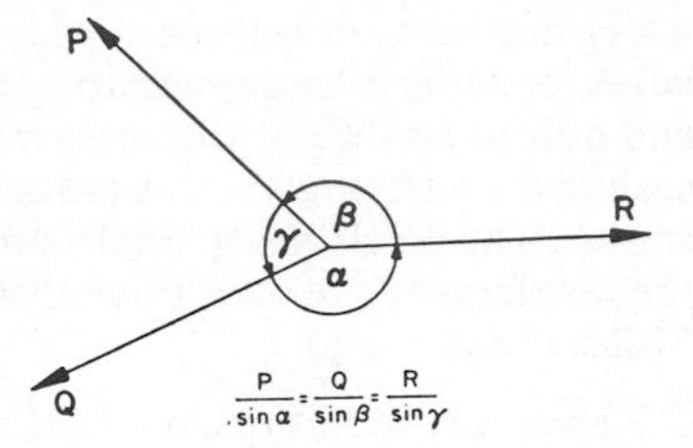

LATIN SQUARE. A square array of n rows and n columns developed from n different elements, n being the *order* of the Latin square. The defining property of such a square is that each row and each column contains the n elements in some arrangement.

LATITUDE. See **Parallels of Latitude**; **Spherical Coordinates in Geodesy.**

LATUS RECTUM. If through S, the focus of a **conic section,** an ordinate is drawn to cut the section at L and L', then LL' is the *latus rectum*, and $LS = SL'$. Either half is a *semi-latus rectum.*

LAWS OF IMPACT IN ELASTICITY. After impact the relative velocity of two balls is less than the relative velocity before. This loss depends on the materials of the balls. If u_1 and v_1 are the velocities in one direction of the faster moving ball, before and after impact, u_2 and v_2, of the slower, then

$$v_1 - v_2 = -e(u_1 - u_2),$$

where e is called the *coefficient of restitution.* When e has a value near zero the bodies are said to be inelastic: when e has a value near unity the bodies are said to be highly elastic.

LAWS OF IMPACT IN MOMENTUM. If m_1 and m_2 are the masses of the balls considered, then, for direct central impact—the least complicated case—

$$m_1u_1 + m_2u_2 = m_1v_1 + m_2v_2$$

The total **momentum** after impact equals the total momentum before. If the velocities are not in one straight line, the law is true of the four velocities resolved in a common direction. See **Conservation of Momentum.**

LAWS OF IMPACT IN REBOUND. If a ball is allowed to fall from a height h_0 on to a horizontal plane, and it rebounds to a height h_1, then $h_1 = e^2h_0$, where e is the *coefficient of restitution.*

LAWS OF MOTION. See **Newton's Laws of Motion.**

LEAST (LOWEST) COMMON MULTIPLE (L.C.M.) IN INTEGERS. The smallest positive **common multiple.**

LEAST (LOWEST) COMMON MULTIPLE (L.C.M.) in POLYNOMIALS. The smallest **common multiple** of least degree.

LEAST SQUARES. See **Method of Least Squares.**

LEAST UPPER BOUND (l.u.b.) See **Bound of Function; Bound of Set.**

LEIBNIZ' THEOREM. A rule for finding the nth **derivative** of the product of two **functions.** It is expressed as a series of $n+1$ terms involving the successive derivatives of the factors and the **binomial coefficients,** as follows:

$$D^n(u\ v) = D^n u\ v + n\ D^{n-1} u\ D\ v + \frac{n(n-1)}{1\cdot 2} D^{n-2} u\ D^2\ v$$

$$+ \frac{n(n-1)(n-2)}{1\cdot 2\cdot 3} D^{n-3} u\ D^3\ v + \ldots + u\ D^n\ v.$$

LEMMA. A **proposition** which is proved for use in the proof of another proposition.

LEMNISCATE OF BERNOULLI. Let O be a point at a distance $a\sqrt{2}$ from the centre C of a fixed circle of radius a, and any variable secant through O cut the circumference at Q and Q'. The loci of P and P' on OQ such that $PO = OP' = QQ'$ are the two branches of the curve called the *lemniscate*. It is also the locus of the foot of the perpendicular from the centre of a **rectangular hyperbola** to a variable tangent. Its **polar equation** is $r^2 = a^2 \cos 2\theta$; its **Cartesian equation** is $(x^2+y^2)^2 = a^2(x^2-y^2)$

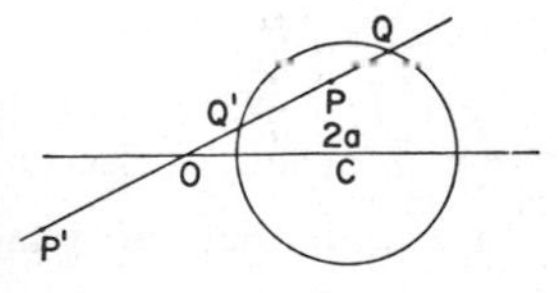

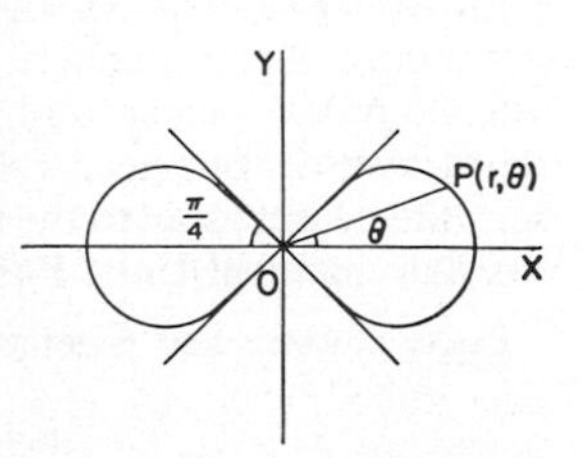

LENGTH. Extension in one dimension along a straight line or curve. See **Dimension of Ordinary Space.**

LENGTH IN BRITISH UNITS. The **yard** is divided into three feet, each of twelve inches. The **statute mile** is divided into eight furlongs, each of ten chains. The chain is twenty-two yards in length.

LENGTH IN METRIC UNITS. The **metric system** of length based on the **metre.** See **Angstrom.**

LEVEL. An arc of a great circle of the earth, measured accurately to serve as a base for a trigonometrical **survey**. The Salisbury Plain (England) base of the Ordnance Survey was seven miles long.

LEVEL LINE. If the shape of the earth, the geoid, is defined by sea-level, then level lines are in fact level curves drawn on shells concentric with the earth's figure. See **Contours.**

LEVELLING. The process of determining differences in **elevation** of points on the earth's surface for the purpose of **survey**. There are three methods of procedure: (1) *gravitational,* making use of the spirit level and the plumb-line; (2) **angular levelling,** which employs trigonometrical measurements; (3) *hypsometrical,* which measures variations in **atmospheric pressure.**

LEVER. A rigid rod used to lift weights by placing it against a fixed support called the *fulcrum.* Levers fall into three types according to the relative positions of the fulcrum and the two points at which the weight and the applied force act. These are (1) fulcrum below rod and between weight and applied force, (2) fulcrum at end and under rod, (3) fulcrum at end and above rod. In equilibrium, the algebraic sum of three parallel forces will be zero and the algebraic sum of three moments about a point is zero. See **Moment of Force.**

LEVY'S CONJECTURE. Levy (1889–) postulated in 1964 that every odd number can be written as $2P+Q$, where P and Q are **prime numbers**. See **Goldbach's Conjecture.**

Lg *x*. *lg x*. This symbol for a **logarithm** to a base 10 is called log *x*, synonym of $\log_{10}x$. See **Ln *x*.**

LIFT. See **Angle of Attack.**

LIGHT YEAR. One of the units of length measurement appropriate to astronomy. Assuming light to travel 300,000 kilometres (a little more than 186,000 miles) per second and a year to be 3.15576×10^7 seconds then light travels in one year 9.46728×10^{12} Km (5.86971×10^{12} miles). In this unit, the nearest star to the solar system is more than four light years away. See **Astronomical Unit; Parsec.**

LIKE TERMS. See **Similar (Like) Terms.**

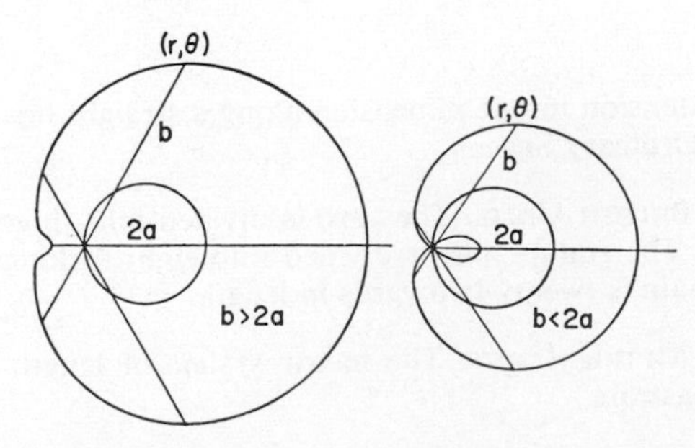

LIMAÇON OF PASCAL. A special case of a **conchoid**. The locus on a plane of a point on a line at a fixed distance from the point of intersection of the line and a fixed circle, as the line rotates about a fixed point on the circle. If the radius of the fixed circle is a, and the fixed distance is b, the **polar equation** of the limaçon is $r = 2a \cos \theta + b$. If $b < 2a$ there are two loops. If $b > 2a$ there is one loop which approaches the shape of a circle as b increases. When $b = 2a$, the curve is a **cardioid** with a **cusp** at the pole.

LIMIT. In general, the ultimate value towards which a **variable** tends.

LIMIT OF CLASS INTERVAL. See **Class Limits.**

LIMIT OF FUNCTION. A **function** $f(x)$ approaches a limit L when x approaches a, written $\lim_{x \to a} f(x)$, if, for every positive number ϵ there is a number δ, such that when $|x - a| < \delta$, $|f(x) - L| < \epsilon$. The limits of two functions can be combined according to the *fundamental theorem on limits*: if u and v are functions of x with limits U and V respectively, then, as x approaches a,

$$\lim(u + v) = U + V; \lim(u - v) = U - V,$$
$$\lim(u \times v) = U \times V; \lim(u/v) = U/V; (V \neq 0).$$

If k is a constant, $\lim ku = kU$, $\lim kv = kV$.

LIMIT OF SEQUENCE. The infinite **sequence** of numbers, $a_1, a_2, a_3, \ldots a_n, \ldots$ approaches a limit L, if an interval having L as its centre contains all the members of the sequence except for a finite number depending upon the size of the interval. It is written: $\lim_{n \to \infty} a_n = L$. For any positive number ϵ there is a number N such that $[L - a_n] < \epsilon$, for all n greater than N. A sequence which has a limit is convergent; one which has no limit is divergent. See **Accumulation Point; Convergence, Divergence of Sequence.**

LIMIT POINT. Synonym of **Accumulation Point.**

LIMITING POINTS OF COAXAL CIRCLES. See **Coaxal (Coaxial) Circles.**

LIMITS OF INTEGRATION. See **Definite Integration.**

LINE. Synonym of **straight line** in contrast to **curve.**

LINE AT INFINITY. Synonym of **ideal line.**

LINE, NODAL. See **Nodal Line.**

LINE OF BEST FIT. (1) A **trend line.** (2) A line determined by the **method of least squares.**

LINE OF NODES. See **Nodes in Astronomy.**

LINE SEGMENT. A finite portion of a line (**straight line**).

LINES OF LATITUDE. See **Parallels of Latitude.**

LINE SYMMETRY. Synonym of **axial symmetry.**

LINES OF LONGITUDE. See **Meridians of Longitude.**

LINEAR. (1) Having the properties of, or referring to, a **straight line.** (2) Sometimes the first definition is extended to include a **curve.** (3) One-dimensional. (4) Of first order or degree. See **Degree in Algebra; Order of Equation.**

LINEAR ALGEBRA. The study of the algebraic properties of **vector spaces.** The basis of the theories is a set of vectors (**vector quantities**) over a field of scalars (**scalar quantities**). Given a set of vectors, $\mathbf{V}=\{\mathbf{v}_1, \mathbf{v}_2, \mathbf{v}_3, \ldots\}$, and a set of scalars, $F=\{a_1, a_2, a_3, \ldots\}$, $\mathbf{V}$ can be subjected to the operation of addition ($\mathbf{V}_1+\mathbf{V}_2$, etc.), and combined with the set of scalars in **scalar multiplication** ($a_1\mathbf{v}_1$, $a_2\mathbf{v}_2$, etc.) or both ($a_1\mathbf{v}_1+a_2\mathbf{v}_2+$, etc.). The results of such operations are vectors of a *linear form.*

LINEAR COMBINATION. The sum $AQ_1+BQ_2+CQ_3+\ldots$ where Q_1, Q_2, $Q_3, \ldots$ are a set of quantities and $A, B, C, \ldots$ are a set of constants of which at least one is not zero. Thus, if $S=0$ is the equation of a circle and $L=0$ the equation of a line, then the linear combination $aS+bL=0$ is the equation of another circle which passes through the intersection of S and L.

LINEAR CONGRUENCE. A **congruence,** such as $ax+by+c=0 \pmod{n}$ in which all the variable terms have first degree. See **Order of Equation.**

LINEAR DEPENDENCE. Set of quantities Q_1, Q_2, $Q_3, \ldots$ which may be **polynomial expressions, matrices, vectors,** etc., are linearly dependent if there is a **linear combination** $AQ_1+BQ_2+CQ_3+\ \ldots\ =0$. $A, B, C, \ldots$ must belong to a specified field (e.g. real complex numbers) and at least one be nonzero.

LINEAR DIFFERENTIAL EQUATION. A **differential equation** of the first degree in y and its **derivatives** where the coefficient of y and its derivatives are **functions** of x only.

LINEAR DISPLACEMENT. Synonym of **translation.**

LINEAR EQUATION. An equation of first order, of form $ax+b=0$ $(a\neq 0)$. See **Order of Equation; Root of Equation.**

LINEAR EXPANSION. Expansion in one direction. See **Coefficient of Linear Expansion.**

LINEAR EXPRESSION. An expression of first degree, of form $ax+b$ $(a\neq 0)$. See **Degree in Algebra.**

LINEAR INTERPOLATION. See **Interpolation.**

LINEAR MEASURE. The measurement of **length** in any system.

LINEAR PROGRAMMING. **Mathematical programming** which deals with **functions** and conditions which are linear in form. It is the problem of solving linear **inequalities.** The problem of *maximizing* the function $x+y$

when it is subject to the restraints $x \geqslant 1$, $2y \geqslant 1$, $5x + 6y \leqslant 30$, $3x - y \leqslant 3$, is the problem of finding the point within a region of possible solutions which is farthest away from the line $x + y = 0$.

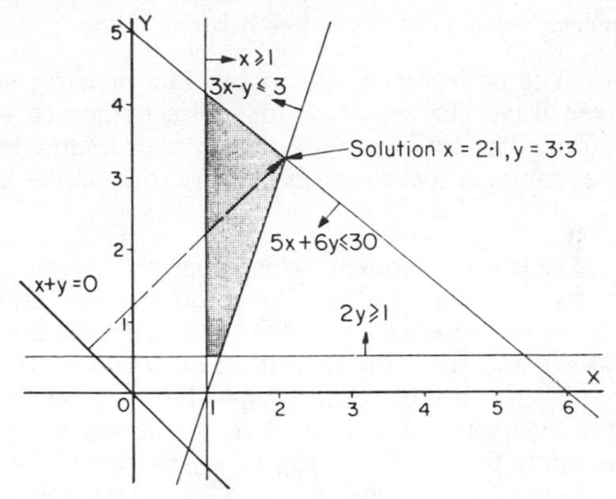

LINEAR TRANSFORMATION. A **transformation** produced by the use of algebraic **linear equations** in terms of old and new variables, for example, **transformation of axes, transformation of similitudes**. A linear transformation can also be used in n-dimensional spaces and **vector spaces**. See **Dimensions of Space**.

LINK. A hundredth part of a surveyor's or Gunter's chain, 0.66 feet or 7.92 inches. A hundredth part of an engineer's chain, one foot.

LINK POLYGON. Synonym of **funicular polygon.**

LITERAL. Composed of letters in contrast to numbers.

LITERAL COEFFICIENTS. **Coefficients** which are **literal constants.**

LITERAL CONSTANT. A letter used to denote a constant, in contrast to a numerical constant such as 1, 2, 3, etc.

LITERAL EQUATION. An equation in which constants are represented by letters. Thus, while $x^2 - 3x - 7 = 0$ has, as its roots, $\frac{1}{2}(3 \pm \sqrt{37})$, the literal equation $x^2 + px + q = 0$ has, as its roots, $\frac{1}{2}(-p \pm \sqrt{p^2 - 4q})$, which have any values depending upon the values assigned to p and q.

LITERAL EXPRESSION. An expression in which constants are represented by letters.

LITRE. The unit of **volume** in the **metric system**. It is equal to a volume of 1,000 cubic centimetres or 1 cubic decimetre. It is equivalent to 1.760 pints approximately. The term *litre* is not used for volume measurement involving high precision and should be reserved for commercial and not scientific purposes.

Ln x, ln x. This symbol for a **logarithm** to a base e is called lon x, synonym of $\log_e x$. See **Lg x**.

LOCAL TIME. The time at any place determined by the position of the sun. See **Apparent Solar Time**; **Greenwich Mean Time**.

LOCUS, LOCI. The path traced by a point, line or curve constrained to move in a given way. The set of points, lines or curves which satisfies certain conditions. The set of points whose coordinates satisfy a given equation or inequality. A locus may be a point, line, curve, plane, surface or solid.

LOGARITHM. The **index** (exponent) which changes a given number, called the *base*, into any required number. The solution of the equation $b^x = N$, where b and N are known, is a logarithm. The equation can be written in the form $\log_b N = x$ and read 'the logarithm of N to the base b is x'. The inverse relation is 'the antilogarithm of x to the base b is N'. The operational rules for logarithms are derived from the **theory of indices**. The following laws apply for any base: $\log (A \times B) = \log A + \log B$, $\log (A/B) = \log A - \log B$, $\log (A^n) = n.\log A$, $\log \sqrt[n]{A} = (1/n).\log A$. Logarithms provide a convenient way of converting computations involving multiplication and division into addition and subtraction. Logarithms can be changed from one base to another by the following relation: $\log_a N = \log_b N/\log_b a$. Two bases are in normal use. The incommensurable e, (2.718 28 . . .), used in **analysis**, is the base of *natural* (*hyperbolic* or *Naperian*) logarithms. The early concept of a logarithm is credited to Napier, in the early part of the seventeenth century; but his ideas did not involve indices. For computation the number 10 is used as base of *common* (*Briggs*') logarithms, introduced by Briggs, a friend of Napier. The conversion of natural into common logarithms is achieved by the conversion relation above, written in the form $\log_{10} N = \log_e N \times \log_{10} e$, where $\log_{10} e$ is 0.434 29 . . . and is called the *modulus of common logarithms*. A modern notation for this is lg N = ln $N \times$ lg e, or log N = lon $N \times$ log e. The conversion of common into natural logarithms is achieved by a reverse relation: lon N = log $N \times$ lon 10, where lon 10 is 2.302 58 . . . and is called the *modulus of natural logarithms*. See **Characteristic**; **Mantissa**.

LOGARITHMIC CURVE. The curve having equation $y = \log_e x$. Its image in the line $y = x$ is the **exponential curve** $y = e^x$.

LOGARITHMIC EQUATION. One in which the unknown quantity appears in the form of a **logarithm**. It can often be solved by transforming it into an **exponential equation**.

LOGARITHMIC FUNCTION. If two variables, x and y, are such that their relation can be expressed in the form of an **exponential function**, $x = a^y$, where a is positive and not unity, then y is called the logarithm of x and is written $y = \log_a x$. It is one of the **transcendental functions**.

LOGARITHMIC SCALE. If n is the *number* associated with the *point* N on a logarithmic scale, then the distance ON of N from the origin is proportional to the **logarithm** of n. The farther the point N is from the origin the closer it is to the point associated with the number $n+1$. The slide rule is a logarithmic scale. Such a scale is an excellent device for showing relative heights above the earth.

10^6;	1,000,000 miles	
10^5;	100,000 miles	Inter-planetary space begins
10^4;	10,000 miles	Outer } radiation limits
10^3;	1 000 miles	Inner } radiation limits
10^2;	100 miles	Aurora
10^1;	10 miles	Troposphere

LOGARITHMIC SERIES. Series used in the computation of natural **logarithms.** The basic series is

$$\ln(1+x)=x-\tfrac{1}{2}x^2+\tfrac{1}{3}x^3-\tfrac{1}{4}x^4+\ldots.$$

where $-1<x\leqslant+1$. This series converges very slowly. It can be converted into a rapidly converging series, true for all positive values of N:

$$\ln(N+1)-\ln N$$
$$=2\left[\frac{1}{2N-1}+\frac{1}{3(2N-1)^3}+\frac{1}{5(2N-1)^5}+\frac{1}{7(2N-1)^7}+\ldots\right].$$

LOGARITHMIC SPIRAL. Synonym of **equiangular spiral.**

LOGARITHMIC TABLES. A tabulated set of mantissae, the fractional parts of logarithms. See **Theory of Logarithms.**

LOGISTIC SPIRAL. Synonym of **equiangular spiral.**

LONG RADIUS. Synonym of radius of **circumscribed circle of polygon.**

LONG TON. 2 240 pounds **avoirdupois.**

LONGITUDE. See **Meridians of Longitude.**

LOSS. See **Profit, Loss.**

LOWER BOUND. See **Bound of Function; Bound of Set.**

LOXODROMIC SPIRAL. Synonymous with loxodrome, **rhumb-line,** spherical **helix.** A curve which cuts the meridians (generating lines of a sphere) at a constant angle.

LUNAR MONTH. See **Calendar; Month.**

LUNE. If two **great circle** planes are not **perpendicular planes** their circumferences will divide a spherical surface into two major and two minor lunes.

M

m. Abr. **metre.**

M.K.S. System. The system of measurements based on the metre, the kilogramme and the second as units of length, mass and time respectively. See **Absolute Units, Metric System.**

Machine. Any device of suitably arranged ropes, pulleys, levers, screws, gears, etc., which permits a **force**, called the effort, to be applied to one part of the device to produce at another part of the device an effective force of different magnitude. The effective force is applied to a load. The *efficiency* of the machine is measured by the ratio *work done on load/work done by effort.* Since some of the **work** is utilized in overcoming resistance (e.g., friction) which is independent of the load, the efficiency is, in practice, less than unity or 100 per cent. The *velocity ratio* is the ratio *distance moved by point of application of effort in unit time/distance moved by point of application of load in same time.* The *mechanical advantage* is the ratio *load/effort.* The three ratios are related: *efficiency = mechanical advantage/velocity ratio.*

Maclaurin's Theorem (Formula). The substitution of 0 for a in **Taylor's theorem** gives Maclaurin's: $f(x) = f(0) + f'(0)\,x + f''(0)(x^2/2!) + \ldots + f^{(n-1)}(0)\{x^{(n-1)}/(n-1)\,!\} + R_n$.

Maclaurin's Trisectrix. See **Trisectrix of Maclaurin.**

Magic Square. A square array of n rows and n columns, developed from the integers 1, 2, 3, . . . n^2, n being the *order* of the magic square. The defining property of such a square is that the sum of the elements in any row, column or diagonal is the same. The study of such squares originated in China about 4000 years ago. The three basic ones are shown. The

2	9	4
7	5	3
6	1	8

12	7	9	6
13	2	16	3
8	11	5	10
1	14	4	15

9	2	25	18	11
3	21	19	12	10
22	20	13	6	4
16	14	7	5	23
15	8	1	24	27

elements of such arrays can be rearranged; there is one type of order 3, 880 of 4 and an unknown number of 5. Other magic squares can be produced from the basic ones by applying to every element one of the operations $+a$, $-b$, $\times c$, $\div d$, where a, b, c and d are any **rational numbers**. The resulting elements can always be arranged in an **arithmetic progression**. The sum of the elements in row, column or diagonal of a basic square is $\frac{1}{2}n(n^2+1)$.

MAGNETIC BEARING. See **Bearing.**

MAGNETIC DECLINATION. At any point on the earth's surface, the angle between the **magnetic meridian** and the geographical **meridian of longitude.** This angle is subject to various changes (*magnetic variation*). See **Agonic Line; Magnetic Poles.**

MAGNETIC MERIDIAN. The direction of the horizontal component of the earth's magnetic field at any point on the earth's surface. The compass direction at a point. See **Magnetic Declination.**

MAGNETIC POLES. The earth behaves as a bi-polar magnet with a north pole lying below a point N.N.W. of Boothia Peninsula, North America, and a south pole lying below a point in Victoria Land, Antarctica. The positions of the magnetic poles relative to the earth are constantly varying. See **Magnetic Declination.**

MAGNETIC VARIATION. See **Magnetic Declination.**

MAGNIFICATION. One aspect of **multiplication.**

MAGNIFICATION TRANSFORMATION. Synonym of **dilation.**

MAGNITUDE. The **absolute value** of a real number, complex number or vector.

MAGNITUDE IN ASTRONOMY. The relative apparent brightness of the stars; a star of one magnitude of brightness being $\sqrt[5]{100}$ or about 2.512 or roughly $2\frac{1}{2}$ times as bright as a star of the next lower order of brightness.

MAJOR, MINOR ARC OF CIRCLE. An **arc** greater or less than half the circumference.

MAJOR, MINOR AXIS OF ELLIPSE. See **Axes of Ellipse.**

MAJOR, MINOR AXIS OF ELLIPSOID. See **Axes of Ellipsoid.**

MAJOR, MINOR SECTOR OF CIRCLE. A **sector** of a circle, greater or less than a **semicircle,** respectively.

MAJOR, MINOR SECTOR OF SPHERE. A **sector** of a sphere greater or less than a **hemisphere,** respectively.

MAJOR, MINOR SEGMENT OF CIRCLE. A **segment** of a circle greater or less than a **semicircle,** respectively.

MAJOR, MINOR SEGMENT OF SPHERE. A **segment** of a sphere greater or less than a **hemisphere,** respectively.

MANIFOLD. A class with sub-classes. A plane is a two-dimensional manifold of points, because it is the class of all its points, and each is specified by two coordinates, (x, y). **Euclidean space** is a three-dimensional manifold of points or a four-dimensional manifold of straight lines. If n numbers are required to specify the individual members of a class, the class is called an n-dimensional manifold.

MANTISSA. This Latin word for make-weight is given to the positive fractional part of a **logarithm**. Thus, if $n = e^{-1.4165}$ it is written conventionally as $n = e^{\bar{2}.5835}$, the negative sign being written above the **characteristic** to which alone it applies. The positive portion 0.5835 is the weight needed to make -2 into -1.4165.

MANY-VALUED FUNCTION. Synonym of **multi-valued function.**

MAP PROJECTION. The representation on a plane surface of the **parallels of latitude** and the **meridians of longitude** of the globe. This may consist of a strictly **geometric projection** or it may be an arrangement which distorts some or all of the shapes, relative sizes and distances, the distortions being minimized by various techniques of drawing or calculation.

homalographic (equal area) projection. Any projection in which the ratio of any small area on the earth and its corresponding area on the map is constant.

orthomorphic projection. Any projection in which the shapes of any small region on the earth is preserved on the map. This can only be approximately true.

orthographic projection. One in which parallel rays project a hemisphere on to its equatorial plane, the rays being perpendicular to the plane. Distortion increases towards the perimeter of the map.

cylindrical projection. One in which the equator and the meridians are projected on to the surface of a cylinder circumscribing the globe along the equator. The meridians appear as parallel straight lines perpendicular to the equator. The parallels of latitude are drawn parallel to the equator at intervals chosen to counteract distortion which would be produced by their geometric projection.

conical projection. One in which an arbitrarily chosen parallel of latitude, the *standard parallel*, and the meridians are projected on to the surface of a cone making contact with the globe along the standard parallel. The developed surface shows the standard parallel as an arc of a circle with centre at a point which shows also the vertex of the cone. This parallel alone shows true length. Other parallels are arcs of concentric circles with arbitrarily chosen radii. The meridians appear as rays from the vertex of the cone, dividing the standard parallel into equal parts.

secant conical projection. A conical projection in which the cone cuts the globe along two standard parallels.

zenithal projection. One in which the parallels of latitude and the meridians of longitude are projected on to a plane tangential to the globe. The projection is described as *polar*, *normal* or *oblique* according to whether the plane touches the globe at either of the poles, a point on the equator or any other point, respectively. Also they are described as *gnomonic* (*central*), *stereographic* or *orthographic* according to whether the intersection of the rays of projection (**centre of projection**) is at the centre of the globe, at the opposite end of the diameter or at an infinite distance, respectively.
See **Pictorial Projection.**

MAPPING, MAP. Synonym of **correspondence, function, transformation.** If to each element x of set X there is a unique corresponding element $f(x)$ of set Y, then set X is said to be mapped into set Y, and $f(x)$ is the *image* of x.

MASS. The **weight** of a body varies with its distance from the centre of the earth. Newton calculated that a body weighing 194 pounds at sea-level at the equator would weigh 195 pounds at the North Pole. (The actual weight does not differ from this by half an ounce.) Weight is a **force** which can be measured by the extension it causes in a spring. If a force acts on a body it accelerates it in such a way that the ratio of the force to the **acceleration** is a constant. It is this constant which is called the mass of the body and is frequently defined as the 'quantity of matter' in the body. If a body equal to ten such bodies is acted on by the same force, it will be accelerated to one-tenth the amount. In general, acceleration produced by a force on a body is inversely proportional to the mass of the body.

According to Einstein's theory of **relativity**, the mass of a body is not independent of its speed, and is actually $m_0/(1 - v^2/c^2)^{\frac{1}{2}}$, where m_0 is the mass of the body at relative rest, v its speed and c the speed of light. (This adds, maybe, a milligramme to each kilogramme in the weight of an artificial satellite.) From this it can be shown that, at speeds near the speed of light, the laws of **conservation of momentum** and mass, which hold for any changing process, are not different from the laws of **conservation of energy**; and that any destruction of mass is necessarily accompanied by a creation of energy through Einstein's *mass-energy equation*, $E = mc^2$, where c is the speed of light in empty space. See **Newton's Laws of Motion.**

MASS-ENERGY EQUATION. $E = mc^2$. See **Mass.**

MATHEMATICAL INDUCTION. See **Induction.**

MATHEMATICAL PROBABILITY. See **Probability.**

MATHEMATICAL PROGRAMMING. The process of finding an optimum value (**maximum, minimum value**) of a **function** when the variables are subjected to certain restraints which usually take the form of **equations** or inequalities. The digital **computer** is normally used for problems of programming which usually arise in operational research. Programming may be **linear programming** or non-linear when the function and restraints may be quadratic, concave or convex. See **Computer Programming.**

MATHEMATICS. A vast system of organized thinking of an analytic and synthetic nature that has developed since the Golden Ages of Greece and the earlier Babylonian civilization. Its power rests on the discovery that it was possible to represent abstract concepts such as those of number and shape by means of concrete symbols; and through the physical arrangement of these symbols with respect to each other to express relations between these concepts. Thus, permissible rules for the changing of the arrangements of the symbols reflected permissible steps in expressing logical relations between the original concepts. The great contribution of the Greeks was to show how this approach could be expressed in an axiomatic form so that from a number of primitive **assumptions**, **axioms** and **postulates** it was possible to *deduce* a succession of propositions of a meaningful nature that were not self-evident initially. Thus mathematics is at one and the same time a field of symbolic representation of ideas and relations between them that have been extracted from that part of their context considered irrelevant for the purpose in hand, and an instrument for effecting a logical examination of the implications of these ideas.

Mathematics as a whole can be divided into three main branches, each of which has its own history: **geometry**, **astronomy** and **chronology** which are concerned with spatial and temporal concepts developed from primitive needs to **measure**; **algebra**, which is concerned with **number** and the abstractions which arise from the study of **arithmetic computation** and **number theory**; **analysis**, which is concerned with the concepts of continuity and limits, the eventual comprehension of which gave rise in the sixteenth and seventeenth centuries to the invention of the **calculus**. These three branches often overlap. Algebra is used in geometry and analysis; analysis is used in **analytic geometry**; the **algebra of sets**, the **algebra of propositions** and **order relations** often use concepts and language which originally belonged to the field of geometry and its sub-branch **topology**; **trigonometry**, growing out of simple mensuration, geometry and survey, in its modern form utilizes concepts from algebra and analysis. In the traditional branches of applied mathematics, **astronavigation**, astronomy, **mechanics**, **statistics**, **survey**, concepts are used which belong to most of the branches of mathematics.

Today, the mathematician works in a world of intense scientific investigation aided by a revolution in methods of computation and means of communication. His thinking is part of the whole climate of intellectual thought in which distinction between pure and applied, abstract and practical, are too subtle to be of much use.

MATRIX. An array of numbers or letters, called elements, in rows and columns, used as a form of tabulation in problems where the relations between the elements are of fundamental importance. For example, **transformations in geometry** and the study of **simultaneous equations** lend themselves to this treatment. The algebra of matrices was developed by Cayley in the middle of the nineteenth century.

square matrix. A matrix with m rows and m columns, referred to as an m by m ($m \times m$) array of m^2 elements and of order m.

rectangular matrix. A matrix with m rows and n columns, referred to as an m by n ($m \times n$) array of mn elements.

identity (unit) matrix. The matrix $\begin{pmatrix} 1 & 0 \\ 0 & 1 \end{pmatrix}$. A matrix is unaltered if it is multiplied by the unit matrix.

product of matrices. $\begin{pmatrix} a & b \\ c & d \end{pmatrix} \times \begin{pmatrix} e & f \\ g & h \end{pmatrix} = \begin{pmatrix} ae+bg & af+bh \\ ce+dg & cf+dh \end{pmatrix}$.
The **commutative law** does not apply.

sum of matrices. $\begin{pmatrix} a & b \\ c & d \end{pmatrix} + \begin{pmatrix} e & f \\ g & h \end{pmatrix} = \begin{pmatrix} a+e & b+f \\ c+g & d+h \end{pmatrix}$.

inverse matrix. In general, the inverse of $\begin{pmatrix} a & b \\ c & d \end{pmatrix}$ is $\begin{pmatrix} d/\triangle & -b/\triangle \\ -c/\triangle & a/\triangle \end{pmatrix}$, where $\triangle$ is the **determinant** of the matrix.

determinant of matrix. Determinant of $\begin{pmatrix} a & b \\ c & d \end{pmatrix}$ is $\triangle = \begin{vmatrix} a & b \\ c & d \end{vmatrix} = ad - cb$.

MAXIMUM, MINIMUM VALUES. The values of a **function** which occur at the *turning points* (*stationary points*) of the graph of the function. A maximum value of the function $y=f(x)$ occurs when $dy/dx=0$ and $d^2y/dx^2<0$; a minimum value occurs when $dy/dx=0$ and $d^2y/dx^2>0$. When both dy/dx and d^2y/dx^2 are zero, the point is not a turning point but a special case of a **point of inflexion**. Maximum and minimum values are sometimes referred to as *stationary values* or *turning values*. See **Derivatives; Gradient of Curve.**

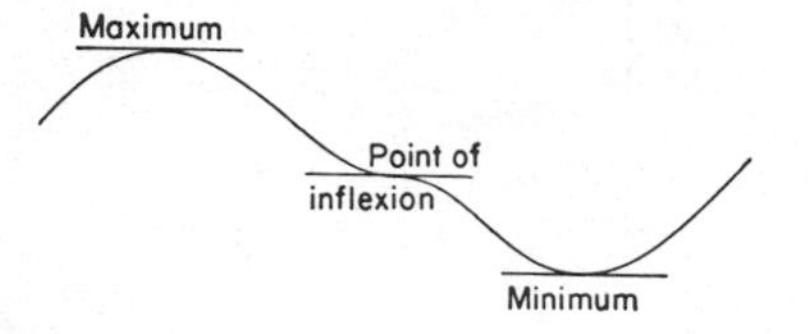

MEAN AXIS OF ELLIPSOID. See **Axes of Ellipsoid.**

MEAN CURVATURE. See **Curvature.**

MEAN DEVIATION. See **Deviation.**

MEAN IN STATISTICS. The **arithmetic mean** of a set of measurements. If in a set of n measurements, measurement a_1 occurs f_1 times, measurement a_2, f_2 times, and so on, the statistical mean of the set is $(f_1a_1+f_2a_2+\ldots)/n$. See **Median in Statistics; Mode.**

MEAN ORDINATE. Synonym of **mean value of function.**

MEAN POINT. Given three points A, (a,b), B, (c,d), C, (e,f), the mean point is that point which has coordinates $x=(a+c+e)/3$, $y=(b+d+f)/3$. If G is this point, it is the **centroid** of triangle ABC. The mean point of any number of points is found in a like manner.

MEAN SOLAR DAY. The day of **apparent solar time** is not of fixed length. The mean solar day is the average of all the solar days in the year and is used as the standard day for the purposes of time measurement. It is subdivided into 24 hours; each hour is divided into 60 minute parts (60 minutes); each minute into 60 second minute parts (60 seconds). See **Calendar; Second of Time-Interval; Sexagesimal.**

MEAN SOLAR NOON. The noon of **mean solar day.**

MEAN TERMS OF PROPORTION. See **Arithmetic Means; Geometric Means; Harmonic Means.**

MEAN VALUE OF FUNCTION. The mean value of function $f(x)$ for $a \leqslant x \leqslant b$ is $\{1/(b-a)\}\int_a^b f(x)dx$. Thus, the mean value of the ordinates under a curve or the mean ordinate is the quotient of the area under the curve and the interval $(b-a)$, where the area is bounded by the curve, the x-axis and the ordinates $x=a$, $x=b$. See **Area under Curve.**

MEAN VALUE THEOREM IN DIFFERENTIAL CALCULUS. If $f(x)$ and $f'(x)$ are continuous and single-valued over the range $a \leqslant x \leqslant b$, then $f(b)-f(a) = (b-a)f'(c)$ for some c in the range of x. This states that every arc of a continuous single-valued curve has at least one tangent parallel to its secant. When the secant is on the x-axis this theorem is known as **Rolle's theorem.**

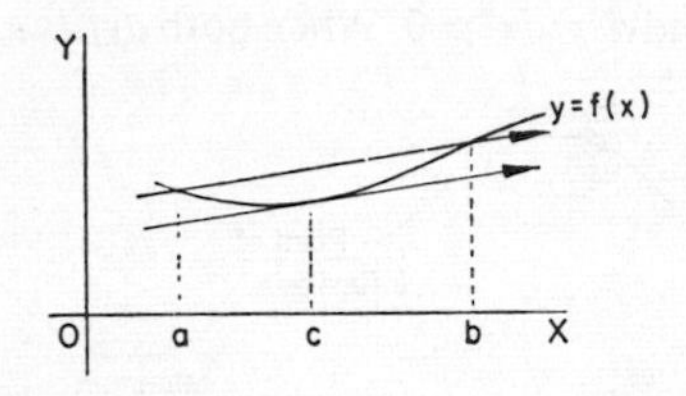

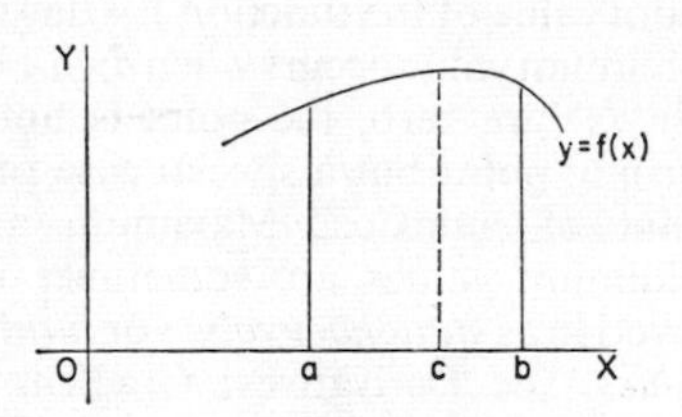

MEAN VALUE THEOREM IN INTEGRAL CALCULUS. If $f(x)$ is continuous and single-valued over the range $a \leqslant x \leqslant b$, then $\int_a^b f(x)dx = (b-a)f(c)$ for some c in the range of x. This states that the area below a continuous single-valued curve over a given interval is the product of the width of the interval and some ordinate within the interval.

MEASURE. The size of something expressed in terms of a standard unit. A term sometimes used to include all the concepts and techniques involved in measuring (measurement, **mensuration**). Historically the concept evolved from that of direct comparison. Where this was impossible two things were compared separately with a convenient movable third (e.g. length, in terms of part of the body, later standardized). In certain measures (e.g., heat, light, sound, etc.) the effect of the phenomenon is measured and its size (intensity) assumed to be proportional.

MEASUREMENT. The process of determining a **measure.**

MECHANICAL ADVANTAGE. See **Machine.**

MECHANICS. The science which studies the effects of **forces** on bodies at rest or in **motion. Dynamics** is often used to describe the theoretical treatment and mechanics is reserved for the practical application in the fields of building and machinery.

MEDIAL. Pertaining to the mean, middle, average.

MEDIAN DEVIATION. Variation of a set of measurements about their median. See **Deviation**; **Median in Statistics.**

MEDIAN IN STATISTICS. The middle element of a set of measurements when arranged in order of size. If the number of elements is even, the median is taken to be the average of the middle two elements. See **Mean in Statistics; Mode.**

MEDIAN OF TRAPEZIUM, TRAPEZOID. The line joining the midpoints of the non-parallel sides.

MEDIAN OF TRIANGLE. The line joining a vertex to the midpoint of the opposite side. See **Apollonius's Theorem.**

MEDIAN POINT OF TRIANGLE. The point of intersection of the **medians of a triangle.**

MEGA. See **Metric System.**

MENELAUS'S THEOREM. If a straight line cuts the sides AB, BC, CA of a triangle ABC in P_1, P_2, P_3 respectively, then $(AP_1/P_1B)\,(BP_2/P_2C)\,.\,(CP_3/P_3A) = -1$. Conversely, if this equality can be established for any three points P_1, P_2, P_3 on the sides AB, BC, CA of a triangle ABC, then P_1, P_2 and P_3 are collinear. In this theorem all line segments are directed line segments. Of the points P_1, P_2, P_3, either one is external with one consequent negative segment or all three are external with three consequent negative segments. Written alternatively,

$$\frac{(AP_1\ BP_2\ CP_3)}{(P_1B\ P_2C\ P_3A)} = -1.$$

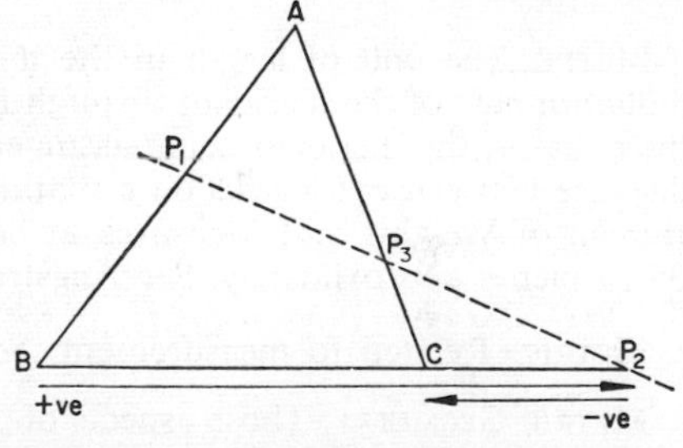

MENSURATION. The branch of arithmetic which deals with the measurement of geometrical shapes and consequent calculations for determining *length*, *area* and *volume*. Only certain lengths can be measured directly, using one physical object to measure another physical object. Lengths of curves, areas of surfaces and volumes of solids, defined in mathematical terms, are determined from formulae derived from mathematical analysis. These entities are abstractions which approximate to the physical realities. For example πr^2 is an approximation to the area of a real circle.

MERIDIAN OF CELESTIAL SPHERE. The projection from the centre of the earth of any one of its **meridians of longitude**. The intersections of any two fix the celestial poles. Such a projection passes through an observer's **zenith** and **nadir** and crosses his horizon and the celestial equator perpendicularly. See **Spherical Coordinates in Astronomy.**

MERIDIANS OF LONGITUDE. **Great circles** on a sphere whose planes contain the polar axis. Thus the geographical meridians of longitude all pass through the **geographical poles**. The word comes from Latin *meridies*, midday: all points on a meridian have midday (noon) at the same time. See **Spherical Coordinates in Geodesy; Spherical Coordinates on Sphere.**

METACENTRE. When a floating body is in a state of equilibrium the **centre of gravity** G of the body and the centre of buoyancy H (centre of gravity of the displaced liquid) are in one vertical line with G below H. If the body is subjected to a small angular displacement θ the centre of buoyancy is displaced to H' due to the change of shape of the displaced liquid. The point of intersection of the line through HG and a vertical line through H', meet at M. The limiting position of M as the angle of rotation θ is made infinitely small is called the *metacentre* and the height GM the *metacentre height*. The relative positions of G and M determine the stability of floating vessels.

METHOD OF LEAST SQUARES. A method for determining the best representative value for a set of data based on the assumption that this occurs when the sum of the squares of the **deviations** is a minimum. The same principle can be used to determine a **line of best fit**. Thus, if $y = mx$ is the equation of a line assumed to fit best the set of points (1, 3), (2, 5), (3, 10), (4, 13), m will be determined when $(m-3)^2 + (2m-5)^2 + (3m-10)^2 + (4m-13)^2$ is a minimum. This gives $m = 3\frac{1}{6}$ and the equation $6y = 19x$.

METRE. The unit of **length** in the **metric system**. Originally, one ten-millionth part of the meridian through Paris between the North Pole and the equator, the meridian being assumed to be at sea-level. It is now the distance between two marks on a platinum alloy bar at the International Bureau of Weights and Measures at Sèvres, France. It is equivalent to 39.37 inches approximately. See **Angstrom**.

METRIC. Related to measurement. See **Measure**.

METRIC GEOMETRY. Those aspects of **geometry** which utilize concepts of magnitude and therefore of measure, in contrast to non-metric geometry or **topology**.

METRIC SYSTEM. The system of measurement based on the units **are** (land area), **gramme (gram)** (mass and weight), **litre** (volume), **metre** (length). With these are used some or all of the prefixes Mega- (1,000,000), Myria- (10,000) Kilo- (1,000), Hecto- (100), Deca- (10), deci- (1/10), centi- (1/100), milli- (1/1,000). micro- (1/1,000,000). Other units of the metric system include **bar** (atmospheric pressure), **bel** (sound), **farad** (electrostatic capacity), **watt** (electric power), **volt** (electric potential).

METRIC TON. See **Tonne**.

METRIC WEIGHT. A system of measurement based on the **gramme (gram) weight**, the prefixes of the **metric system** together with Millier (1 000 Kg). Quintal (100 Kg), Myriagramme (10 Kg = 22.046 lb).

MHO. Reciprocal of **ohm**. Unit of conductance; the conductance of a conductor is the ratio of the current in **amperes** and the **potential difference** in **volts** between the ends of the conductor.

MICRO-. See **Metric System.**

MICRON. The millionth part of one **metre.**

MILE. See **Geographical Mile; Nautical Mile; Statute Mile.**

MILLI-. See **Metric System.**

MILLIBAR (mb.). The unit of **atmospheric pressure** shown on weather charts. 1.450×10^{-2} lb in^2. 1 mb supports 0.029 53 in or 0.750 1 mm of mercury at 0°C.

MILLIER. See **Metric Weights.**

MINOR ARC OF CIRCLE. See **Major, Minor Arc of Circle.**

MINOR AXIS OF ELLIPSE. See **Axes of Ellipse.**

MINOR AXIS OF ELLIPSOID. See **Axes of Ellipsoid.**

MINOR IN DETERMINANTS. The minor of any element in a **determinant** is that determinant of next lower order obtained by deleting the row and column in which the element occurs. If it is given a positive or negative sign, for reasons below, it is called a *signed minor* or a *cofactor* of the element. The position number of the row of the element, counted downwards, and the position number of the column counted from the left are added, and if the sum is even the signed minor is positive; if odd, negative.

MINOR SECTOR. See **Major, Minor Sector of Circle, Sphere.**

MINOR SEGMENT. See **Major, Minor Segment of Circle, Sphere.**

MINUS. This is the Latin word for less. The original commercial sign for a deficiency, in contrast with **plus** for a superfluity. It is now the symbol $-$ which implies that the term or collection of terms in brackets immediately following it must be subtracted from any terms preceding it. It has come to signify a negative quantity as one 'subtracted' from zero. For example, absolute zero, -273°C, is really 0°C $-$ 273°C.

MINUS NUMBERS. See **Directed Numbers.**

MINUTES, SECONDS. The subdivisions of **angle (plane-angle)** and **mean solar day** based on the **sexagesimal** system of notation. See **Second of Time-Interval.**

MÖBIUS BAND (STRIP). If a rectangular strip *ABCDA* is joined so that *D* fits *A* and *C* fits *B*, an ordinary cylindrical surface is obtained on the outside and another on the inside of the continuous strip so formed. If, however, *C* is joined to *A* and *D* to *B* by giving the strip a single twist, a Möbius strip is obtained with only one, continuous surface. Any point on the original rectangular strip can now be joined to any point on the other side without crossing an edge. If a Möbius strip is cut along its medial line it falls into a single two-sided strip. The study of the properties of such a surface falls into the field of **topology**. See **Klein Bottle**.

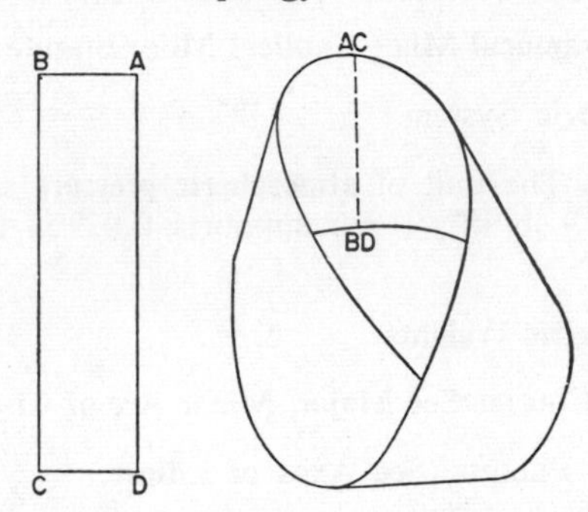

MODE. The most frequently occurring value in a series of measurements or observations. A peak in a **frequency curve**. See **Mean in Statistics**; **Median in Statistics**.

MODULO (MODULAR) ARITHMETIC. The application of fundamental operations to number systems which involve the use of numbers of the systems only. Thus, the modulo n or mod n system uses only the numbers 0, 1, 2, . . . $(n-1)$. The fundamental operations are the same as those of ordinary arithmetic except that if the number is greater than $(n-1)$ it is divided by n and the remainder is used in place of the ordinary result. There are no negative numbers: thus

$$x+3=0 \text{ (real numbers)} \rightarrow x=-3;$$
$$x+3\equiv 0 \pmod 7 \rightarrow x=4, \text{ since } 4+3\equiv 0, \pmod 7.$$

There are no fractions: thus

$$3x=5 \text{ (real numbers)} \rightarrow x=\tfrac{5}{3}=1\tfrac{2}{3};$$
$$3x\equiv 5 \pmod 7 \rightarrow x=4, \text{ since } 3\times 4\equiv 5 \pmod 7.$$

Some equations in real numbers have no solutions, e.g., $x^2=-6$: some equations in modulo arithmetic have no solutions, e.g., $x^2\equiv 6 \pmod 7$. Modulo arithmetic is used in **casting out nines**. See **Congruent Integers**; **Residue Class**.

MODULUS, BULK. See **Bulk Modulus**.

MODULUS OF COMPLEX NUMBER. Synonym of magnitude, or absolute value of complex number. See **Polar Form of Complex Number**.

MODULUS OF ELASTICITY. See **Elasticity**.

MODULUS OF LOGARITHMS. The number by which the **logarithms** of one system are multiplied to give the logarithms of another system. Given a set of logarithms to the base a, the corresponding set of logarithms to the base b may be obtained by multiplying each logarithm of the first set by $\log_b a$, the appropriate modulus.

MODULUS, RIGIDITY. See **Rigidity Modulus.**

MODULUS, YOUNG'S. See **Young's Modulus.**

MOMENT IN STATISTICS. A quantity given by $\Sigma(x)^n/N$, where x is the deviation of any item and N the number of items considered. See **Deviation in Statistics.**

MOMENT OF COUPLE. See **Couple in Mechanics.**

MOMENT OF FORCE. Synonym of **torque.** The measure of the turning effect of a **force** acting on a body about a point. If this is O, and the point of application of the force is P, the moment of the force $\mathbf{F}$ through P about O is the **vector product** of the **position vector r,** and the force $\mathbf{F}$, i.e. $\mathbf{L}=\mathbf{r}\times\mathbf{F}$. Its magnitude is $L=Fa=Fr\sin\theta$.

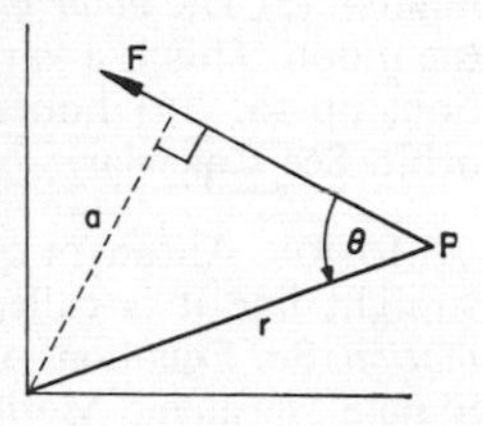

MOMENT OF INERTIA. The product mk^2, of the **mass** of a body and the square of a length called the **radius of gyration.**

MOMENT OF MASS. Synonym of *static moment.* If m is the mass of a particle and x its distance from a point, line or plane, the moment of the mass of the particle about the point, line or plane is the product mx. See **Centre of Mass.**

MOMENT OF MOMENTUM. Synonym of *angular momentum.* If a particle of **mass** m with a **position vector r** relative to a fixed point is moving with a velocity $\mathbf{v}$, the moment of momentum of the particle is the **vector product** $\mathbf{r}m\mathbf{v}$.

MOMENTUM. For a particle of **mass** m having a **velocity v**, the momentum is $m\mathbf{v}$. If a body is moving so that all its particles are moving in parallel lines, $\Sigma m=M$ and the momentum of the body is $M\mathbf{v}$. The rate of change of momentum in the direction of motion is proportional to the **force** applied in this direction to cause the change.

MONOMIAL EXPRESSION. An algebraic expression which contains one term. A simple expression. See **Multinomial Expression.**

MONOTONE. Synonym of **monotonic.**

MONOTONIC DECREASING QUANTITY. A **function, sequence,** etc., which either decreases or remains the same, but never increases.

MONOTONIC INCREASING QUANTITY. A **function, sequence,** etc., which either increases or remains the same, but never decreases.

MONOTONIC MAPPING. A **mapping** of **ordered set** X on to ordered set Y such that if element a_x precedes or equals element b_x in set X, then image a_y precedes or equals image b_y in set Y.

MONOTONIC SETS. A system of sets in which, for any two sets, one is contained in the other. See **Set Theory**.

MONTH. In Anglo-Saxon, *monath*, the moon-length of time. The word has three connotations. (1) The *calendar month*, one of twelve civil divisions of the year comprising 28, 29, 30 or 31 days. (2) Any four consecutive weeks or any 28 consecutive days; or two fortnights (i.e. fourteen days and nights). (3) The *lunar month*, the time interval between recurrent phases of the moon. This is a varying quantity averaging 29.530 59 days. The variation, up to $\pm 6\frac{1}{2}$ hours is due mainly to the eccentricity of the moon's orbit. See **Calendar**.

MOTION. Antonym of rest. If the change of position involved is along a straight line it is called *rectilinear motion*; if along a curve, *curvilinear motion*. See **Equations of Motion**; **Newton's Laws of Motion**; **Rigid Motion**; **Simple Harmonic Motion**.

MOTION GEOMETRY. A term which describes a dynamic approach to the study of geometry rather than a branch of geometry. It involves the study of those properties of space which remain invariant under transformations such as **rigid motions**. This is in contrast to the classical approach of the Greek geometers which was axiomatic, logical and in a sense static. See **Topological Transformations**; **Transformation in Geometry**.

MOTION IN CIRCLE. See **Circular Motion**.

MULTIFOIL. A plane figure bounded by congruent arcs of a circle. The centres of the arcs are the vertices of a regular polygon. Such figures are called *trefoil*, *quatrefoil*, *pentafoil*, *hexafoil*, etc., according to whether the polygon is a triangle, square, pentagon, hexagon, etc.

MULTINOMIAL EXPRESSION. An algebraic expression which is the sum of two or more terms. It is binomial, trinomial, quadrinomial, etc., when it has two, three, four, etc., terms. If an expression has only one term it is called a monomial. See **Polynomial Expression**.

MULTINOMIAL THEOREM. A theorem for the expansion of the expression $(x_1 + x_2 + \ldots + x_p)^n$ derived from successive applications of the **binomial theorem**.

MULTIPLE. Any member of an infinite set of integers or **polynomial expressions** which has a given integer or polynomial expression as a **factor** is called a multiple of that factor.

MULTIPLE INTEGRAL. A generalization from the **definite integral** of a function of one variable. The concept can be extended to cover integrals over two (double), three (triple) and higher dimensional regions.

MULTIPLE POINT. A point through which pass two or more branches of a curve. It may be double (**node**), triple, . . . n-tuple. At an n-tuple point there will be n tangents to the curve either all distinct, all coincident or some distinct and some coincident.

MULTIPLE ROOTS. Synonym of **repeated roots.**

MULTIPLE-VALUED FUNCTION. Synonym of **multi-valued function.**

MULTIPLICAND, MULTIPLIER. In a product of two factors, in which, say, A is 'multiplied by' B, A is called the multiplicand and B the multiplier.

MULTIPLICATION. A general term for the **binary operation** equivalent to that of combining equal groups into a whole; the reverse operation of **division.** It is expressed in the forms $a \times b = c$, $a \cdot b = c$, $ab = c$, where a and b are factors of the result of multiplication and c is called the *product*. Fundamentally the operation is an abstraction from different concrete experiences. *Magnification* corresponds to partition (reduction) in division and is the process of sharing objects to produce a given number of equal piles. For example, the placing of objects into four boxes produces the pattern: 4 ones are 4, 4 twos are 8, 4 threes are 12, ... *Successive addition* (extended count) corresponds to quotition (grouping) in division and is the process of adding equal amounts. For example, counting table legs produces the pattern 4, 4 and 4, 4 and 4 and 4, . . . or 1 four is 4, 2 fours are 8, 3 fours are 12, ... The **ratio** aspect of multiplication is based on direct comparison of objects and involves the concept of a unit of measure.

The concept of multiplication is developed by extending the number concept to embrace **rational numbers** (fractions). When applied to **directed numbers** it is referred to as **algebraic multiplication**. It is further developed by its application to the field of **complex numbers.**

The term multiplication can be applied to fields outside traditional arithmetic. Two sets of things are often combined in a multiplicative sense as in intersection (or product) in **set theory**. See **Cartesian Product**; **Operations in Ordinary Algebra**; **Operator.**

MULTIPLICATION, DIVISION OF COMPLEX NUMBERS. See **Complex Numbers.**

MULTIPLICATION, DIVISION OF VECTORS. See **Scalar Multiplication**; **Scalar Product**; **Vector (Cross) Product.**

MULTIPLIER. See **Multiplicand, Multiplier.**

MULTI-VALUED FUNCTION. Synonym of multiple-valued function, many-valued function. A correspondence between two sets A and B such that at least one element of set A is matched with two or more elements of set B, e.g., $y^2=4ax$. A multi-valued function is sometimes referred to as a **relation** when the term **function** is reserved for a *single-valued function.*

N

N. **Neighbourhood** of a.

n-DIMENSIONAL SPACE. See **Dimensions of Space.**

NADIR. The point on the **celestial sphere** in the direction downwards of the plumb-line. The **zenith**, the observer and the nadir are in a straight line.

NAPERIAN LOGARITHMS. Synonym of natural **logarithms** to base e.

NAPPE. A **conical surface** and the surface of a pyramid extend both ways from the apex. Each part of such a surface is a nappe of the surface.

NATURAL LOGARITHMS. Synonym of Naperian **logarithms** to base e.

NATURAL NUMBERS. Positive integers; whole **numbers** which may be **cardinal numbers** or **ordinal numbers.**

NAUTICAL MILE. The fundamental unit of distance used in navigation. The length of a minute of arc on a **great circle** drawn on the surface of a sphere equal in area to that of the earth. It is 6 080.27 feet. A minute of arc along a meridian varies with latitude owing to the flattening at the poles. Its *average* length is 6 076.82 feet. For practical purposes the minute of arc along the equator (the **geographical mile**), or along any meridian, and the nautical mile are approximated to 6 080 feet. See **Statute Mile.**

NEGATION OF PROPOSITION. A denial or contradiction of a given proposition formed by prefixing 'It is not true that' or by interpolation at an appropriate point of 'not'. Thus, the negation of the proposition: *he is alive*, is either: *it is not true that he is alive*, or more simply: *he is not alive.* This is not necessarily the same as: *he is dead*, which is a distinctly different positive proposition. See **Algebra of Propositions.**

NEGATIVE. See **Positive, Negative.**

NEGATIVE ANGLE. An **angle (plane angle)** measured in a clockwise direction.

NEGATIVE NUMBER. See **Directed Numbers.**

NEGATIVE QUANTITY. A quantity of length, time, temperature, etc., measured from zero in the direction opposite to that chosen as positive. See **Positive, Negative.**

NEIGHBOURHOOD. Part of a line which contains a given point. Thus, 0, on a number line can be described as being in the neighbourhood $(-1, +1)$ or the neighbourhood $(-\frac{1}{2}, +\frac{1}{2})$ or $(-\frac{1}{4}, +\frac{1}{4})$, etc. Any point on the line is contained in an infinity of neighbourhoods. A neighbourhood N of real number α is an **interval** $[a, b]$ containing α within it $(a<\alpha<b)$. See **Nested Intervals.**

NEPER. (From Napier.) A unit measuring intensity of sound on a **logarithmic scale**, and equal to 6.686 **decibels.**

NEST. Synonym of **nested set.**

NESTED INTERVALS. A sequence of intervals $[a_1, b_1], [a_2, b_2], \ldots$ which has the properties: (1) each interval except the first is included in the preceding one; (2) the width of the interval shrinks to zero; (3) two different numbers cannot be in the same nest since they are separated by a distance greater than zero; (4) a nest of intervals represents a number if that number is attached to the only point, called the *final residue*, which lies inside all the intervals of the sequence. The number 3.142 857, for example, lies between 3 and 4 and is said to be in interval [3, 4] or interval 3. It also lies in the intervals [3.1, 3.2], [3.14, 3.15], [3.142, 3.143], etc., or intervals 3.1, 3.14, 3.142, etc. See **Neighbourhood.**

NESTED SETS. Synonym of *nest*. Any collection of sets in which one set is contained in another.

NET PREMIUM. A **premium** which does not include the expenses of a company.

NET PROCEEDS. The selling price less the expenses of trading (except the original cost).

NET PROFIT, LOSS. See **Profit, Loss.**

NETS. See **Development of Solids.**

NEUTRAL EQUILIBRIUM. See **Equilibrium.**

NEWTON. (From Sir Isaac Newton.) The **force** required to produce an **acceleration** of one metre per second per second in a **mass** of one kilogramme. The unit is part of the **m.k.s. system** and corresponds to the **dyne** of the **c.g.s. system** or the **poundal** of the **f.p.s. system**. It is 100,000 dynes.

NEWTON'S LAW OF GRAVITATION. Two **masses** attract one another with a **force** directly proportional to the product of their masses and inversely proportional to the square of their distance apart. See **Constant of Gravitation.**

NEWTON'S LAWS OF MOTION. (1) Every body remains at rest or continues to move uniformly in a straight line unless compelled by external forces to change this condition (the law of **inertia**). (2) The rate of change of **momentum** of a body is proportional to the **force** applied to it and the change takes place along the line of the applied force. (3) **Action and reaction** are equal and opposite.

NINE POINTS CIRCLE. The unique circle passing through nine points fixed in the plane of triangle ABC: the midpoints of the sides, M_a, M_b, M_c; the feet of the perpendiculars, P_a, P_b, P_c; AP_a being an altitude of the triangle; the midpoints of the segments of the altitudes between the **orthocentre,** O, and the vertices, S_a, S_b, S_c. The centre of the nine points circle, N, lies midway between the orthocentre and the **circumcentre,** T. The radius of the circle is given by $abc/8\triangle$, where $\triangle = [s(s-a)(s-b)(s-c)]^{\frac{1}{2}}$, s being the semi-perimeter, $\frac{1}{2}(a+b+c)$. Alternatively,

$$r = \tfrac{1}{2}abc[(a+b+c)(b+c-a)(c+a-b)(a+b-c)]^{-\frac{1}{2}}.$$

See **Feuerbach's Theorem.**

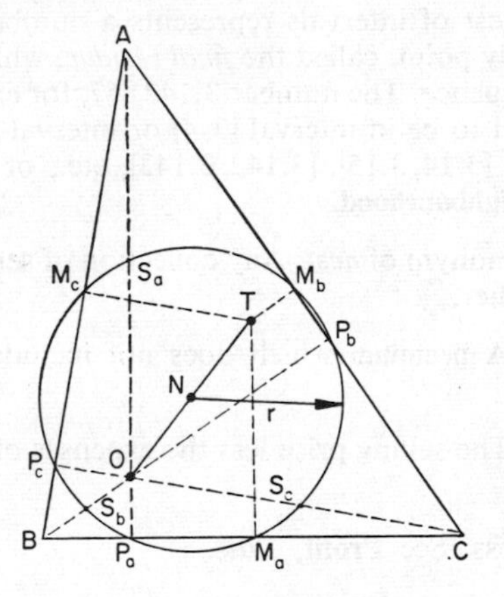

NODAL LINE. The line in any **configuration** which remains fixed while the configuration is subjected to any **rotation** or deformation.

NODAL POINT. A vertex common to three or more polygons which are part of a **tessellation.**

NODE OF CURVE. A **double point** common to two branches of the same curve at which the two tangents are real and distinct. It is sometimes called a *crunode.*

NODES IN ASTRONOMY. The points where the line of intersection of any two planes in space, known as the *line of nodes,* meets the **celestial sphere.** The **equinoxes** are examples of such points.

NODES IN PHYSICS. See **Standing (Stationary) Waves.**

NOMOGRAM. Synonym of time chart. A graphic device designed so that when any two variables out of three involved in a relation are known, the third can be determined directly. Thus the relation $3a+c=6b$ can be represented as three scales. A line will cross the scales at points representing values of a, b and c which satisfy the relation.

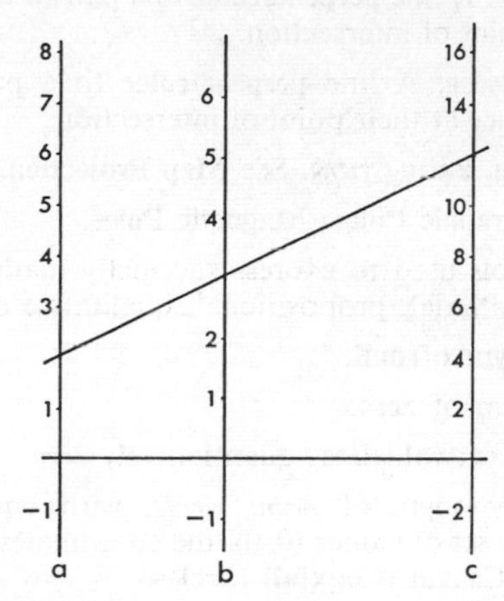

NONAGON. A **polygon** with nine sides.

NONARY. Associated with nine, as in nonary **system of notation.**

NON-EUCLIDEAN SPACE. A space which is not based on the parallel postulate of **Euclidean space**. There are two possibilities. Through a point external to a given line (1) there is no line parallel to the given line, (2) there are several lines parallel to the given line. The first possibility gives rise to **elliptic space**; the second to **hyperbolic space** in contrast to *parabolic space*, a term used to describe Euclidean space.

NON-METRIC GEOMETRY. Those aspects of **geometry** which do not utilize any concepts of magnitude and therefore of measure. Synonymous with **topology.**

NON-TRANSITIVE RELATION. See **Transitive Relation.**

NORMAL. Synonym of **perpendicular.** Sometimes used for usual, as in *normal* distribution. See **Error Curve.**

NORMAL DISTRIBUTION. See **Distribution.**

NORMAL DISTRIBUTION CURVE. See **Error Curve.**

NORMAL EQUATION OF LINE. If p is the line segment from the origin perpendicular to a straight line, and it makes an angle θ with the positive direction of the x-axis, then the equation of the line may be written: $x \cos \theta + y \sin \theta - p = 0$.

NORMAL TO CURVE. The line at right angles to a tangent to a curve at the point of contact of the tangent. Its equation is: $x - x' + (y - y')f'(x') = 0$, where $y = f(x)$ is the equation of the curve, (x', y') the point of contact and $f'(x')$ the **gradient of curve** at that point.

NORMAL TO PLANE. A line **perpendicular** to a pair of intersecting lines in the plane at their point of intersection.

NORMAL TO SURFACE. A line **perpendicular** to a pair of intersecting tangents to the surface at their point of intersection.

NORMAL ZENITHAL PROJECTION. See **Map Projection.**

NORTH. See **Geographic Poles; Magnetic Poles.**

NOTATION. Symbols used to express succinctly mathematical concepts which may be operational, propositional, qualitative or quantitative.

NOTHING. Synonym of **null.**

NOUGHT. Synonym of **zero.**

NULL. Non-existent; valueless; quantitatively **zero.**

NULL CIRCLE. Synonym of *point circle*, with equation $x^2 + y^2 = 0$, satisfied by only one set of values (0, 0), the coordinates of one point. The radius is zero. See **Coaxal (Coaxial) Circles.**

NULL ELLIPSE. Synonym of *point ellipse*, with equation $x^2/a^2 + y^2/b^2 = 0$, satisfied by only one set of values (0, 0), the coordinates of one point.

NULL MATRIX. A **matrix** with elements all zero.

NULL SEQUENCE. A **sequence** having **zero** as its **limit.**

NULL SET. The set with no elements. See **Set Theory.**

NULL VECTOR. A **vector quantity** with **zero** magnitude (zero length, zero absolute value).

NUMBER. A term with several interpretations. The earliest concept was that of the **counting** numbers or **natural numbers**. They suffice for counting, denoting the sizes of collections (**cardinal numbers**) or order (**ordinal numbers**). The invention of zero as a **place holder** facilitated computation with the natural numbers. The addition or multiplication of any two natural numbers produces a natural number and the system is said to be closed under addition and multiplication. When numbers are understood as **directed numbers** the concept is developed to embrace positive and negative integers and zero as a separator. The system is then closed under subtraction also. The system of **rational numbers**, defined as ratios of integers, enables quotients in the form of fractions to be included and the system is then closed under the arithmetic operations of addition, subtraction, multiplication and division. This is the fundamental concept of what constitutes a **number field**. However, some pairs of lengths cannot be expressed as a rational number; e.g., lengths of side and diagonal of a square, lengths of diameter and circumference of a circle. The set of real numbers includes rational numbers and the **irrational numbers** called **surds** which can be ordered and represented by the points on a line. This is the basis of the concept of a **continuum**. The real numbers can be subdivided into *algebraic numbers* (the solutions of polynomial equations with rational coefficients; (e.g., $x^2 = 5$) and non-algebraic or **transcendental numbers** (such as π and e). **Complex numbers** embrace all the real and **imaginary numbers** which permit the solution of equations which have no real roots (such as $x^2 = -1$). See **Amicable Numbers; Cubical Numbers; Perfect Numbers; Polygonal Numbers; Prime Numbers; Square Numbers.**

NUMBER FIELD. A set of **numbers** which is arithmetically closed. If the operations of addition, subtraction, multiplication and non-zero division are applied to any two elements, a third element of the same set is produced.

NUMBER LINE. An infinite line on which **directed numbers** are represented by points on both sides of a chosen zero point; conventionally, positive ones to the right, negative ones to the left or positive ones upwards, negative ones downwards.

NUMBER SCALE. (1) The **scale** formed by placing the directed numbers $\ldots -3, -2, -1, 0, +1, +2, +3, \ldots$ at equal intervals along a line. (2) A scale of notation. See **System of Notation.**

NUMBER SYSTEM. (1) A mathematical system such as **complex number** system, **real number** system. (2) A **system of notation.**

NUMBER THEORY. The study of integers (whole numbers) and the relationships between them. See **Number.**

NUMERAL. Any symbol used to express a **number**. Those derived from Hindu-Arabic sources, 0, 1, 2, 3, . . . 9, are the most commonly used. The Roman, I, V, X, L, C, D, M are rarely used and never in computation. See **Hindu-Arabic Numerals.**

NUMERATOR. The term N in the common **fraction,** N/D.

NUMERICAL COEFFICIENTS. See **Coefficients.**

NUMERICAL VALUE. Synonym of **absolute value**. The value of a number or quantity without reference to direction or sign. See **Absolute Value of Complex Number; Real Number; Vector.**

O

OBLATE. See **Ellipsoid.**

OBLIQUE. Neither **parallel** nor **perpendicular** to a given direction.

OBLIQUE COORDINATES. **Cartesian coordinates** with respect to axes which are not mutually **perpendicular**.

OBLIQUE PROJECTION. See **Pictorial Projection; Geometric Projection.**

OBTUSE ANGLE. An **angle (plane angle)** greater than a right angle (90° or $\pi/2$ radians) and less than a straight angle (180° or π radians).

OBTUSE (OBTUSE-ANGLED) TRIANGLE. A **triangle** with one **obtuse angle.**

OCTAGON. A **polygon** with eight sides.

OCTAHEDRON. A **polyhedron** with eight faces. It is one of the **regular solids** when its faces are regular trigons (equilateral, equiangular triangles).

OCTAVE. The difference in pitch between two tones with frequencies in the ratio of 2 : 1. See **Pitch in Music.**

OCTONAL, OCTONARY. Synonymous words associated with eight as in octonal (octonary) **system of notation.**

ODD FUNCTION. A **function,** $f(x)$, of a **variable,** x, which changes its sign but not its **absolute value** when the sign of x is changed. Its graph is symmetric with respect to the origin of coordinates; e.g., $y=x^3$, $y=\sin x$. See **Even Function.**

ODD NUMBER. An **integer** (whole number) which is not even, and therefore has remainder 1 when divided by 2. Any number of the form $2n+1$, where n is an integer (including zero).

OFFSET. A measurement usually taken at right angles to survey lines in **traversing** and **triangulation.** Such a measurement is usually limited to about 50 **links.**

OHM (Ω). Unit of electrical resistance. The resistance through which a **potential difference** of 1 **volt** will produce a current of 1 **ampere.** It is defined internationally as the resistance of a column of mercury with uniform cross-section, having length 106.3 cm and weight 14.452 1 g, at 0°C. Other units of electrical resistance are megohm ($M\Omega$) $=10^6$ Ω and micro-ohm ($\mu\Omega$) $=10^{-6}\Omega$.

ONE. See **Unity.**

ONE-ONE CORRESPONDENCE. A **correspondence** between two **sets** A and B in which each element of A is paired with one element only of B and conversely. It is written symbolically, $A \sim B$. A and B are said to be *equinumerous* and possess the same **cardinal number.** They can be described as $A=\{a_1, a_2, a_3, \ldots a_n\}$, $B=\{b_1, b_2, b_3, \ldots b_n\}$ where n is their cardinal number. Such sets are called *equipollent*. See **Inverse Function.**

ONE-ONE TRANSFORMATION. See **Transformation in Geometry.**

OPEN INTERVAL. See **Interval.**

OPEN TRAVERSE. See **Traversing.**

OPERATION. See **Algebraic Operation; Inverse Operation; Operations in Ordinary Algebra; Operation in Set; Operator; Transcendental Operation.**

Operation in Set. In any **set** an **operation** which combines two or more elements into a single element of the same set. A **binary operation** involves a pair, a *ternary operation*, a triple, etc.

Opposite Angles, Sides, Vertices of Polygon. Two angles, sides, vertices which have an equal number of sides between them irrespective of the direction round the polygon.

Opposite Angles, Sides, Vertices of Triangle. In any triangle each side is opposite a vertex or angle and vice versa. In Euler's notation for a **standard triangle**, $AcBaCbA$, there are three pairs of opposites, (a, A), (b, B), (c, C).

Operations in Ordinary Algebra. For the set of **rational numbers** $R = \{a, b, c, \ldots\}$ subject to two **binary operations**, addition $(+)$ and multiplication $(\times)$, the following laws hold:

closure. $a + b$ and $a \times b$ are rational numbers.

associative law. $a + (b + c) = (a + b) + c$; $a \times (b \times c) = (a \times b) \times c$.

commutative law. $a + b = b + a$; $a \times b = b \times a$.

identity. Zero under addition, $a + 0 = 0 + a = a$; unity under multiplication, $b \times 1 = 1 \times b = b$.

inverse. Negative $(-a)$ under addition, $a + (-a) = (-a) + a = 0$; reciprocal (a^{-1}) under multiplication, $a \times a^{-1} = a^{-1} \times a = 1$ $(a \neq 0)$.

cancellation. $a + b = a + c$, then $b = c$; $a \times b = a \times c$ $(a \neq 0)$, $b = c$.

distributive law. $a \times (b + c) = (a \times b) + (a \times c)$; $(a + b) \times c = (a \times c) + (b \times c)$.

Operator. (1) The number, symbol or expression associated with a sign of operation, which may be implied. For example: in $n(x + 1)$, n is the operator associated with an implied *multiplication*; in 3×4, 3 is the operator, though 4 is often incorrectly made the operator, possibly in attempting to make the notation consistent with addition, subtraction and division. However, in general, the operator in mathematics appears at the front of an expression. (2) A symbol used to represent an operation. For example: $(D^2 + xD + 2)\, y = d^2y/dx^2 + x(dy/dx) + 2y$, D being the **differential operator.** See **Binary Operation.**

Oppositely Congruent. See **Congruent Figures in Space.**

Orbit. The path through space taken by any moving body. The orbits of planets and satellites are **ellipses** with small **eccentricity,** those of comets are very elongated ellipses, where e nearly equals 1. The **centre of mass** of the Earth–Moon is 1,000 miles below the earth's surface. The earth and the moon rotate about this point monthly; and it is *this* point that completes what is called the earth's orbit, annually.

Order of Derivative. See **Derivative.**

Order of Determinant. See **Determinant.**

Order of Differential Equation. The order, first, second, third, . . . , of a differential equation expresses the fact that the first, second, third, . . .

differential coefficient is the highest one involved in the equation. Thus, $d^2y/dx^2 + py^2 = 0$ is of second order.

ORDER OF EQUATION. The degree of the highest power represented. Synonym of **dimension of equation.** See **Degree in Algebra; Power of Quantity.**

ORDER OF GROUP. The number of elements in a finite group. See **Group Theory.**

ORDER OF MATRIX. If a set of numbers is represented by a square **matrix** (array) of m rows and m columns, the number m is the order of the matrix.

ORDER OF NODAL POINT. The number of polygons which share a common vertex (nodal point) in a **tessellation.**

ORDER OF RADICAL. If a **radical expression** is written as $\sqrt[n]{a}$ (the nth root of a), then n is the order of the radical.

ORDER OF ROOT. The number of times a **root of equation** is repeated. In the cubic equation $x^3 - 7x^2 + 15x - 9 = 0$, which can be factorized as $(x-3)^2(x-1) = 0$, the root 3 is of order 2; root 1, of order 1.

ORDER OF SYMMETRY. A geometrical figure has a **centre of symmetry** of the nth order, if, on rotating the figure about the centre by an angle of $360°/n$, it coincides with its original position.

ORDER RELATIONS. For a set $R = \{a, b, c, \ldots\}$ of **rational numbers:**

trichotomy. One and only one of $a<b$, $a=b$, $a>b$ is true.

transitivity. If $a<b$ and $b<c$, then $a<c$.

density. If $a<b$, then there is an element c such that $a<c<b$.

extension. For any element a there exist elements b and c such that $b<a<c$.

consistency. If $a<b$, then $(a+c)<(b+c)$ for any c; and $(a \times c)<(b \times c)$ for any positive c.

ORDERED FIELD. A **field** which includes a set of positive elements to which the following rules apply: (1) the sum and the product of any two elements is positive (the subset is closed); (2) for any element x either (i) x is positive, (ii) x is zero, or (iii) $-x$ is positive.

ORDERED PAIR, TRIPLE, n-TUPLE. A set of elements written in the order in which they are to be considered. Thus (a, b) in **Cartesian coordinates** can represent a point in a plane; (a, b, c) a point in ordinary space. Such sets can also be considered as **vector quantities,** e.g. (a, b, c, d) can represent a vector in four-dimensional space.

ORDERED SET. A set in which (1) any two elements a and b satisfy one only of the relations $a<b$, $a=b$, $a>b$; (2) if $a <$ or $> b$ and $b<$ or $> c$ then also $a <$ or $> c$. If these conditions apply only to some of the elements, it is a *partially ordered set.*

ORDINAL. Expressing order or succession.

ORDINAL NUMBER. A number considered as a place in the ordered sequence of whole **numbers**. Used in counting as first, second, third, fourth, etc., to *n*th in a set of *n* elements. See **Cardinal Number**; **Order Relations.**

ORDINARY FRACTION. Synonym of **common fraction.**

ORDINATE. The *y* coordinate of a point. See **Cartesian coordinates.**

ORIENTATION. The determination of **direction** in terms of accepted standard directions, e.g., earth's axis in astronomy; Cartesian coordinate axes; vertical and horizontal lines through a point. See **Sense of Orientation.**

ORIGIN. A fixed point from which a measurement is made; and especially the point at the intersection of axes in **Cartesian coordinates.** Also the pole of **polar coordinates in plane.**

ORIGIN OF RAY. The initial point of a **ray.**

ORTHOCENTRE. The point of **concurrency** of the three **altitudes of triangle.**

ORTHOGONAL. Right-angled. Referring to right angles and the use of them in projection.

ORTHOGONAL BASIS VECTORS. See **Basis Vectors.**

ORTHOGONAL CIRCLES. A pair of circles which intersect at right angles, i.e. at each common point the tangents are perpendicular. The tangents to one circle pass through the centre of the other.

ORTHOGONAL PROJECTION. Synonym of orthographic projection. See **Geometric Projection; Map Projection.**

ORTHOGONAL TRAJECTORY. A member of a family of curves, all members of which cut at right angles *every* member of another family of curves.

ORTHOGONAL VECTORS. Any **vector quantities** which operate along two perpendicular lines in a plane or three mutually perpendicular lines in space. The **scalar product** of any pair is zero.

ORTHOGRAPHIC PROJECTION. Synonym of orthogonal projection. See **Geometric Projection; Map Projection.**

ORTHOMORPHIC PROJECTION. See **Map Projection.**

ORTHONORMAL BASIS VECTORS. See **Basis Vectors.**

ORTHOTOMIC. If P is a fixed point and P' its image in any tangent to a curve, the set of points P' is the orthotomic of the curve with respect to the point P.

OSCILLATION. In general a synonym of vibration. Technically, it is the total movement in any vibratory motion which occurs during one period of oscillation, the time taken by a particle to move from one point to the same point, returning in the same direction with the same velocity. Many types of oscillation approximate to **simple harmonic motion**. When the **amplitude** of an oscillation is continuously diminished by any force resisting the motion the oscillation is referred to as a *damped oscillation*.

OSCULATING CIRCLE. That which passes through three close-together points on a curve. It is the circle of closest contact, and the innermost point gives equal values of dy/dx for the circle and curve, and equal values of d^2y/dx^2. It is sometimes called the circle of **curvature**.

OSCULATION. See **Point of Osculation.**

OUTER PRODUCT. Synonym of **Cartesian product.**

OVERLAPPING SETS. Sets having one or more members in common. If not overlapping they are called *disjoint*. Thus the even numbers and the odd numbers form disjoint sets; the set of perfect squares and the set of perfect cubes have 1 and 64 in common and are overlapping. See **Set Theory**.

P

P (*a*). Probability of statement (*a*). See **Probability in Logic.**

PAPPUS'S THEOREMS. Various theorems bear the name of Pappus, a mathematician of the third century A D. One is associated with the area of the surface of a **solid of revolution,** one with the volume of such a solid and one with a basic idea of projective geometry.

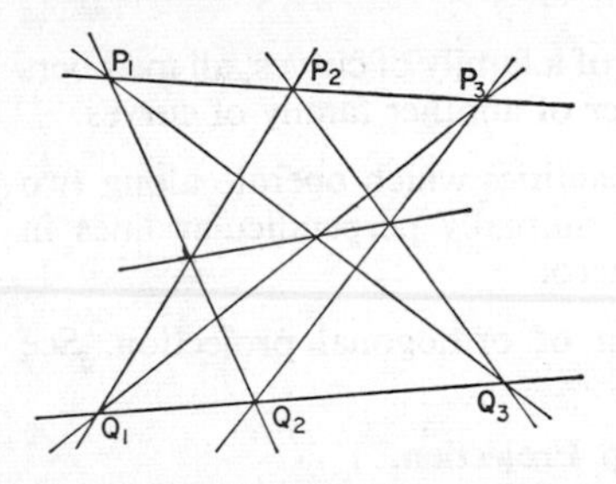

(1) When an arc of a plane curve is rotated about an axis in its plane not cutting the curve, the area generated by the arc is the product of the length of this arc and the segment of a circular path travelled by the **centroid** of the arc. If the path of the centroid is the circumference of a circle, the area generated is a surface of revolution.

(2) When a plane surface is rotated about an axis in its plane not cutting the surface, the volume generated by the surface is the product of the area of the surface and the segment of a circular path travelled by the centroid of the surface. If the path of the centroid is the circumference of a circle, the volume generated is a volume of revolution.

(3) If P_1, P_2, P_3 are points in order on a straight line, and Q_1, Q_2, Q_3 are three points in order on another straight line, then the pairs of joins: P_1Q_2, P_2Q_1; P_2Q_3, P_3Q_2; P_3Q_1, P_1Q_3 will define at their points of intersection, points on a third straight line.

PARABOLA. The **conic section** made by a plane parallel to a **generating line** of a cone. The set of points which satisfy the equation $y^2 = 4ax$. The focus is the point $(a, 0)$ and the directrix is the line $x = -a$. See **Focus-Directrix Definition of Conic; Parametric Coordinates.**

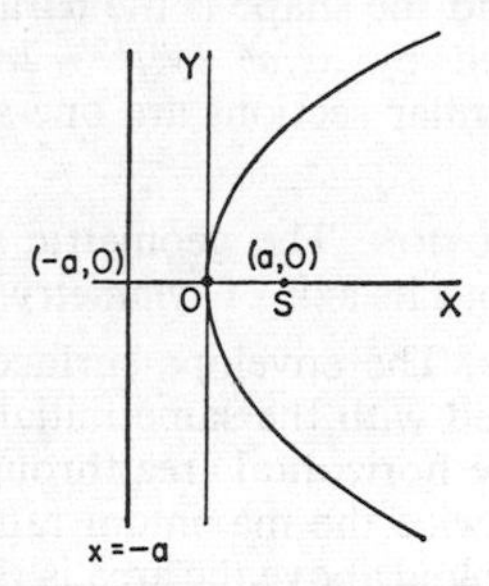

PARABOLA OF SAFETY. A vertical plane section through the vertex of the **paraboloid of safety.**

PARABOLIC MOTION. The motion of a projectile in a vertical plane. Near the earth's surface for speeds not much more than that of sound (760 m.p.h., 332 m/s), with g, the **acceleration due to gravity,** a constant and negligible air resistance, the **parametric equations** of the parabola of the trajectory are $x = v_0 t \cos \alpha$ and $y = v_0 t \sin \alpha - \frac{1}{2}gt^2$, where v_0 is the initial velocity and α the initial **angle of inclination** of the projectile with respect to the horizontal plane. The maximum height reached is $(v_0 \sin \alpha)^2/2g$ and the range on a horizontal plane, $(v_0^2/g) \sin 2\alpha$. The greatest range occurs when $\alpha = 45°$, the parametric equations being: $x = \frac{1}{2}\sqrt{2}\, v_0 t, y = \frac{1}{2}\sqrt{2}\, v_0 t - \frac{1}{2}gt^2$ and the parabola of the trajectory is $y = x - \frac{1}{2}gt^2$. Maximum height is $\frac{1}{4}v_0^2/g$ and the range, v_0^2/g.

PARABOLIC SEGMENT. Section of a plane bounded by a chord of a **parabola** perpendicular to the axis of the parabola and the arc of the parabola cut off by the chord. Its area is $2ca/3$, where c is the length of the chord and a the perpendicular distance from the vertex of the parabola to the chord.

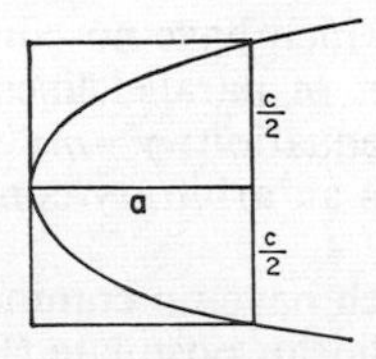

PARABOLIC SPACE. Synonym of **Euclidean space.**

PARABOLIC SPIRAL. Synonym of Fermat's spiral; the **spiral** whose **polar equation** is $r^2 = a\theta$.

PARABOLOID. A geometric surface in which certain sections are **parabolas.** The surface represented in rectangular coordinates by $x^2/a^2 + y^2/b^2 = 2cz$ is called an *elliptic paraboloid* because its sections parallel to the co-ordinate planes are one set of **ellipses** and two sets of parabolas. If $a = b$, the ellipses are circular and the shape is the **paraboloid of revolution.**

The surface represented by $x^2/a^2 - y^2/b^2 = 2cz$ is called a *hyperbolic paraboloid* because the similar sections are one set of **hyperbolas** and two sets of parabolas.

PARABOLOID of REVOLUTION. The geometric surface generated by the rotation of a **parabola** about its axis of **symmetry.**

PARABOLOID OF SAFETY. The envelope surface touching all the trajectories of particles projected with the same initial velocity in all directions from one point. It cuts the horizontal area through the point in a circle of diameter $2v^2/g$, which is twice the maximum range of a particle: and the highest point of the **paraboloid** above the area is v^2/g. See **Parabolic Motion.**

PARADOX. An apparent contradiction between two conclusions each of which seems to be supported by fact or reason. Paradoxes have arisen throughout the history of mathematics and have stimulated greater comprehension of the logical foundations of mathematics.

PARADOX OF GALILEO. See **Countably Infinite Set.**

PARALLAX. If E, S and A suggest the positions of the earth, the sun and a star, when the angle ESA is a right angle, then the extremely small angle at A is the measure of the parallax of the star. There is no star with a parallax as great as one second of arc.

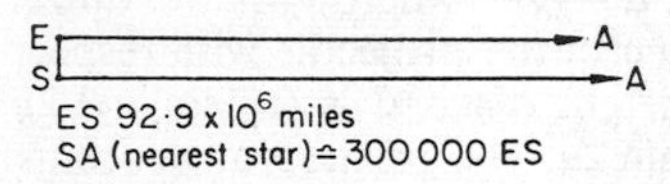

PARALLEL. From the Greek, *para allelos*, beside one another, the word which describes the common property of **parallel curves, lines, planes.**

PARALLEL CURVES, LINES, PLANES.

Curves in a plane, points of which have a **one-one correspondence** such that any pair share a common **normal** and centre of **curvature.** The tangents at these points are parallel lines, and their distance apart is the same for all pairs.

Straight lines in a plane which have no common point no matter how far they are extended. A set of parallel lines is represented in **Cartesian coordinates** by the general equation: $y = mx + c$, where m is the constant gradient of the lines and c is an arbitrary constant which is the intercept of the line on the y-axis.

Planes lying in space which have no common points no matter how far they are extended. The Euclidean postulate (**Playfair's axiom**), applied to parallel lines, can be extended to cover parallel planes. All perpendiculars to parallel planes are parallel lines; intercepts between a pair of planes being equal.
See **Anti-parallel; Skew Lines.**

PARALLEL (CYLINDRICAL) PROJECTION. See **Geometric Projection; Map Projection.**

PARALLEL FORCES. **Forces** which act in lines **parallel** to one another, but not necessarily directed in the same way. Their resultant is the single force parallel to themselves, which has the same **moment**, about an arbitrarily chosen point. See **Couple in Mechanics.**

PARALLEL LINES. See **Parallel Curves, Lines, Planes.**

PARALLEL PLANES. See **Parallel Curves, Lines, Planes.**

PARALLELS OF LATITUDE. Small circles on a sphere, having the polar axis as a common axis. When the sphere is the earth, the axis is that diameter through the geographical North and South Poles. See **Spherical Coordinates in Geodesy; Spherical Coordinates on Sphere.**

PARALLELS POSTULATE OF EUCLID. See **Playfair's Axiom.**

PARALLELEPIPED. A solid bounded by six faces in three pairs of similar, parallel, equal **parallelograms.**

PARALLELEPIPEDON LAW. In a rectangular **prism,** one of the simple **parallelepipeds,** any one of the diagonals has a length d given in terms of the length, breadth and height of the prism by the equation: $d=\sqrt{(l^2+b^2+h^2)}$. If one diagonal makes angles α, β, γ with the edges of length, breadth and height, then the directions of that diagonal through the prism are $\alpha=\cos^{-1}l/d$, $\beta=\cos^{-1}b/d$ and $\gamma=\cos^{-1}h/d$. See **Velocity.**

PARALLELOGRAM. A **quadrilateral** with two pairs of **parallel** sides. Its opposite sides are equal in length; its opposite angles are equal; its diagonals bisect one another.

PARALLELOGRAM OF FORCES, VELOCITIES, ETC. See **Parallelogram of Vectors.**

PARALLELOGRAM OF VECTORS. A graphical method for the **composition of vectors.** If **vector quantities P** and **Q** acting at a point O are represented in magnitude and direction by the line segments $\overline{OA}$ and $\overline{OB}$, then their resultant **R** is similarly represented by $\overline{OC}$ along the diagonal of the parallelogram $OACB$. This resultant is the **vector sum** of **P** and **Q**. The other diagonal represents in magnitude and direction the vector difference of **P** and **Q**; its sense being $\overline{AB}$ if $OB>OA$ and $\overline{BA}$ if $OA>OB$.

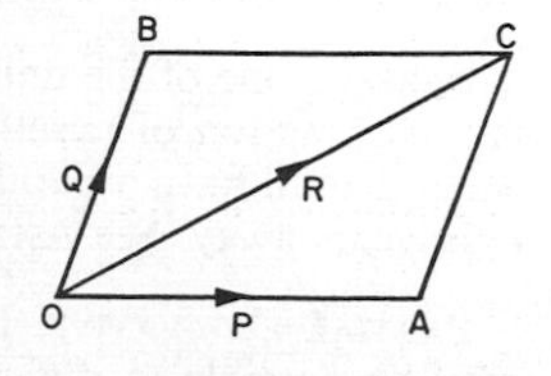

PARAMETER. A family of parallel lines can be represented by the equation $y=mx+c$. The m is a constant: the arbitrary constant c is the parameter. Every value assigned to c gives an equation associated with one line only. A second use of the word is made in describing the auxiliary variables in the **parametric equations** of any configuration. The concept of a parameter can be extended to embrace surfaces, solids and other spaces. See **Dimension of Space.**

PARAMETRIC COORDINATES. The **Cartesian coordinates** of a point, expressed in terms of other variables, called **parameters,** which are fewer in number than the original coordinates. Thus, the point (x, y, z) on the **ellipsoid** with equation $x^2/a^2 + y^2/b^2 + z^2/c^2 = 1$, may be written with two parametric coordinates, u and v, as $(a \cos u \cos v,\ b \cos u \sin v,\ c \sin u)$.

PARAMETRIC COORDINATES OF CONIC SECTIONS. The **parametric coordinates** of a point lying on the following **conic sections**:

circle. $x^2 + y^2 = a^2$, either $(a \cos \theta,\ a \sin \theta)$ or $\{a(t^2 - 1)/(t^2 + 1),\ 2at/(t^2 + 1)\}$.

ellipse. $x^2/a^2 + y^2/b^2 = 1$, either $(a \cos \theta,\ b \sin \theta)$ or $\{a(t^2 - 1)/(t^2 + 1),\ 2bt/(t^2 + 1)\}$.

parabola. $y^2 = 4ax$, $(at^2, 2at)$.

hyperbola. $x^2/a^2 - y^2/b^2 = 1$, either $(a \sec \theta,\ b \tan \theta)$ or $(a \cosh \phi,\ b \sinh \phi)$.

rectangular hyperbola. $xy = c^2$, $(ct, c/t)$.

See **Parametric Equations.**

PARAMETRIC EQUATIONS. Equations of any **configuration** in which the coordinates are separately expressed as functions of **parameters.** For example, $x = a \cos \theta$, $y = a \sin \theta$ represent a circle of radius a with centre at origin. The parameter θ is the angle the radius makes with the positive direction of the x-axis.

PARENTHESIS. See **Aggregation.**

PARITY. Two numbers have the *same parity* when they are both even or both odd. When one is even, the other odd, they have *opposite parity.*

PARITY (PAR). Equality used in a special financial sense to state the fact that the buying or selling price of a share, or a holding of stock is equal to the price at the time of their first being marketed. See **Premium.**

PARSEC. One of the units of length appropriate to astronomy. The word is a contraction of **parallax** and second. A star, one parsec distant from earth, would have a parallax of one second of arc, and be some 3·258 . . . **light years** away. See **Astronomical Units.**

PARTIAL DERIVATIVE. If a **function** contains n independent variables it may be possible to differentiate it in n distinct ways, once in respect of each variable. Let $u = f(x, y, z, \ldots)$ and all the variables except one, x, be held constant. Then if u is a differentiable function of x, its **derived function** in these circumstances is called the *partial differential coefficient* of u or the partial derivative of u with respect to x. It is denoted by the distinguishing symbol $\partial u/\partial x$. Then

$$\frac{\partial u}{\partial x} = \lim_{\delta x \to 0} \frac{f(x+\delta x, y, z, \ldots) - f(x, y, z, \ldots)}{\delta x}.$$

Similarly,

$$\frac{\partial u}{\partial y} = \lim_{\delta y \to 0} \frac{f(x, y+\delta y, z, \ldots) - f(x, y, z, \ldots)}{\delta y}.$$

For example: if $u = 2x^5y^4$, $\partial u/\partial x = 10x^4y^4$ and $\partial u/\partial y = 8x^5y^3$. Also if $pv = kt$, or $p = k(t/v)$, $\partial p/\partial t = k/v$ and $\partial p/\partial v = -kt/v^2$. See **Partial Differentiation.**

PARTIAL DERIVATIVE SYMBOL, ∂. See **Partial Derivative.**

PARTIAL DIFFERENTIAL COEFFICIENT. See **Partial Derivative.**

PARTIAL DIFFERENTIATION. The operation of finding **partial derivatives.** If $u = f(x, y, z, \ldots)$, the infinitesimal change in u due to infinitesimal changes in x, y, z, . . . at the same time is

$$\delta u = (\partial u/\partial x)\delta x + (\partial u/\partial y)\delta y + (\partial u/\partial z)\delta z + \ldots,$$

a sum of terms, each of which is an infinitesimal change in one of the variables multiplied by the partial differential coefficient of the function with respect to that variable. See **Partial Derivative.**

PARTIAL FRACTIONS. Fractions of the form

$$\frac{A}{(ax+b)^n}, \quad \frac{Ax+B}{(ax^2+bx+c)^n},$$

where n is a positive integer. Any given proper **fraction** can be written as a sum of such partial fractions, whose denominators are the factors of the denominator of the given fraction. This is the reverse process of the addition of fractions; it is used in **integration of rational functions.**

PARTIAL SUMS OF SEQUENCE. If $a_1, a_2, a_3, \ldots a_r, \ldots a_n, \ldots$ is a **sequence** with first element a_1; general element a_r; the nth element, usually the last of a finite set, a_n, then $a_1 + a_2$, $a_1 + a_2 + a_3$, . . . $\sum_{r=1}^{n} a_r$, . . . are the partial sums, each an ordered **series** of a specified number of terms.

PARTIALLY ORDERED SET. See **Ordered Set.**

PARTICLE. A minute portion of a body at which **scalar quantities** like mass may be considered as being concentrated, and towards which **vector quantities** such as forces may be directed.

PARTITION. One aspect of **division**, synonym of reduction. See **Theory of Partitions.**

PARTITION OF SET. For any set S, a set of disjoint (non-overlapping) subsets $\{S_a, S_b, S_c, \ldots\}$ such that $S_a + S_b + S_c + \ldots = S$. See **Set Theory**.

PASCAL'S THEOREM. One of the early theorems of **projective geometry,** discovered in the early seventeenth century. It states that the three points of intersection of the three pairs of opposite sides of a hexagon inscribed in a **conic section** lie on a line.

PASCAL'S TRIANGLE. The **binomial theorem** may be written:

$$(1+x)^n = 1 + {}^nC_1\, x + {}^nC_2\, x^2 + {}^nC_3\, x^3 + \ldots + {}^nC_r\, x^r + \ldots$$

If the coefficients are calculated for varying values n, they can be arranged in the following pattern, known as Pascal's triangle:

When $n=0$, the coefficient is 1
When $n=1$, the coefficients are 1 1
2, 1 2 1
3, 1 3 3 1
4, 1 4 6 4 1
5, 1 5 10 10 5 1
6, 1 6 15 20 15 6 1

This array can be extended indefinitely. The numbers are related in certain ways. (1) Each number is the sum of the two nearest numbers in the row above, e.g. $15 = 5 + 10$. (2) The sum of the numbers in each row is a power of 2, (2^n), e.g. $1+4+6+4+1 = 2^4 = 16$. (3) Each number is the sum of all the numbers in the diagonal sequence beginning with 1 and finishing with either of the two nearest numbers in the row above, e.g. $15 = 1+2+3+4+5 = 1+4+10$. (4) Each diagonal sequence gives the coefficients of the expansion of $(1-x)^{-m}$ where m is the number of the diagonal, e.g. $(1-x)^{-3} = 1 + 3x + 6x^2 + 10x^3 + 15x^4 + \ldots$ See **Extension of Pascal's Triangle; Fibonacci Sequence.**

PEAUCELLIER'S LINKAGE. A mechanical contrivance for constructing inverse curves. O is fixed, then P (or Q) traces a curve and Q (or P) traces the inverse curve. See **Inversion.**

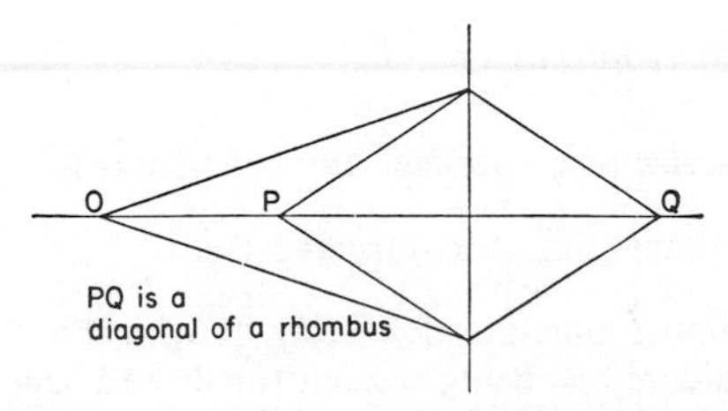

PEDAL CURVE. If P is any point on the curve $y = f(x)$, t the tangent to the curve at P and OT the perpendicular from a fixed point O to t, then the locus of T is the pedal curve, associated with $y = f(x)$. The pedal curve of a

parabola with respect to the focus as fixed point is the tangent at the vertex. The pedal curve of an **ellipse** or a **hyperbola** with respect to a focus is the **auxiliary circle**.

PEDAL EQUATION. Let P be a point on a curve, and O a fixed point called the *pole*. Let the length of the perpendicular from O to the tangent at P, OM, be p; and the length of OP, r. The equation giving the relation between p and r is the pedal equation of the curve. It is sometimes called the *tangent-polar equation*. For example, if a focus of an ellipse $x^2/a^2+y^2/b^2=1$ is taken as pole, the pedal equation of the ellipse is $b^2/p^2=(2a/r)-1$.

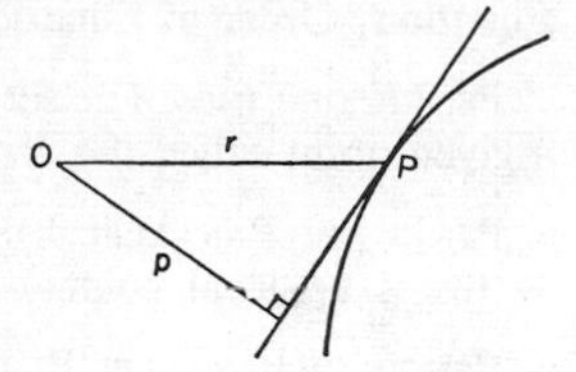

PEDAL LINE (SIMSON'S LINE). See **Pedal Triangle.**

PEDAL TRIANGLE. (1) If from any point P in the plane of a given triangle perpendiculars are drawn to the sides or sides produced, the triangle formed by joining the feet of the perpendiculars is called a *pedal triangle*. (2) If P lies on the circumference of the **circumcircle**, the triangle has zero area because the vertices of the pedal triangle are **collinear**. The line so formed is called a *pedal line* or *Simson's line*. (3) If P is at the **orthocentre** of the given triangle, the perpendiculars bisect the angles of the pedal triangle.

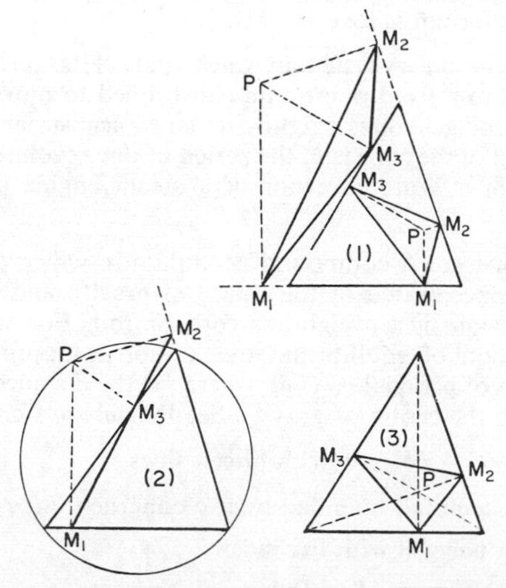

PENCIL, GENERAL EQUATION. If $F_1=0$ and $F_2=0$ are equations of the same order representing any two members of a **pencil of lines**, then $hF_1+kF_2=0$ is the general equation of the pencil, where h and k are arbitrary **parameters** (not simultaneously zero). See **Order of Equation.**

PENCIL OF CIRCLES. The set of all circles in a given plane which pass through two given points which lie on the **radical axis** of the circles. See **Pencil, General Equation.**

PENCIL OF CURVES. The set of all plane algebraic curves of the same order n, which pass through all of a set of n^2 given points. See **Pencil, General Equation**; **Order of Equation.**

PENCIL OF LINES. The set of all lines in a given plane which pass through a given point called the *vertex* of the pencil. See **Pencil, General Equation.**

PENCIL OF PARALLEL LINES. A special case of **pencil of lines** where the vertex is an **ideal point.**

PENCIL OF PARALLEL PLANES. A special case of **pencil of planes** where the axis is an **ideal line.**

PENCIL OF PLANES. The set of all planes which pass through a given line called the *axis* of the pencil. See **Pencil, General Equation.**

PENCIL OF SPHERES. The set of all spheres which pass through a given circle which lies in the **radical plane.**

PENDULUM, COMPOUND. Any rigid body suspended from a horizontal axis passing through it. If k is the **radius of gyration** and h the distance of the **centre of gravity** from the axis, the period of small **oscillations** about the position of equilibrium is $2\pi\sqrt{(k^2/gh)}$.

PENDULUM, CONICAL. A system in which a particle is attached by a weightless cord or rod to a fixed point, and constrained to move in such a way that the cord or rod generates a right-circular **conical surface** with a vertical axis. If the height of the cone is h, the **period** of one revolution is $2\pi\sqrt{(h/g)}$. The operation of a Watt governor of a steam engine depends on this principle.

PENDULUM, SIMPLE. A compound pendulum in which the mass is considered to be concentrated at the **centre of gravity** and attached to the point of suspension by a weightless cord or rod. For small **oscillations** about the position of equilibrium the motion is approximately **simple harmonic motion** of **period** $2\pi\sqrt{(l/g)}$, where l is the distance from the point of suspension to the centre of gravity. See **Pendulum, Compound.**

PENTADECAGON. A **polygon** with fifteen sides.

PENTAFOIL. A **multifoil** bounded by five congruent arcs of a circle.

PENTAGON. A **polygon** with five sides.

PENTAGONAL NUMBERS. See **Polygonal Numbers.**

PENTAGRAM OF PYTHAGORAS. The five-pointed star obtained by drawing the five diagonals of a regular **pentagon.**

PENTAHEDRON. A **polyhedron** with five faces

PENTOMINO. A **polyomino** made of five adjacent squares.

PER ANNUM (p.a.). For each and every year.

PER CENT. Symbol %. Hundredth(s). 100% is unity. 17% is a way of writing seventeen per cent, or 17/100 or 0.17. Every number is 100% of itself, and of two numbers A and B, A is (A/B) 100% of B.

PERCENTAGE. A common **fraction** with 100 as its denominator. It may be proper or improper. Thus, $N/100$ may be expressed as N **per cent** or N%. A percentage is used to express a relation, ratio or rate between two quantities; e.g., profit/price, interest/principal, duty/value, etc.

PERCENTAGE ERROR. **Relative error** expressed as a **percentage.**

PERCENTAGE PROFIT, LOSS. **Relative profit, loss,** expressed as a **percentage** of the cost price, selling price or turnover, according to circumstances.

PERCENTILE. A statistical term for a value which divides the range of a set in such a way that a given **percentage** of it lies before the point of division.

PERFECT CUBE. See **Perfect Power.**

PERFECT NUMBERS. If $2^n - 1$, n being an integer, is a **prime number**, then $2^{n-1}(2^n - 1)$ is called a *perfect number*. Euclid was interested in these numbers, each of which is equal to the sum of its factors, including 1, but excluding itself. Thus, $6 = 2^{2-1}(2^2 - 1) = 2 \times 3 = 1 + 2 + 3$; $28 = 2^{3-1}(2^3 - 1) = 4 \times 7 = 1 + 2 + 4 + 7 + 14$. The next perfect number is 496. When the sum of the factors of a number is greater than the number, the number is said to be *defective* or *deficient*; when the number itself is greater than the sum of its factors, it is said to be *abundant*, *excessive* or *redundant*.

PERFECT POWER. A number or a polynomial which is the nth power of a number or a polynomial respectively. Thus 9 is a perfect square, (3^2); 27 is a perfect cube, (3^3); $x^2 + 2x + 1$ is a perfect square, $(x+1)^2$; $x^3 - 3x^2 + 3x - 1$ is a perfect cube, $(x-1)^3$. If x is fractional, the possibility arises of having a mixed number as a perfect power; thus $2\frac{1}{4}$ is a perfect power of $1\frac{1}{2}$.

PERFECT SQUARE. See **Perfect Power.**

PERFECT TRINOMIAL SQUARE. An expression of the form $a^2 \pm 2ab + b^2$. It is equivalent to a **square of binomial.**

PERICYCLOID. The path traced by a point on the circumference of a circle which rolls without slipping on the outside of a fixed circle. The fixed circle lies within the larger, rolling circle. See **Epicycloid; Hypocycloid.**

PERIGAL'S DISSECTION. A proof, by fitting, of **Pythagoras's theorem** due to Henry Perigal in 1830. The middle point of the medium-sized square is found and through it are drawn lines at right angles to and parallel to the hypotenuse. The four equal portions, *a*, *b*, *c*, *d* of this square are translated to occupy corner positions of the square on the hypotenuse. The small square, *e*, can be translated to fill the uncovered central part.

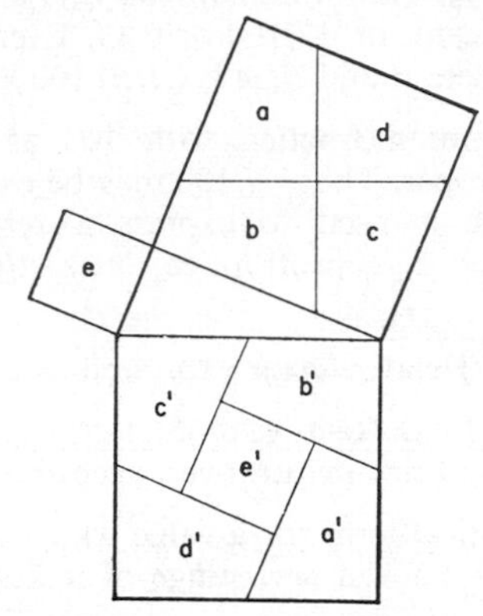

Perigee. The point on the orbit of the moon, or of any planet, or of the apparent orbit of the sun, which is nearest to the earth. See **Apogee**.

Perigon. 2π radians. 360°. 1 revolution. See **Angle (Plane Angle)**.

Perihelion. The point on the orbit of the earth or of any heavenly body, which is nearest to the sun. See **Aphelion**.

Perimeter. The length of a **closed curve**. The length of the boundary of a plane figure.

Period in Astronomy. The time taken to complete an **orbit**.

Period in Involution. In the arithmetic process of extracting the roots of a number, the number of digits repeatedly marked off to the left and right of the decimal point, equal to the **index** of the root required. See **Root of Real Number**.

Period in Notation. The number of digits bounded by thin spaces in the writing of a number for the purpose of easier reading. Thus: 1,684,327 or 3.141 592 6. The use of commas is obsolescent.

Period in Repeating Decimals. The set of digits which repeat, indicated by dots over the first and last digits of the period. Thus $1/7 = 0.\dot{1}4285\dot{7}$ $= 0.142\,857\,142\,857\,142\ldots$

Period of Function. See **Periodic Function**.

Period of Oscillation. The time taken to complete one oscillation. See **Pendulum**; **Simple Harmonic Motion**.

Periodic Continued Fraction. See **Continued Fraction**.

Periodic Curve. The graph of a **periodic function**.

PERIODIC FUNCTION. A **function** that repeats the same value at equal intervals of the independent **variable.** If $f(x)=f(x+p)$ for all x, then $f(x)$ is periodic, with period p. The graph of such a function is a *periodic curve* with amplitude equal to the largest absolute value of $f(x)$, and frequency $1/p$, the number of repetitions of the pattern of the curve in unit length. Typical periodic functions, sine and cosine, with period 2π radians, are associated with **simple harmonic motion**. Any periodic function can be represented as an infinite **trigonometric series.** See **Fourier Series.**

PERMUTATION. If from a given number, say n, of different things, an arrangement of r things is to be made, the number of possible arrangements is represented by ${}^nP_r$, in which P implies that type of an arrangement known as a permutation. The value of ${}^nP_r$ is $n!/(n-r)!$ These arrangements enter into the study of theories of equations, group theory and statistics. See **Combination; Factorial n.**

PERPENDICULAR. From Latin *pendo*, I hang, comes the fundamental concept of a vertical line at right angles to a horizontal plane. The word now describes any line which is at right angles to any other line or plane. A perpendicular line is said to intersect such other line or plane normally. The symbol is $\perp$. See **Normal.**

PERPENDICULAR BISECTOR OF LINE SEGMENT. The line at right angles to and passing through the midpoint of a segment.

PERPENDICULAR CURVES. Curves which intersect in such a way that tangents drawn at the point of intersection are **perpendicular lines.**

PERPENDICULAR LINE TO PLANE. A line at right angles to any two distinct lines in the plane which pass through the point of penetration.

PERPENDICULAR LINES. Lines which intersect at right angles. In **analytic geometry** the product of the **gradients** of such lines is -1.

PERPENDICULAR PLANES. Planes having a **dihedral angle** of 90°.

PERSPECTIVE PROJECTION. See **Pictorial Projection.**

PHASE DIFFERENCE. See **Phase in Simple Harmonic Motion.**

PHASE IN ASTRONOMY. A particular stage in the appearance of the illuminated half of a sphere as the respective positions of the sphere, the illuminant and the observer change. The moon and Venus best show phases.

PHASE IN COMPLEX NUMBER. See **Polar Form of Complex Number.**

PHASE IN SIMPLE HARMONIC MOTION. A stage in a **periodic function.** From phase to repeated phase a vibrating object completes a **period.** If $y=a \sin(\omega t+\epsilon)$ represents a periodic function, the angle $(\omega t+\epsilon)$ is the *phase*, and, when $t=0$, ϵ is called the *initial phase*. If two **simple harmonic motions** are represented by $y=a \sin(\omega t+\epsilon_1)$ and $y=a \sin(\omega t+\epsilon_2)$, then the *phase difference* is $\epsilon_1-\epsilon_2$.

PI; Π; π. Archimedes' constant. The letter *P*, *p*, of the Greek alphabet. Π is used as the symbol for product: π is the first letter of *perimetron*, perimeter, the measure of the circumference of a circle of unit diameter. For all circles, π is the ratio of the length of the circumference to the length of the diameter. It is both an **irrational number** and a **transcendental number**. In Hebrew times it was thought to have the value 3. Very roughly it can be thought of as $3\frac{1}{7}$. A closer approximation, $3\frac{10}{71}$ was known to Archimedes, who knew that the true value lay between 220/70 and 223/71. Calculation from series yields the sequence of figures: 3.141 592 653 6 . . . The common fraction 355/113 yields the first seven of these figures. Ptolemy used the value 3° 8′ 30″, standing for $3+8/60+30/60^2$, or $3.141\dot{6}$. The value of π may be expressed in the following ways:

$$\tfrac{1}{2}\pi = \frac{2\cdot2}{1\cdot3}\cdot\frac{4\cdot4}{3\cdot5}\cdot\frac{6\cdot6}{5\cdot7}\cdot\frac{8\cdot8}{7\cdot9}\ldots \text{ the limit of a product;}$$

$$\tfrac{1}{2}\pi = 1 + \frac{1}{3}\left(\frac{1}{2}\right) + \frac{1}{5}\left(\frac{1\cdot3}{2\cdot4}\right) + \frac{1}{7}\left(\frac{1\cdot3\cdot5}{2\cdot4\cdot6}\right) + \ldots \text{ the limit of a } \textbf{series;}$$

$$\tfrac{1}{4}\pi = 1 - 1/3 + 1/5 - 1/7 + 1/9 - 1/11 + \ldots \text{ the limit of a series;}$$

$$4/\pi = 1 + \frac{1^2}{2\,+}\,\frac{3^2}{2\,+}\,\frac{5^2}{2\,+}\,\frac{7^2}{2\,+}\ldots \text{ the limit of a } \textbf{continued fraction.}$$

PICTORIAL PROJECTION. A pictorial representation of a solid which shows the three dimensions on one diagram. If such a projection is a **scaled drawing** it is called a *metric projection*.

axonometric projection. One in which all vertical lines in an object are drawn as upright lines on the page; all those following the main horizontal direction of the object are drawn at a convenient angle (usually 45° or 30°) to the uprights. The lines following the other horizontal directions at right angles to the main ones are drawn on the diagram at right angles. This projection has the advantage of displaying the plan of the object. All vertical and horizontal distances are drawn to the same scale.

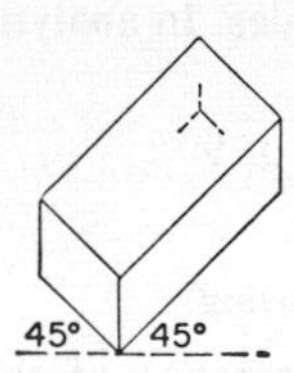

isometric projection. A phrase coined by Professor Farish of Cambridge (England) for a variety of conventional perspective used in technical drawing. All vertical lines in an object are drawn as upright lines on the page. All those following one main direction of the plan are drawn at 60° to the uprights on one side; all those following the other main direction (at right angles to the first) at 60° to the uprights on the other side.

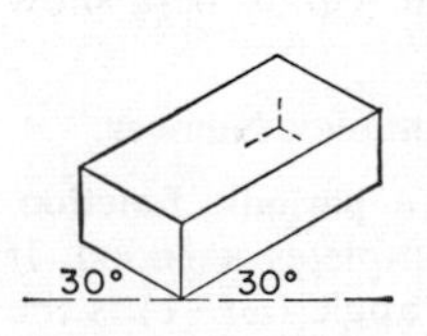

oblique projection. One in which all vertical lines in an object are drawn as upright lines on the page, and all those following one main horizontal direction, at right angles to the uprights. Those lines following the other main horizontal direction are drawn at 45° to the upright and horizontal lines already drawn. All vertical distances and those drawn horizontally in the diagram are drawn to the same scale. Sometimes the distances along the 45° direction are drawn to half scale, thus making the projection appear more like a perspective projection.

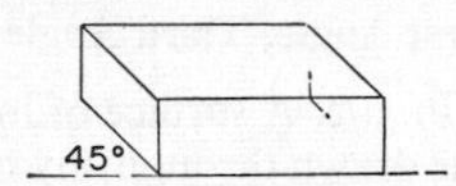

perspective projection. One portraying on a plane the features of an object occupying three-dimensional space. The eye of the observer is at the vertex of a cone of rays, called the centre of projection, each ray fixed by some point on the object viewed. A transparent picture plane is imagined to be between the viewer and the object, normal to the axis of the cone of rays. Each point of penetration of this picture plane by a ray is marked. The points on the picture plane and the appropriate lines joining them give a near photographic image of the object. The location of the *vanishing points* on the *eye-level* (the **true horizon**) towards which parallel lines are directed and the deciding of the relative positions of the observer and the **ground line,** etc., are techniques of architectural drawing. One diagram shows an object with two principal vanishing points and the apparent directions taken by two sets of parallel lines. All vertical lines are drawn upright on the page and all rectangular side elevations appear as **trapeziums.** One **invariant property** is: the point of intersection of the diagonals of a trapezium lies on the upright median of the rectangle the trapezium represents.

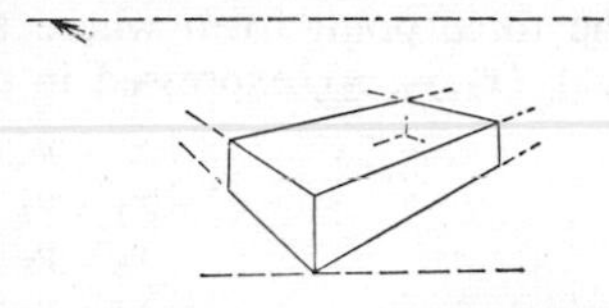

See **Geometric Projection**; **Map Projection**; **Plan, Elevation, Section.**

PIERCING POINT. See **Trace of Line.**

PITCH IN MUSIC. The term for the characteristic chiefly responsible for our being able to distinguish one musical sound from another. It depends primarily on the **frequency** of the principal sound wave causing the **tone.**

PITCH OF ROOF. The **dihedral angle** between the roof and the horizontal plane of the wall plates.

PLACE HOLDER. Zero, when used in any **system of notation,** to give place value to another digit. Thus in 60, the zero gives place value to the 6.

PLACE VALUE. See **System of Notation.**

PLAN, ELEVATION, SECTION. Related diagrams which show the construction of a solid object in two-dimensional **scaled drawings**, by means of **geometric projection** in which the rays of projection are perpendicular to the plane of projection (*orthographic projection*). The plan, elevation and **section** are projections of the object on a horizontal, a vertical and any convenient intersecting plane, respectively. If the plane of intersection is horizontal, the section is sometimes referred to as a sectional plan. See **First Angle, Third Angle Projections.**

PLANE. A surface of infinite extent that contains the whole of a straight line drawn through any two points in it. It can be represented by the general linear equation $Ax+By+Cz+D=0$, where (x, y, z) are the coordinates of any point in the plane. The equation can take different forms:

(1) intercept form, $x/a+y/b+z/c=1$, where a, b and c are intercepts on three rectangular axes;

(2) perpendicular (normal) form, $x \cos \alpha+y \cos \beta+z \cos \gamma=p$, where p is the perpendicular distance from the origin to the plane and α, β and γ are the angles between the perpendicular and the axes;

(3) the three point form where the plane passes through (x_1, y_1, z_1), (x_2, y_2, z_2), (x_3, y_3, z_3), expressed in **determinant** form,

$$\begin{vmatrix} x & y & z & 1 \\ x_1 & y_1 & z_1 & 1 \\ x_2 & y_2 & z_2 & 1 \\ x_3 & y_3 & z_3 & 1 \end{vmatrix} = 0.$$

PLANE COORDINATES. **Cartesian coordinates** or **polar coordinates in plane** which determine the position of a point in a **plane.**

PLANE FIGURE. A geometric figure which lies entirely in a **plane.**

PLANE GEOMETRY. The study of geometric configurations in a **plane.**

PLANE OF PROJECTION (IMAGE PLANE). See **Geometric Projection.**

PLANE SECTION. See **Section.**

PLANE TRIGONOMETRY. The **trigonometry** of **plane figures.**

PLAYFAIR'S AXIOM. Through any point, one line only can be drawn parallel to a fixed line. This assumption replaced Euclid's twelfth axiom: if a straight line, falling on two other straight lines, makes the two interior angles on the same side of it together less than two right angles, these two lines will meet, if continually produced, on that side on which the angles are together less than two right angles. See **Non-Euclidean Space.**

PLOTTING. The process of placing in a coordinate system selected points of a curve. See **Curve Tracing.**

PLUMB-LINE. A **vertical line** determined by the position assumed by a string, rigidly supported and carrying a plummet or plumb-bob (from Latin, *plumbus*, the metal lead). Its direction differs imperceptibly from that of a radius of the earth.

PLUS, +. Latin for more. The sign was first used by the German, Widman, in 1489, to indicate overweight in commercial consignments. In 1544, Stifel used the same sign to replace p, used by mathematicians to denote **addition.** It was generally accepted by 1630. Today it serves a second purpose—to distinguish positive numbers. See **Directed Numbers.**

POINT. Euclid's definition of this most fundamental idea was: *that which has position but no magnitude.* A point exists as a concept rather than as something which can be defined. Sometimes it is thought of as a position, as in the phrase, two lines meet at a point; sometimes as one of a set of points which determine a geometric figure, as when a specific triangle is described as the triangle ABC.

POINT CIRCLE. See **Null Circle**; **Coaxal (Coaxial) Circles.**

POINT ELLIPSE. See **Null Ellipse.**

POINT OF CONTACT. Synonym of point of tangency. See **Tangent Curves**; **Tangent Line**; **Tangent Plane.**

POINT OF DIVISION. If (a, b), (c, d) are the **Cartesian coordinates** of two points, then the point dividing their join in the ratio $e:f$ will have co-ordinates $\{(ec \pm fa)/(e \pm f),\ (ed \pm fb)/(e \pm f)\}$, the four additions giving a point of internal division, the four subtractions, a point of external division. See **Division of Line**; **Points at Infinity.**

POINT OF INFLEXION. See **Curvature**; **Inflexion**; **Maximum, Minimum Values.**

POINT OF OSCILLATION. See **Centre of Oscillation.**

POINT OF OSCULATION. Synonym of **cusp.** A point on a curve at which two branches of the curve share a common tangent.

POINT SYMMETRY. Synonym of **central symmetry.**

POINTS AT INFINITY. If A, B, C are **collinear** points in order and the ratio $AB:BC$ is taken as positive, then the ratio $AB':B'C$ will be negative if B' lies in the extension of AC. The positions of B' if the ratio becomes -1 are suggested in the diagram: they are called **ideal points.** They divide the line into equal parts externally. The sequence of ratios from $+1$ to -1 may pass imperceptibly through zero or infinity. See **Ideal Line.**

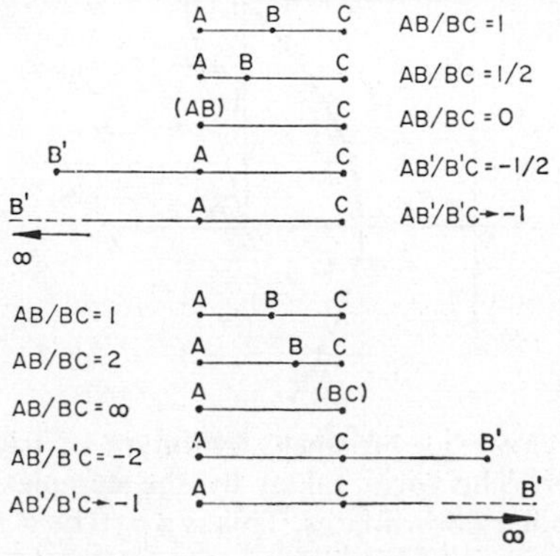

A B C AB/BC = 1
A B C AB/BC = 1/2
(AB) C AB/BC = 0
B' A C AB'/B'C = −1/2
B' A C AB'/B'C → −1
∞

AB/BC = 1 A B C
AB/BC = 2 A B C
AB/BC = ∞ A (BC)
AB'/B'C = −2 A C B'
AB'/B'C → −1 A C B'
∞

POINTS, CARDINAL. See **Cardinal Points.**

POISSON'S RATIO. In an elastic body, the ratio of the transverse **strain** to the longitudinal strain, when it is subjected to a longitudinal **stress,** T. If e_1 is the linear contraction in its cross-sectional dimensions and e_2 is the longitudinal extension, then Poisson's ratio, σ, is the numerical value $|e_1/e_2|$. It is related to **Young's Modulus** in tension by the formula $\sigma = -e_1E/T$ (from **Hooke's Law**: $T = Ee_2$).

POLAR ANGLE. See **Polar Coordinates in Plane.**

POLAR AXIS. The initial line in **polar coordinates in plane** and **spherical coordinates in space.**

POLAR COORDINATES IN PLANE. The system of coordinates in which the position of a point P is determined by its distance OP from a fixed point O and the angle which OP makes with an initial line OX. The point O is the *pole*; OP is the *radius vector* of P; OX is the *polar axis*; angle XOP, θ, measured anti-clockwise, is the *polar angle*, *vectorial angle*, *azimuth* or the *amplitude* of P. If the pole and the origin are the same point O, the polar coordinates (r, θ) and the Cartesian coordinates (x, y) of P are related by the equations: $x = r\cos\theta$, $y = r\sin\theta$; $r = \sqrt{(x^2 + y^2)}$, $\theta = \tan^{-1}(y/x)$.

POLAR EQUATION. An equation of a **configuration** when the variables are **polar coordinates in plane.**

POLAR EQUATION OF CONIC SECTION. The equation expressed in terms of the polar coordinates of a point on the curve. Usually a focus S is taken as pole and the positive direction of the major axis or the transverse axis is taken as the initial line. If e is the eccentricity, l the length of the semi-**latus rectum,** θ the angle XSP, then the **conic section** has equation $l/r = 1 - e\cos\theta$. See **Polar Coordinates in Plane.**

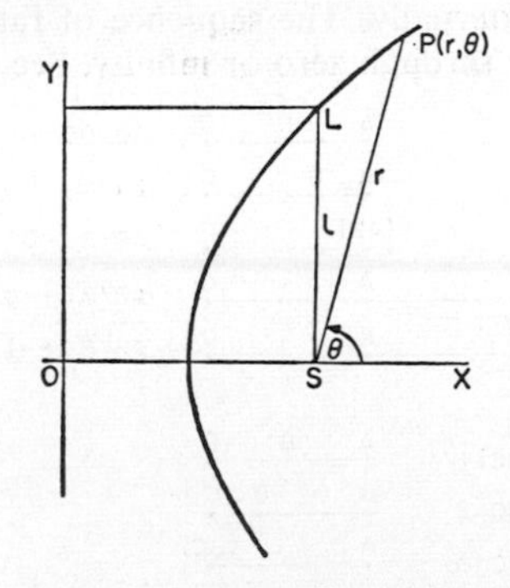

POLAR FORM OF COMPLEX NUMBER. Synonym of trigonometrical form of complex number. The form taken by the **complex number** $z = x + iy$ when expressed in polar coordinates. This is $z = r(\cos\theta + i\sin\theta)$, where r is the radius vector and θ the vectorial angle of the point representing the complex number. The radius vector, r, in value $\sqrt{(x^2+y^2)}$, is called the *modulus* (*magnitude* or *absolute value*) and the vectorial angle, θ, the *amplitude* (*argument* or *phase*). See **Polar Coordinates in Plane.**

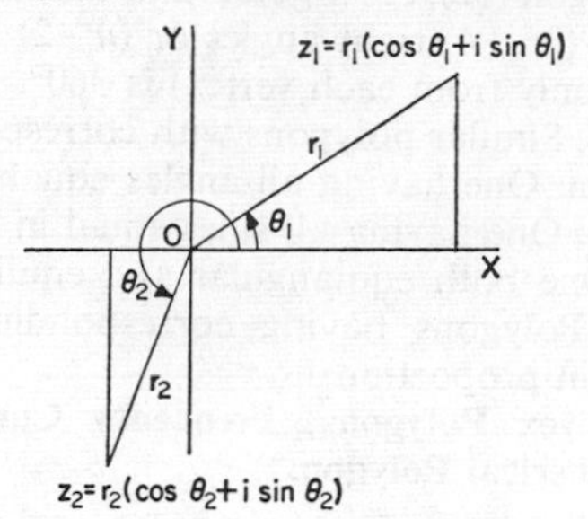

Polar Spherical Triangle. If A', B', C' are those poles of the sides of a spherical triangle *abc* nearest to the vertices A, B, C respectively, then $A'B'C'$ is the polar spherical triangle of *abc*. See **Pole of Circle on Sphere.**

Polar Zenithal Projection. See **Map Projection.**

Pole. See **Geographical Poles; Magnetic Poles; Pole and Polar of Conic; Pole in Coordinates; Pole of Circle on Sphere.**

Pole and Polar of Conic. If A and B are the points of intersection with a conic of a variable secant through a given point P, there is a set of points Q, each of which with P are **harmonic conjugates** with respect to A and B. This set of points is called the *polar* of P. It is a straight line and P is called its *pole*. If tangents to the conic are drawn at A and B to intersect in T, T lies on the extension of the polar chord. If tangents PC, PD can be drawn from P to the conic, the secant CD is the polar of P. P and T are *conjugate points* and AB and CD are *conjugate lines* (chords) of the conic. The equation of the polar of any point is obtained by replacing the coo:dinates of a point of contact of a tangent to the conic by the coordinates of the pole. See **Conic Section.**

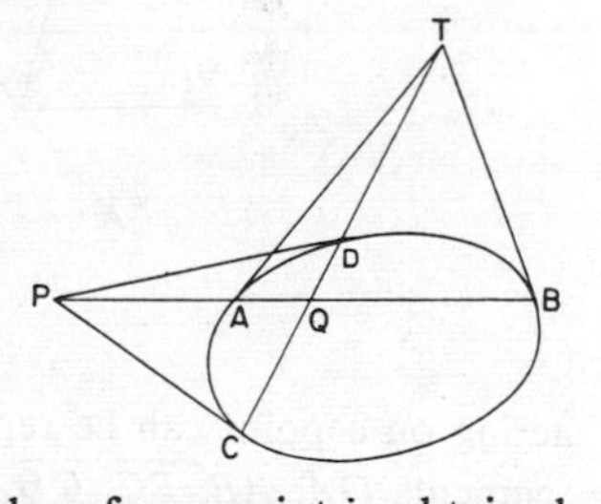

Pole in Coordinates. The fixed point in **polar coordinates in plane** and **spherical coordinates in space.**

Pole of Circle on Sphere. The two extremities of a diameter of a sphere which is perpendicular to the plane of the circle are the near and far poles of the circle. When the circle is a **great circle,** the poles are equidistant from it. On the earth, the poles of the equator, called the North and South (geographical) Poles, are at the extremities of the diameter which is the axis of rotation.

POLYGON. A plane **rectilinear figure**. It is named variously by the number of its sides or angles: triangle (3), quadrilateral (4), pentagon (5), hexagon (6), heptagon (7), octagon (8), nonagon (9), decagon (10), undecagon (11), dodecagon (12), . . . n-gon (n). In the convex n-gon the sum of the interior angles is $(2n-4)$ right angles or $(n-2)\ 180°$. The sum of the exterior angles (one only from each vertex) is 360°.

congruent polygons. Similar polygons with corresponding sides equal.

equiangular polygon. One having all angles equal.

equilateral polygon. One having all sides equal in length.

regular polygon. One both equiangular and equilateral.

similar polygons. Polygons having corresponding angles equal and corresponding sides in proportion.

See **Concave, Convex Polygons**; **Frequency Curve, Graph, Polygon**; **Regular Polygon**; **Spherical Polygon.**

POLYGON OF FORCES. See **Polygon of Vectors.**

POLYGON OF VECTORS. The graphical method for the **composition of vectors.** It is a convenient development of the **triangle of vectors.** If **vector quantities** $\mathbf{v}_1$, $\mathbf{v}_2$, $\mathbf{v}_3$, $\mathbf{v}_4$, acting at a point are represented in magnitude and direction by the line segments $\overline{OA}$, $\overline{AB}$, $\overline{BC}$, $\overline{CD}$, then the *resultant* $\mathbf{r}$ is represented in magnitude and direction by the segment $\overline{OD}$ which is needed to complete the polygon $OABCD$. It follows that if five vectors

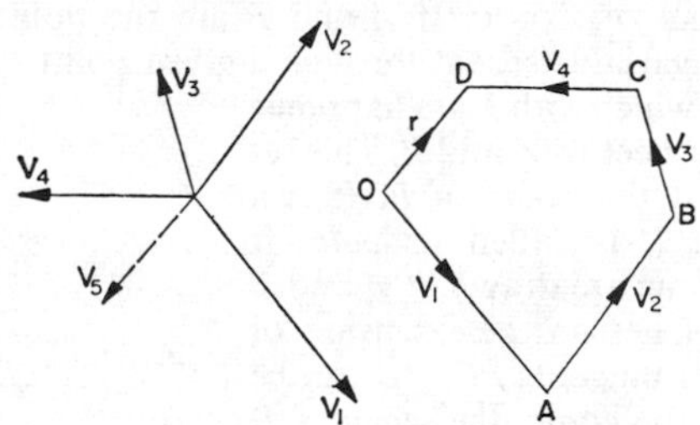

acting on a point can be represented in magnitude and direction by line segments $\overline{OA}$, $\overline{AB}$, $\overline{BC}$, $\overline{CD}$, $\overline{DO}$ along the sides of a polygon $OABCD$, their resultant is zero. Such a system is in equilibrium.

POLYGONAL NUMBERS. **Numbers** which can be arranged in a polygonal array. Each one is the sum of an **arithmetic progression** with first term 1 and common difference $n-2$, where n is the number of sides of the **polygon.** Synonym of *figurate numbers*.

triangular. 1, 3, 6, 10, . . . $\frac{1}{2}(r+1)(r+2)$, . . . from 1, $1+2$, $1+2+3$, . . . $1+2+3+\ldots+(r+1)$, . . .

square. 1, 4, 9, 16, . . . $\frac{1}{2}(r+1)(2r+2)$, . . . from 1, $1+3$, $1+3+5$, . . . $1+3+5+\ldots+(2r+1)$, . . .

pentagonal. 1, 5, 12, 22, . . . $\frac{1}{2}(r+1)(3r+2)$, . . . from 1, $1+4$, $1+4+7$, . . $1+4+7+\ldots+(3r+1)$, . . .

hexagonal. $1, 6, 15, 28, \ldots \frac{1}{2}(r+1)(4r+2), \ldots$ from $1, 1+5, 1+5+9, \ldots 1+5+9+ \ldots +(4r+1), \ldots$

***n*-gonal.** $1, n, 3n-3, 6n-8, \ldots \frac{1}{2}(r+1)(\overline{n-2}.r+2), \ldots$ from $1, 1+(n-1), 1+(n-1)+(2n-3), \ldots 1+(n-1)+(2n-3)+ \ldots +\{(n-2)r+1\}, \ldots$

Since the linear dimensions of each polygon is *doubled* at the *third* term, it is r-tupled at the $(r+1)$th term, and $\frac{1}{2}(r+1)\{r(n-2)+2\}$ gives every polygonal number.

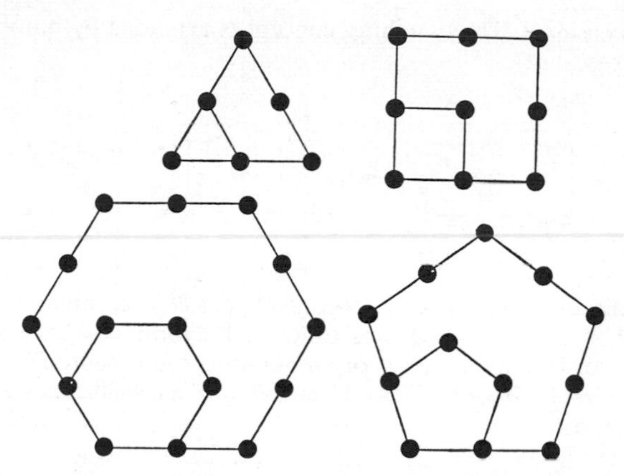

POLYHEDRAL ANGLE. The geometrical configuration formed in a **polyhedron** by the faces and edges which share a common vertex. See **Solid Angle.**

POLYHEDRON. A solid bounded by plane **polygons** called the *faces*; the intersections of three or more edges called the *vertices*. The numbers of faces, edges and vertices in polyhedra in general obey Euler's law: $f+v=e+2$. A *simple polyhedron* is topologically equivalent to a sphere, i.e. it has **genus** 0. See **Concave, Convex Polyhedron; Regular Polyhedron; Topological Transformation.**

POLYNOMIAL EXPRESSION. A rational, integral, algebraic expression of the general form

$$a_0x^n + a_1x^{n-1} + a_2x^{n-2} + \ldots + a_{n-1}x + a_n,$$

where n is the *degree* of the expression. It is linear, quadratic, cubic, quartic, etc., when its degree is 1, 2, 3, 4, etc. See **Integral Expression; Multinomial Expression; Rational Expression.**

POLYOMINO. In general, the term for a rectilinear plane figure constructed from congruent squares arranged to share some sides. For example the one dimensional domino is a double-square rectangle. These figures form the basis of various number games involving the matching of squares numbered in the same way, or the fitting of them into various shapes. The term *polyomino* was first used in 1954 by Golomb of the Californian Institute of Technology.

one dimensional. These may be dominoes consisting of two squares; trominoes, (3); tetrominoes, (4); pentominoes, (5); etc. The domino is unique. There are two trominoes, five tetrominoes, twelve pentominoes, thirty-five hexominoes, etc. Each square bears one number.

two dimensional. The square bears four numbers permitting matching in two directions.

three dimensional. The matching concept is extended by numbering the faces of cubes (dice).

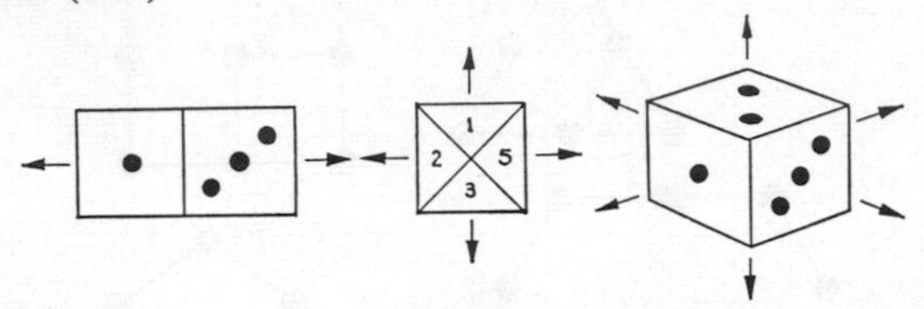

A set of polyominoes which bears all possible combinations of the numbers 0, 1, 2, . . . $(n-1)$, has order n. Polyominoes have interesting properties; for example, the set of all non-congruent pentominoes (12) fit into rectangles 3×20, 4×15, 5×12 and 6×10, each 30 times as big as a domino, 2×1.

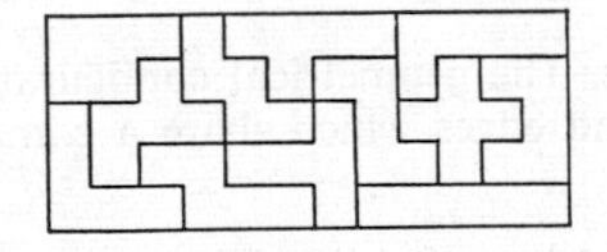

POPULATION. The whole set of items which have a common characteristic, which is the subject of some **sampling** in the process of statistical analysis.

POSITION VECTOR. A **vector quantity** represented by the line segment from the origin of coordinates to a point under consideration. For point (a, b, c) the position vector is $\mathbf{R} = a\mathbf{i} + b\mathbf{j} + c\mathbf{k}$, where **i, j, k** are unit vectors (magnitude 1) in the directions of the x, y, z axes.

POSITIVE, NEGATIVE. An arbitrarily chosen direction or order of counting which has become familiar and accepted. The signs used are borrowed from arithmetic practice, + and −, and need careful handling to avoid confusing their older use of operation and their newer one of description. See **Directed Line Segment**; **Directed Numbers.**

POSITIVE, NEGATIVE NUMBERS. See **Directed Numbers.**

POSTULATE. An assumption upon which a logical argument is based. It is accepted as a self-evident truth which can or cannot be proved. A term which has superseded the term **axiom.** See **Validity.**

POTENTIAL. At any point, the potential is the work done against a conservative field in bringing a unit quantity from infinity to the point. See **Conservative Forces**; **Potential Difference.**

POTENTIAL DIFFERENCE (P.D.). If between any two points in a circuit, electrical energy may be converted into any other form of energy, then a potential difference is said to be established between the two points. P.D. always involves the concept of an electrical pressure difference between two points in contrast to an **electromotive force** which refers to the electric pressure generated in the interior of any source. The direction of the P.D. depends on that of the current. For convenience, earth is taken as zero potential. The electric potential of a point relative to the earth is defined as being positive or negative according to whether the flow of electrons is from the earth to the point or from the point to the earth. The standard unit of P.D. is the **volt.**

POTENTIAL ENERGY (P.E.). The **energy** of a body which it possesses by virtue of its position only. The term can be applied only to fields of **conservative forces.** The P.E. of a body is the negative value of the work done in displacing the body from its standard position to any other position. See **Conservation of Energy**; **Kinetic Energy.**

POUND FORCE, lbf. The absolute unit of **weight** approximately equal to 32.173 9 **poundals,** the weight of a **pound mass** under standard gravity, 32.173 9 ft/s^2.

POUND MASS. The unit of **mass** defined by a block of platinum alloy at the Standards Office in London, England.

POUNDAL. A unit of force in the **f.p.s. system.** The **force** which produces an **acceleration** of one foot per second per second (1 ft/s^2) in a **mass** of one pound. A **weight** of 1 lb is equivalent to g poundals, and 1 poundal is equivalent to a weight of 0.497 . . . oz. The **c.g.s. system** equivalent is 13 825.49 **dynes.**

POWER IN MECHANICS. The rate of doing **work.** It is measured in foot-pounds per second or centimetre-grammes per second. The British unit, 1 **horse-power** (hp) is 550 ft lb/s.

POWER OF POINT WITH RESPECT TO CIRCLE. The product $PA \cdot PB$ where PA and PB are line segments on a secant passing through point P and intersecting a circle in A and B. The power is regarded as positive or negative if the point P lies outside or inside the circle respectively. If P lies outside the power is equal to the square on the tangent from P. If P lies on the circumference its power is zero. The power of a point P, (x', y'), with respect to the circle $(x-a)^2+(y-b)^2=r^2$ is $(x'-a)^2+(y'-b)^2-r^2$.

POWER OF POINT WITH RESPECT TO SPHERE. The **power of point with respect to circle** of section of the sphere and a plane containing the point and the centre of the sphere. The power of the point P (x', y', z') with respect to the sphere $(x-a)^2+(y-b)^2+(z-c)^2=r^2$ is $(x'-a)^2+(y'-b)^2+(z'-c)^2-r^2$.

Power of Quantity. The continued product of a number of factors, each equal to a given quantity, gives a power of the quantity specified by the number of factors employed. If a is the quantity, n the number of factors, the nth power of a, briefly, a to the n, is written a^n, and stands for $a \cdot a \cdot a \ldots a$, where a appears n times in the product. See **Theory of Indices.**

Power Series. A **series** of the form $a_0 + a_1x + a_2x^2 + a_3x^3 + \ldots + a_nx^n$. Such a series may be finite, $1 + 4x + 6x^2 + 4x^3 + x^4 = (1 + x)^4$; or infinite, $x - \frac{1}{2}x^2 + \frac{1}{3}x^3 - \frac{1}{4}x^4 + \ldots = \log_e(1 + x)$.

Practice. The arithmetic process by which the cost of a number of things is found if the price of one is known. Also the process by which the cost of a quantity of material is found if the price of a unit quantity is known. The first process is called *simple* practice; the second, *compound.*

Premium. (1) The difference between the selling price and the face value (par) of bonds, stock, shares, when the selling price is greater than the par value. When the selling price is less than the par value, the difference is called a *discount*. In the first case the bonds, etc., are said to be *at a premium* or *above par*; in the second *at a discount* or *below par*. (2) The difference between the selling price of one currency (in terms of another) and its face value, when the selling price is greater than the face value. (3) The amount paid for a loan or for insurance.

Present Worth. The value today of a bill due to be paid at some time in the future. The process of finding it is the reverse of the one for finding compound amounts. Thus,

$$\text{Present Worth} = \text{Value of Bill}/(1 + r/100)^n$$

where r is the current rate of **interest** and n the number of years the bill has to run before payment is due.

Pressure. The **force** exerted on a unit area of surface. See **Centre of Pressure.**

Prime Direction. The initial direction given by a **directed line,** to which other directions are referred.

Prime Factor. A **factor** of a quantity which is a **prime number** or a **prime polynomial.**

Prime Field. See **Subfield.**

Prime Meridian. The **meridian of longitude** chosen to represent zero longitude from North Pole to South Pole. The Greenwich, England, meridian is used internationally.

Prime Number. Any integral number, 2, 3, 5, 7, 11, etc., which has no integral factor other than 1 and itself. It is usual to exclude 1 from the list and sometimes 2 is excluded since it is the only even prime number. These numbers were first defined by Euclid (about 300 B.C.), who proved that there was an infinite number of primes. Assume there is some greatest prime, P, then consider the number $(2\cdot3\cdot5\cdot7\cdot11\ldots P)+1$. This number is not a multiple of any prime from 2 to P, as there will always be a remainder 1. It is therefore either (1) prime and greater than P or (2) composite and divisible by a prime greater than P. Either conclusion establishes the fact that P is not the greatest prime. There can be no greatest prime: the number of primes is infinite. Two consecutive integers n and $n+1$ can be divided only by 1. See **Fermat's Numbers**; **Fermat's Theorems**; **Goldbach's Conjecture**; **Levy's Conjecture.**

PRIME NUMBER THEOREM. If $P(n)$ is the number of prime numbers not greater than n, the ratio $P(n)/n$ is the ratio of prime numbers to all integers from 1 to n. The theorem, proved about 1900, states that this ratio, when n is large, approximates to $1/\ln n$. For example, if $n=10^9$, $1/\ln n \doteqdot 0.05$; thus about 5 per cent of the natural numbers up to 10^9 are prime. The theorem can be stated in terms of limits, thus: $\lim\limits_{n\to\infty} \dfrac{P(n)}{n}\Big/\dfrac{1}{\ln n}=1.$

PRIME POLYNOMIAL. A **polynomial expression** that has no polynomial factors other than constants and itself. For example, $2(x-1)$ and x^2+y^2.

PRIME, SECONDARY SYMBOLS. The symbols $'$ and $''$ placed to the right of and above another symbol to distinguish it from the same symbol with a different meaning. Thus: (1) (x',y') and (x'',y'') are particular points in the set of points (x,y); (2) y' and y'' are the first and second **derivatives** of y with respect to x; (3) 3′ 3″ stands for three feet and three inches; (4) 3′ 3″ stands for three minutes and three seconds of time; (5) 3′ 3″ stands for three minutes and three seconds of rotation; (6) the prime symbol is often used to indicate an inverse, A' is the inverse of A. See **Subscripts, Superscripts.**

PRIMITIVE (UNDEFINED). In the axiomatic development of a theory certain concepts are undefined. On these primitive (undefined) concepts are based definitions and **axioms**. Thus, in **set theory**, 'set' and 'is an element of' can be undefined and yet stated symbolically $a\in A$ (a is an element of set A). A consequent definition is that A is a subset of B, written $A\subseteq B$, if $a\in A$ implies $a\in B$ for all a.

PRINCIPAL. (1) Of greater importance. (2) The sum of money invested, as in **simple interest** and **compound interest.**

PRINCIPAL PLANES OF QUADRIC SURFACE. The planes of **symmetry** of a **quadric surface.**

PRINCIPAL ROOT OF NUMBER. The positive real root of a positive number, e.g., $\sqrt{4}=+2$. And the negative real root of a negative number in the case of an odd root, e.g., $\sqrt[3]{-8}=-2$. See **Radical Expression.**

PRINCIPAL, SECONDARY DIAGONALS. These occur in the arrays of elements in **determinants**. The diagonal passing through the element in the top row and left-hand column is the principal one. In an array of n^2 elements this passes through $a_1{}^1$ and $a_n{}^n$. The secondary one passes through $a_n{}^1$ and $a_1{}^n$.

PRINCIPLE. A fundamental truth which may be a **postulate** or a **proposition** from which other truths are derived.

PRINCIPLE OF UNDETERMINED COEFFICIENTS. If two **polynomial expressions** are identically equal, the coefficient of a term of any power in one may be equated to the coefficient of the term of the same power in the other. The principle is used in determining the numerators of **partial fractions.**

PRISM. A solid bounded by congruent parallel faces (called the *bases*) and a set of parallelograms (called the *lateral faces*) formed by joining corresponding vertices of the bases. It is called a *right prism* if the lateral faces are rectangles.

PRISMATIC SURFACE. Any number of adjacent lateral faces of a **prism.** If all the lateral faces are included the surface is a closed prismatic surface.

PRISMATOID. A **polyhedron** with some vertices forming a base on one plane and the rest of the vertices forming a second base on a parallel plane. The lateral faces are triangles or quadrilaterals.

PRISMOID. A **prismatoid** with an equal number of sides in each base, the lateral faces being quadrilaterals.

PROBABILITY CURVE. The Gauss-Laplace curve of *normal frequency* distribution. See **Error curve.**

PROBABILITY IN LOGIC. Probability theory is used in assessing the chance of statements being true by studying the properties of the corresponding **truth sets**. The *probability* $P(a)$ of statement a is equivalent to a *measure* $\mu\ (a)$ of the truth set A corresponding to statement a. The measure is made according to the definition of **probability in statistics.**

PROBABILITY IN STATISTICS. The numerical measure of the likeliness of an event occurring. If, under certain conditions, there are n alternative ways, exhaustive and mutually exclusive, in which an event can be conceived as occurring, and m of these ways are considered as of class A, then the *mathematical probability* of an event of class A occurring is m/n. If, in a random sequence of p trials of an event, q are considered favourable, the *probability* of the event q is the limit of the ratio q/p as p increases without bound. When the probability of an event occurring is based on previous observation, it is known as *empirical probability.* Thus, if in a number of trials r are considered favourable and s unfavourable, the probability of a favourable event occurring in the next trial is $r/(r+s)$.

PROBLEM. A question offered for solution or a **proposition** requiring something to be done.

PRODUCT. In general, the result of **multiplication.**

PRODUCT OF SETS. See **Algebra of Sets; Cartesian Product; Set Theory.**

PROFIT, LOSS. The difference between the cost price and the selling price of an article when the selling price is the greater or smaller respectively. If the cost price is the original cost the profit or loss is described as *gross*; if it includes the cost of marketing, the profit or loss is described as *net*.

PROGRAMMING. See **Computer Programming; Linear Programming; Mathematical Programming.**

PROGRESSION. A series of values in which every term is obtained from its predecessor in the same way. See **Arithmetic Progression; Geometric Progression; Harmonic Progression.**

PROJECTION. See **Geometric Projection; Map Projection; Pictorial Projection.**

PROJECTIVE GEOMETRY. The study of those properties of geometric **configurations** which are unaltered by *central* **geometric projection**. Straight lines are projected as straight lines on to the projection plane but the length of a line segment, the size of an angle and congruence are not maintained except in special cases of *parallel* projection. All **conic sections** are projected as conic sections but not necessarily of the same type. **Harmonic section** and **anharmonic ratio** are invariant. **Pascal's theorem** and **Desargues' theorem** are theorems of projective geometry. The founding of this branch of mathematics is attributed to Desargues in the early part of the seventeenth century. See **Affine Geometry.**

PROLATE. See **Ellipsoid.**

PROOF. The logical presentation of the necessary **data, postulates** and arguments, with or without the visual aid of a diagram, by means of which some **proposition** is established. Sometimes the term is used loosely in such an expression as 'proved experimentally' to mean that a proposition has been tested experimentally and no exception (to date) found. See **Unsolved Problems.**

PROPORTION. A statement of the equality of two **ratios.** When four quantities, a, b, c, d, have such comparable magnitudes that $a/b = c/d$, they are called proportionals and the relationship is often expressed as 'a is to b as c is to d'. (Obsolete presentation: $a : b :: c : d$; obsolescent, $a : b = c : d$.) The terms a and d are the *extremes*, b and c the *means*. Operations known as *componendo*, *dividendo* and *convertendo* can be performed on the equality to produce the following useful forms:

$$\frac{a \pm b}{b} = \frac{c \pm d}{d}, \quad \frac{a}{a \pm b} = \frac{c}{c \pm d}, \quad \frac{a}{b} = \frac{c}{d} = \frac{a \pm c}{b \pm d}.$$

See **Variation.**

PROPORTIONAL DIVISION. If any line segment AC is divided internally or externally at B in such a way that the ratio of AB to BC is some given amount, another line segment AY can be divided internally and externally in the same ratio. It is effected by drawing AC and AY in different directions, joining CY and drawing BX parallel to CY to find X on AY or on AY produced.

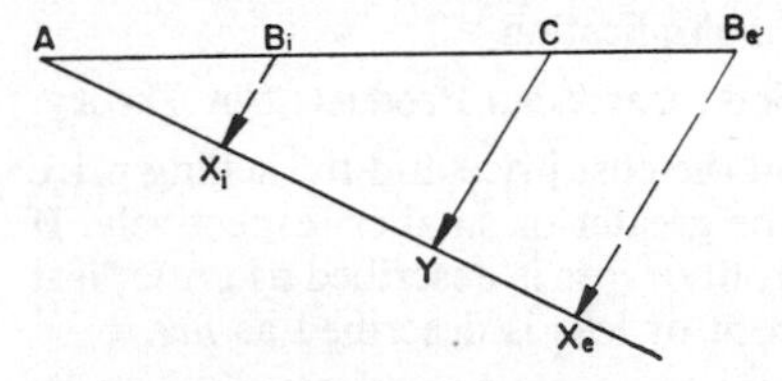

PROPOSITION. Modern name for **theorem.** A term used in logic as in **algebra of propositions.**

PTOLEMY'S THEOREM. This states the relationship between the lengths of the sides and the diagonals of a **cyclic quadrilateral** $ABCD$:

$$AB \cdot CD + BC \cdot DA = AC \cdot BD.$$

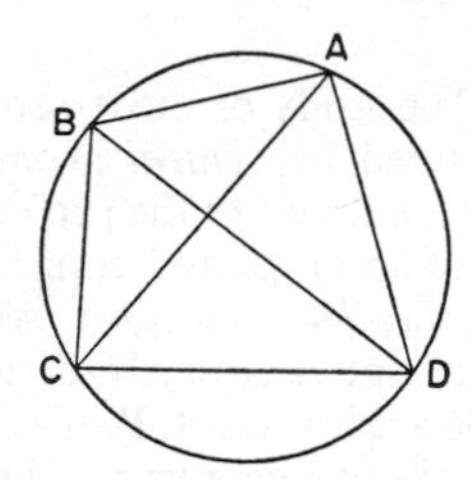

PURE MATHEMATICS. One of the traditional aspects of **mathematics** when contrasted with **applied mathematics.** Abstract mathematics, which did not necessarily have any immediate relevance to the problems of the time or environment, depended for its growth largely on the mathematical history and influences at a particular time.

PURE TRIANGULATION. See **Triangulation.**

PYRAMID. A special case of a **prismatoid** where all but one of the vertices form a base polygon. The remaining one is called the apex. It is a **polyhedron** with one polygonal face (base) and all lateral faces triangles with a common vertex. If the base also is a triangle the pyramid is usually referred to as a **tetrahedron.** A *regular pyramid* has a regular polygonal base and congruent isosceles triangles as lateral faces. A regular tetrahedron is one of the five **regular solids.** The Egyptian pyramids are regular square based.

PYTHAGORAS'S THEOREM. In any right-angled triangle the square on the hypotenuse is equal to the sum of the squares on the other sides. Euclid's proof of this best-known geometric theorem is the substance of his *Elements I 47.* This theorem, dealing with squares, is a special case. If the sides of a right-angled triangle are corresponding sides of three similar figures, the area of the one on the hypotenuse will be equal to the sum of the areas of the other two. An analytical proof is shown. See **Perigal's Dissection.**

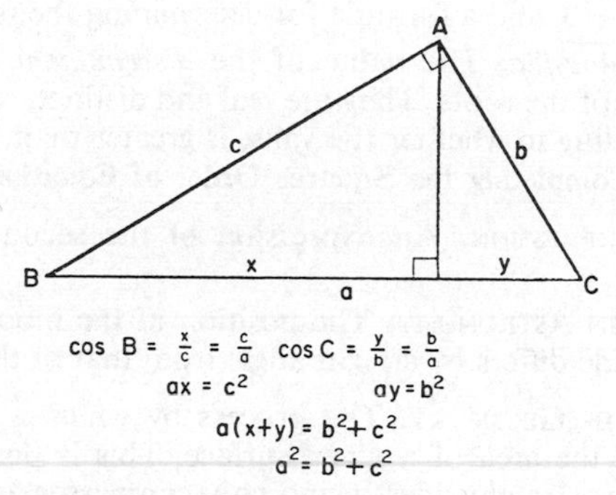

PYTHAGOREAN TRIANGLE. A right-angled triangle with sides of integral lengths. If m and n are any two integers, m^2-n^2, $2mn$ and m^2+n^2 will be a triple of integrals satisfying the equation, $x^2+y^2=z^2$, and a triangle with sides of lengths x, y and z will have a right angle. The **rope-stretchers' triangle** (3, 4, 5), is the simplest example.

PYTHAGOREAN TRIPLES. Any three integers (whole numbers) which are relatively prime in pairs, e.g. (3, 4, 5), (5, 12, 13), (8, 15, 17), that satisfy the equation $x^2+y^2=z^2$, thus forming a **Pythagorean triangle.** They are special cases of **Heronic triples.** See **Prime Number.**

Q

QUADRANGLE. A *simple quadrangle* is a plane geometrical figure consisting of four vertices, A, B, C, D, and their joins AB, BC, CD, DA. If these line segments do not intersect, the figure is a simple **quadrilateral.** A *complete quadrangle* is a plane geometrical figure consisting of four points and the six lines they determine. It is understood that no three of A, B, C, D are **collinear points.** See **Complete Quadrilateral.**

QUADRANT ANGLES. **Angles (plane angles)** having the positive x-axis as their initial line and designated first, second, third or fourth according as their terminal line lies in the first, second, third or fourth **quadrant of plane.**

QUADRANT OF CIRCLE. The part bounded by two perpendicular radii and a quarter of the circumference.

QUADRANT OF PLANE. One of the four quarter planes formed by planar rectangular axes of **Cartesian coordinates.** They are numbered anti-clockwise, the first being the one in which all points have both coordinates positive.

QUADRANTAL ANGLES. Any of the angles $n \cdot \pi/2$ radians ($n \cdot 90$ degrees) where $n = 0, 1, 2, 3, \ldots$ etc.

QUADRATIC EQUATION, FORMULA. An equation of the second order of the form $ax^2 + bx + c = 0$, and a formula for determining the **roots of equation**; $x = (-b \pm \sqrt{b^2 - 4ac})/2a$. The value of the *discriminant*, $b^2 - 4ac$, determines the nature of the roots. They are real and distinct, real and equal, or imaginary according to whether the value is greater than, equal to, or less than zero. See **Completing the Square**; **Order of Equation.**

QUADRATIC EXPRESSION. An expression of the second degree of the form $ax^2 + bx + c$.

QUADRATURE IN ASTRONOMY. The position of the moon or of a planet when the longitude differs by a right angle from that of the sun.

QUADRATURE IN GEOMETRY. The process by which a square is found equal in area to the area of a given surface. This is possible for figures consisting only of straight lines, using only compasses and straight edge. The historic attempts to 'square the circle' proved abortive and were abandoned when π was found to be a **transcendental number. See Unsolved Problems.**

QUADRIC CURVE. Any curve whose equation in **Cartesian coordinates** is algebraic of the second order. See **Order of Equation.**

QUADRIC EXPRESSION. A **homogeneous expression** of the second degree, e.g., $ax^2 + 2hxy + by^2$.

QUADRIC SURFACE. Any surface whose equation in **Cartesian coordinates** is algebraic of the second order. See **Order of Equation**.

QUADRILATERAL. In general, a plane figure with four straight sides. Six special cases may be defined by adding conditions.

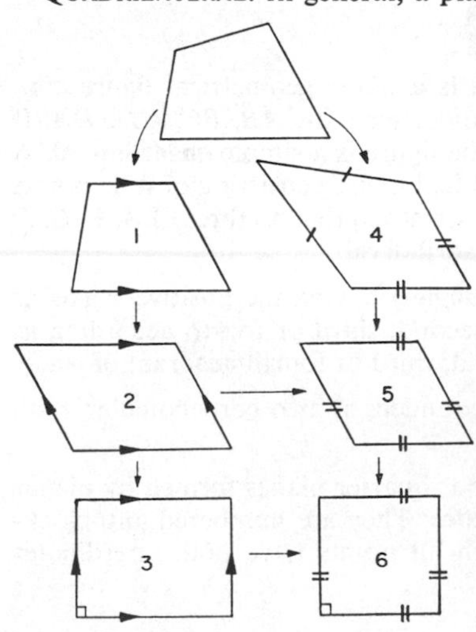

1. **trapezium (trapezoid)** (*q.v.*). One pair of parallel sides.
2. **parallelogram** (*q.v.*). Two pairs of parallel sides.
3. **rectangle** (*q.v.*). Two pairs of parallel sides and four right angles.
4. **kite** (*q.v.*). Two pairs of equal adjacent sides.
5. **rhombus** (*q.v.*). Four equal sides.
6. **square.** Four equal sides and four right angles. See **Square in Geometry**.

The rhombus is a special case of a parallelogram; and a square of a rectangle. See **Complete Quadrilateral**; **Cyclic Quadrilateral.**

QUADRINOMIAL. A **multinomial expression** of four terms.

QUANTIC. An algebraic **polynomial expression** in two or more variables which is a **homogeneous expression.** A rational **integral function** in two or more variables which is a **homogeneous function.** Quantics are classified by the number of **variables** as binary (2), ternary (3), quaternary (4), etc., and by degree as quadric (2), cubic (3), quartic (4), etc. See **Degree in Algebra; Rational Number, Quantity.**

QUART. One quarter of an Imperial **gallon.**

QUARTER IN ARITHMETIC. A common unit **fraction** with 4 as denominator.

QUARTER IN ASTRONOMY. One of the **phases** of the moon.

QUARTER IN CHRONOLOGY. One of the portions of a civil year into which the year is divided by *Lady Day* (Mar. 25), *Midsummer Day* (June 24), *Michaelmas Day* (Sept. 29) and *Christmas Day* (Dec. 25).

QUARTER IN WEIGHT. A quarter of a hundredweight, 28 lb, one eightieth of a **ton.** See **Avoirdupois.**

QUARTIC (BIQUADRATIC) EQUATION. An equation of the fourth order, of form: $a_0x^4+a_1x^3+a_2x^2+a_3x+a_4=0$. See **Order of Equation.**

QUARTIC (BIQUADRATIC) EXPRESSION. An expression of the fourth degree, of form: $a_0x^4+a_1x^3+a_2x^2+a_3x+a_4$. See **Quantic.**

QUARTIC SYMMETRY. **Symmetry** of a plane figure which has four axes of symmetry forming eight angles of 45° at their common point of intersection.

QUARTILE. The 25th, 50th and 75th **percentiles,** are the 1st, 2nd and 3rd quartiles.

QUARTILE DEVIATION, Q. A measure of the scatter or variation of values in a sequence of measurements. It is half the difference between the first and the third **quartiles.** See **Deviation in Statistics; Scatter Diagram.**

QUATERNARY. Associated with four as in quaternary **system of notation** and quaternary **quantic.**

QUATREFOIL. A **multifoil** bounded by four congruent arcs of a circle.

QUINARY. Associated with five as in quinary **system of notation** and quinary **quantic.**

QUINTAL. A term used variously for 100 lb, 112 lb or 100 kg (220.46 lb).

QUINTIC EQUATION. An equation of the fifth order of the form $a_0x^5+a_1x^4+a_2x^3+a_3x^2+a_4x+a_5=0$. See **Order of Equation.**

QUINTIC EXPRESSION. An expression of the fifth degree of the form $a_0x^5+a_1x^4+a_2x^3+a_3x^2+a_4x+a_5$. A **quantic** of fifth degree.

QUOTIENT. The result obtained by **division.**

QUOTIENT FIELD. The field of **rational numbers** is often referred to as the quotient field of *J*, the set of integers, since its elements are ratios of integers in the form p/q. See **Field Theory**.

QUOTIENT GROUP. The group, denoted G/H, whose elements are the **cosets** of *H*, an invariant subgroup of *G*. See **Group Theory.**

QUOTITION. Synonym of grouping, one aspect of **division.**

R

r. Symbol for **coefficient of correlation**; length of radius vector in **polar coordinates in plane**; radius of circle and sphere.

R. **Field** of **rational numbers.**

R*. **Field** of **real numbers.**

R.B. Reduced **bearing.**

RADIALLY RELATED FIGURES. The figures placed such that lines drawn through pairs of corresponding points pass through a common fixed point called the *centre of homothety* or *similitude*, the *homothetic centre* or the *ray centre*. The distances of the two corresponding points from the common point are in a constant ratio, called the *homothetic* or *ray ratio*, the *ratio of homothety* or *similitude*. See **Similitude**; **Transformation of Similitude.**

RADIAN. Unit of measure of **angle (plane angle).** Its size is defined as the angle subtended at the centre of a circle by an arc equal in length to the radius. There are 2π radians in 1 revolution; 1 radian is approximately 57° 17′ 45″.

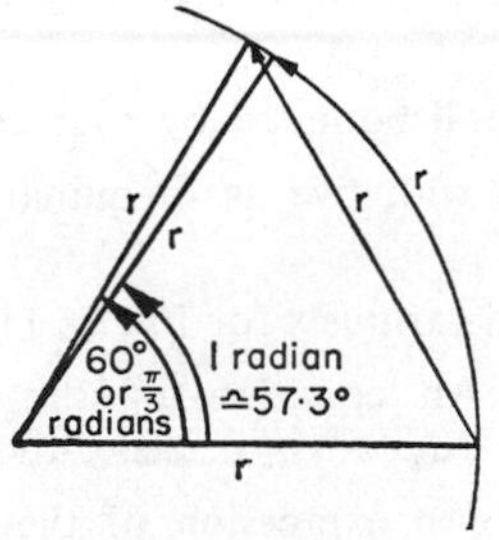

RADICAL. Synonym of **radical expression.**

Radical Axis. The set of points from any one of which four line segments of equal length can be drawn to touch two circles in the plane of the set. If $S_1 = S_2 = 0$ are the equations of two circles, $S_1 - S_2 = 0$ will represent the radical axis.
Thus, for the two circles

$$S_1\colon x^2 + y^2 + 2g_1x + 2f_1y + c_1 = 0$$
$$S_2\colon x^2 + y^2 + 2g_2x + 2f_2y + c_2 = 0$$

the radical axis is

$$2(g_1 - g_2)x + 2(f_1 - f_2)y + (c_1 - c_2) = 0.$$

If the circles intersect, the axis is the common secant; if they touch, the common tangent. In all cases it is perpendicular to the line of centres. See **Coaxal (Coaxial) Circles; Pencil of Circles.**

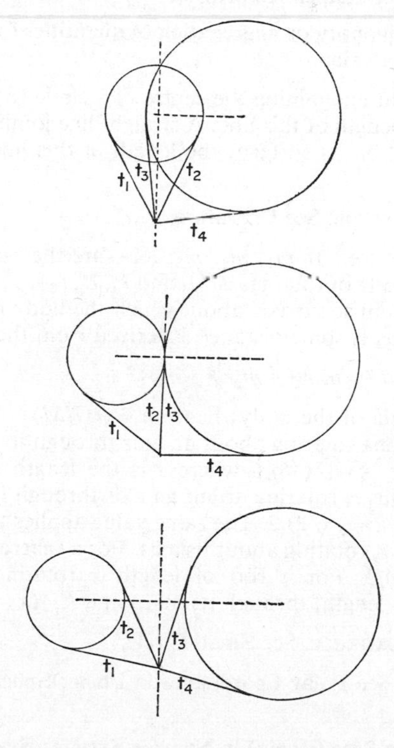

Radical Centre of Four Spheres. The point of intersection of the six **radical planes** of four spheres taken in pairs. If the centres of the spheres are **coplanar** the radical centre is at infinity.

RADICAL CENTRE OF THREE CIRCLES. The point of intersection of the three radical axes of three circles taken in pairs. If the centres of the circles are **collinear** the radical centre is at infinity.

RADICAL EXPRESSION. An indicated root of a quantity, e.g., $\sqrt[2]{3}$ or $\sqrt[3]{x}$. The sign $\sqrt{}$ is the *radical sign*: $\sqrt[2]{}$ represents the square root, usually written without the 2, $\sqrt[3]{}$ the cube root or root with *index* 3, $\sqrt[n]{}$ the *n*th root or root with index *n*. The sign is sometimes combined with a vinculum placed above the *radicand* or preceding some sign of aggregation as in $\sqrt[3]{x-5}$ or $\sqrt{}(b^2-4ac)$. See **Fractional Index; Principal Root of Number; Rational Expression; Rational Function; Surd.**

RADICAL PLANE. The plane represented by the equation obtained by eliminating the second power terms of the equations of two spheres. See **Pencil of Spheres.**

RADICAL SIGN. See **Radical Expression.**

RADICAND. The quantity or **aggregation** of quantities following a radical sign in a **radical expression.**

RADIUS. A straight line joining the centre of a circle to any point on the circumference; the length of this line. A straight line joining the centre of a sphere to any point on its surface; the length of this line. The symbol is usually r.

RADIUS OF CURVATURE. See **Curvature.**

RADIUS OF GYRATION. If m_1, m_2, m_3, . . . are the masses of particles which comprise a body of total **mass** M, and r_1, r_2, r_3, . . . are the distances of these particles from some axis about which the body is revolving, then the *radius of gyration* is some distance k derived from the equation:

$$Mk^2 = m_1r_1^2 + m_2r_2^2 + m_3r_3^2 + \ldots = I,$$

the **moment of inertia** of the body; hence $k = \sqrt{}(I/M)$.

For a square **lamina** rotating about an axis through its **centroid** perpendicular to its plane, $k = (s\sqrt{6})/6$ where s is the length of a side. For a circular disc of radius r, rotating about an axis through its centre perpendicular to its plane, $k = (r\sqrt{2})/2$. The same value applies to a **right circular cylinder**, end radius r, rotating about its axis. For a **sphere** rotating about a diameter, $k = (r\sqrt{6})/3$. For a rod of length x rotating about an axis perpendicular to its length, through its midpoint, $k = (x\sqrt{3})/6$.

RADIUS of SMALL CIRCLE. See **Small Circle.**

RADIUS VECTOR. See **Polar Coordinates in Plane; Spherical Coordinates in Space.**

RADIX (BASE). See **Base (Radix) in Number System; System of Notation.**

RADIX FRACTION. A sum of positive common **fractions** of the form $a/r + b/r^2 + c/r^3 + \ldots$ where r is called the radix.

Any common fraction can be converted into a radix fraction with any radix and expressed in any number system. Thus, the denary fraction $\frac{5}{7}$ can be converted into bicimal, tercimal, . . . decimal fractions:

$$\tfrac{5}{7} \times 2 = \tfrac{10}{7} = 1 + \tfrac{3}{7},$$
$$\tfrac{3}{7} \times 2 = \tfrac{6}{7} = 0 + \tfrac{6}{7},$$
$$\tfrac{6}{7} \times 2 = \tfrac{12}{7} = 1 + \tfrac{5}{7},$$
$$\tfrac{5}{7} \times 2 = \tfrac{10}{7} = 1 + \tfrac{3}{7}, \text{ etc., therefore}$$
$$\tfrac{5}{7} \text{ (denary)} = 1/2^1 + 0/2^2 + 1/2^3 + 1/2^4 + \ldots$$
$$= 0{\cdot}\dot{1}0\dot{1} \text{ (base 2).}$$

Similarly,

$$\tfrac{5}{7} \text{ (denary)} = 2/3^1 + 0/3^2 + 1/3^3 + 0/3^4 + 2/3^5 + 1/3^6 + 2/3^7 + \ldots$$
$$= 0.\dot{2}01\,02\dot{1} \text{ (base 3).}$$

And, $\frac{5}{7}$ (denary) $= 7/10^1 + 1/10^2 + 4/10^3 + 2/10^4 + 8/10^5 + 5/10^6 + 7/10^7 + \ldots$
$$= 0.\dot{7}14\,28\dot{5} \text{ (base 10).}$$

Random. Unordered and unselected. A term used to describe a **sample** when a part is taken to represent the whole; the part being chosen without condition (at random).

Random Sample. See **Sample.**

Range in Statistics. The interval between the greatest and the least values of a set of data.

Range of Function. See **Function.**

Range of Points. A set of points in a straight line.

Range of Variable. The set of all possible values of a **variable.**

Rate. The relation between two sets of quantities. It may be used to express the way in which one **variable** depends upon another, as in **speed** where distance depends on time. It may also be used to express a relative value, as in the price of goods. The rate associated with two quantities of the same kind can be expressed as a **ratio**, e.g., the **scale** of a model would be expressed as a rate 1 ft per 100 ft, or 1 ft ≡ 100 ft or as a ratio 1/100 (1:100). The concept of rate enters into **rates** in taxation, **vital statistics, rate of change of function.**

Rate of Change of Function. For the **function** $y = f(x)$ when $x = x'$, the limit of the ratio of **infinitesimal** increments of the function and the independent variable. It is also the limit of the average rate of change over an interval which includes the point when the interval approaches zero. This is called the *instantaneous rate of change*. Its value is equal to the gradient of the curve of the function and is equivalent to the **derivative** of the function at any point. See **Gradient of Curve.**

Rate of Interest. See **Interest.**

RATES. The money collected by local authorities from ratepayers, the occupiers of properties within the authorities' areas of jurisdiction. The *rate* fixed by an authority is the ratio of the estimated expenditure for the ensuing year to the known total rateable value of all properties in the area. Thus,

current rate = rates paid on one property/rateable value.

It follows: rates paid by an owner equal rateable value of property multiplied by current rate.

RATIO. The relation existing between two quantities of the same kind, say a and b, and expressed as the fraction a/b or b/a. Thus the relationship of 7 cwt to 2 tons is that the first is 7/40 of the second and the second 40/7 ($5\frac{5}{7}$) of the first. See **Proportion.**

RATIO OF HOMOTHETY. Synonym of homothetic ratio. See **Radially Related Figures.**

RATIO OF SIMILITUDE. See **Radially Related Figures; Transformation of Similitude.**

RATIO TESTS. Various tests for determining the convergency or divergency of sequences and series. **Cauchy's test** is the ordinary test and **D'Alembert's test** is the generalized form. See **Comparison Test; Convergence, Divergence of Sequence; Convergence, Divergence of Series.**

RATIONAL EQUATION. An equation containing only **rational expressions.** Any such equation can be reduced to a **polynomial** equation.

RATIONAL EXPRESSION. An algebraic expression in which the variables are not involved in an irreducible **radical expression** (or have a **fractional index**). For example, $1+2x^3$ is rational, but $\sqrt{(1+2x^3)}=(1+2x^3)^{\frac{1}{2}}$ is **irrational.** If a rational expression involves no division by a variable it is a *rational integral expression* or **polynomial.** A rational expression can be written as the quotient of two polynomials.

RATIONAL FUNCTION. A **function** which can be expressed as a **rational expression,** and therefore as the quotient of two **polynomials.** If a rational function contains only integral terms in one or more variables, it is a *rational integral function.* A function can be described as rational and/or integral in part only. Thus, $2x^2+3\sqrt{y}-4/z$ is rational in x and z, irrational in y, integral in x and y, not integral in z.

RATIONAL NUMBER, QUANTITY. A number or quantity which can be expressed as a **ratio** of two integers. All common **fractions** and all terminating decimal fractions are rational. Certain **trigonometric ratios** of certain angles are rational, e.g., $\cos 60° = \frac{1}{2}$. See **Irrational Number, Quantity.**

RATIONALIZATION OF EQUATIONS. The transformation of equations containing **irrational numbers, quantities** in order to simplify their solution. The process involves the separation of all the irrational quantities from the rational quantities by the sign of equality. Both sides of the equality are then raised to a power sufficient to remove the **radicals.** Where there is a sum or difference of radicals the process is repeated. e.g.

$$\begin{aligned}\sqrt{5x-2}-x&=7\\ \sqrt{5x-2}&=7+x\\ 5x-2&=49+14x+x^2\\ x^2+9x+51&=0.\end{aligned}$$

$$\begin{aligned}\sqrt{2x}+\sqrt{x-1}&=2\\ 2x+2\sqrt{2x}\sqrt{x-1}+x-1&=4\\ 2\sqrt{2x}\sqrt{x-1}&=5-3x\\ 8x(x-1)&=25-30x+9x^2\\ x^2-22x+25&=0.\end{aligned}$$

RATIONALIZATION OF FRACTIONS. The transformation of fractions containing **irrational numbers, quantities** in order to simplify their use and evaluation. The identities

$$\frac{a\pm b}{D}\times\frac{a\mp b}{a\mp b}=\frac{a^2-b^2}{D(a\mp b)}$$

$$\frac{N}{a\pm b}\times\frac{a\mp b}{a\mp b}=\frac{N(a\mp b)}{a^2-b^2}$$

can be used to remove **radicals** from the numerator and the denominator respectively. Thus:

$$\frac{N}{\sqrt{5}+\sqrt{3}}\times\frac{\sqrt{5}-\sqrt{3}}{\sqrt{5}-\sqrt{3}}=\frac{N(\sqrt{5}-\sqrt{3})}{2}.$$

In the special case when N or D contain only one **surd** the radical can be removed by multiplying N and D by the radical as often as is needed to remove it. Thus:

$$\frac{\sqrt{3}}{4}\times\frac{\sqrt{3}}{\sqrt{3}}=\frac{3}{4\sqrt{3}};\quad \frac{5}{\sqrt[3]{4}}\times\frac{(\sqrt[3]{4})^2}{(\sqrt[3]{4})^2}=\frac{5(\sqrt[3]{4})^2}{4}.$$

RAY. A **half-line** defined by fixing an initial point (the *origin of ray*) and considering the set of points on only one side of it. Sometimes written $\overrightarrow{AB}$, where A is the origin and B some other point.

RAY CENTRE. The point of intersection of **rays.** In projection, synonymous with *centre of projection.* See **Geometric, Map, Pictorial Projection.**

RAY RATIO. See **Radially Related Figures.**

REACTION. See **Action and Reaction.**

REAL AXIS. A straight line on which **real numbers** are plotted. See **Argand Diagram.**

REAL NUMBER. A **number** which is a **rational number** or an **irrational number.** That part of a **complex number** which is not an **imaginary number.** Thus in $x+iy$, x is a real number. Real numbers form a **continuum.**

REAL PLANE. The set of points in a plane whose coordinates are **ordered pairs** of **real numbers.**

REAL VARIABLE. A variable to which only **real numbers** can be assigned.

RECIPROCAL CURVE. The curve obtained by replacing each ordinate by its reciprocal. The graphs $y=x$ and $y=1/x$ are reciprocal curves. The graphs of $y=\tan x$ and $y=\cot x$ are reciprocal curves.

RECIPROCAL ELEMENT. The **inverse element** in a set when the **binary operation** is defined as **multiplication.**

RECIPROCAL EQUATION. An equation in one variable which is unaltered when the variable is replaced by its reciprocal.

RECIPROCAL FUNCTION. Synonym of **inverse function.**

RECIPROCAL OF MATRIX. Synonym of inverse **matrix.**

RECIPROCAL OF NUMBER. For the number x, the number x^{-1}, such that $x \times x^{-1}=1$.

RECIPROCAL RELATION. Synonym of **inverse relation.**

RECIPROCAL SPIRAL. Synonym of **hyperbolic spiral.**

RECIPROCAL THEOREM. A theorem obtained by interchanging geometric elements such as points for lines. A theorem and its reciprocal are not necessarily simultaneously true or false. See **Duality.**

RECTANGLE. A **quadrilateral** with two pairs of parallel sides, four right angles, opposite sides equal in length, equal diagonals bisecting one another.

RECTANGULAR COORDINATES. **Cartesian coordinates** when the axes of coordinates are perpendicular to one another.

RECTANGULAR FORM OF COMPLEX NUMBER. The form $x+iy$ in contrast to the **polar form of complex number**. See **Argand Diagram.**

RECTANGULAR HYPERBOLA. A **hyperbola** whose **asymptotes** are at right angles. The transverse and conjugate axes are equal. The general equation is $x^2-y^2=a^2$ when the curve is symmetrical to the axes of coordinates (the axes of the hyperbola are the axes of coordinates) and $xy=\frac{1}{2}a^2$ or $xy=c^2$ when the asymptotes are the axes of coordinates. The **parametric coordinates** of any point are $(ct, c/t)$. See **Axes of Hyperbola.**

RECTANGULAR PROPERTIES OF CIRCLE. The relationships between the line segments formed by the intersection of a circle and two intersecting lines. When the points A, B, C, D are concyclic and lines AB, CD intersect at X, then $AX \cdot XB = CX \cdot XD$. Two special cases occur: one where C and D coincide at T, when $AX \cdot XB = XT^2$; and one where the two lines are perpendicular and one is a diameter, when $CX = XD = \sqrt{AX \cdot XB}$.

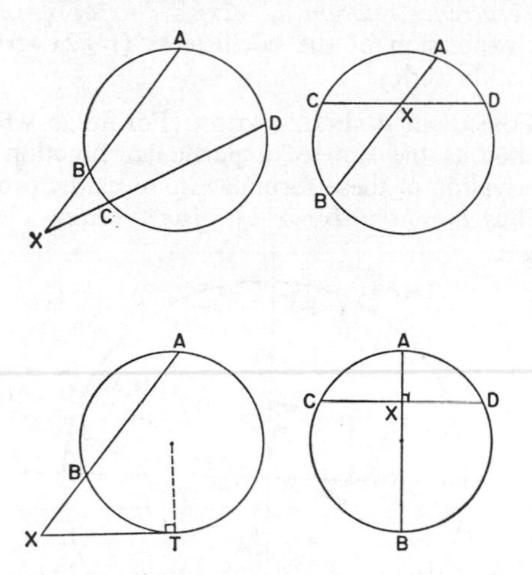

RECTIFIABLE CURVE. A curve of finite length or a curve whose length can be determined by the process of **rectification.**

RECTIFICATION. The process of finding the length of a curve segment. For the plane curve $y = f(x)$, the length of the segment between A (x, y) and B, $(x + \delta x, y + \delta y)$ is δs which equals $\sqrt{(\delta x^2 + \delta y^2)}$. Since $\delta s/\delta x = \{1 + (\delta y/\delta x)^2\}^{\frac{1}{2}}$, the length of the curve between $x = x_1$ and $x = x_2$ is in the limit as $\delta x \to 0$ given by $s = \int_{x_1}^{x_2} \left\{1 + \left(\frac{dy}{dx}\right)^2\right\}^{\frac{1}{2}} dx.$

RECTILINEAR FIGURE. A plane figure bounded by **line segments.** See **Polygon.**

RECTILINEAR GENERATOR. See **Ruled Surface.**

RECTILINEAR MOTION. Motion along a straight line. See **Velocity.**

RECURRENCE RELATION. See **Recurring Sequence, Series.**

RECURRING CONTINUED FRACTION. See **Continued Fraction.**

RECURRING DECIMAL; REPEATING DECIMAL. A decimal **fraction** in which, after a certain decimal place, one digit or a set of digits in the same order, is repeated indefinitely. A dot above one figure or dots above the first and last figures of a set indicate the recurring (repeating) decimal figures. For example: $1/3 = 0.\dot{3}$, $1/7 = 0.\dot{1}4285\dot{7}$. Any recurring decimal can be written as an infinite series of **common fractions**; thus, $0.0\dot{9} = 9/100 + 9/10{,}000 +$ etc.

Recurring Sequence, Series. A **sequence** or **series** in which there is a relation between groups of successive terms. Thus, in the arithmetico-geometric series $1+2x+3x^2+4x^3+\ .\ .\ .$ any three successive terms are related by the *recurrence relation* $u_n=2xu_{n-1}-x^2u_{n-2}$, or $u_n-2xu_{n-1}+x^2u_{n-2}=0$. The expression of the coefficients $(1-2x+x^2)$ is called the *scale of relation* of the series.

Reduction Formulae in Integration. Formulae which express an **integral of function** as the sum of a particular function and a simpler integral. The derivation of these formulae involves the process of **integration by parts.** Thus $I_n=\int x^n e^x dx=x^n e^x-\int nx^{n-1}e^x dx=x^n e^x-nI_{n-1}$.

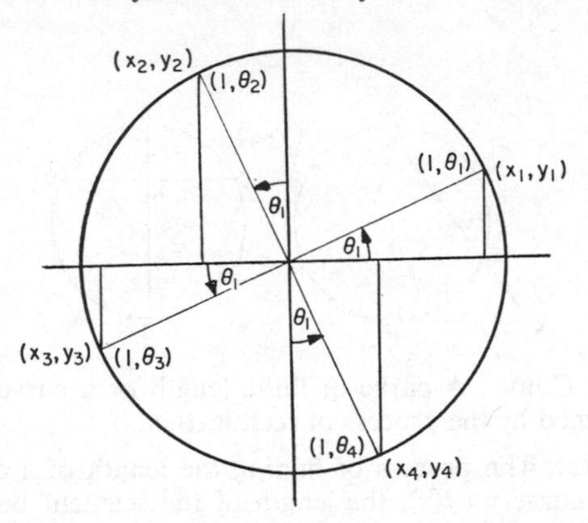

Reduction Formulae in Trigonometry. Identities which express **trigonometric ratios** of any angle in terms of the trigonometric ratios of an acute angle. The identities can be derived from the relation between the **polar coordinates** $(1,\theta)$ and the **Cartesian coordinates** (x,y) of a point on the circumference of a unit circle whose centre is both pole and origin. Thus for the sine ratio, $\sin\theta_1=y_1$; $\sin\theta_2=\sin(90°+\theta_1)=y_2=x_1=\cos\theta_1$; $\sin\theta_3=\sin(180°+\theta_1)=y_3=-y_1=-\sin\theta_1$; $\sin\theta_4=\sin(270+\theta_1)=y_4=-x_1=-\cos\theta_1$. This process is applicable to negative angles and all ratios. See **Circular Functions.**

Reduction in Arithmetic. (1) The process of expressing a quantity given in one unit in terms of another, e.g., 1 hour $\equiv$ 60 minutes; 1000 g $\equiv$ 1 Kg. (2) Simplification of a **common fraction** involving the division of both numerator and denominator by a common factor. The value of the fraction is *not* reduced. (3) Diminution in size, e.g., a selling price reduced by a discount. (4) Synonym of partition, one aspect of **division.**

Reduction in Geometry. A **scaled drawing** smaller than the original.

Redundant Number. Synonym of abundant number, excessive number. See **Perfect Number.**

RE-ENTRANT ANGLE. An **interior angle** of a geometrical plane figure which is a **reflex angle. See Concave, Convex Polygons.**

REFLECTION. A **transformation in geometry** in which each point P has a corresponding point P' such that the line PP' is bisected by a plane to which it is normal, a line to which it is perpendicular, or a point. See **Isometry; Symmetry.**

REFLEX ANGLE. An **angle (plane angle)** greater than half a revolution (straight angle, 180°, π radians) and less than a revolution (360°, 2π radians)

REFLEXIVE RELATION. A **relation** such that any element x, bears the relation to itself. An **equality** is a *reflexive* relation, since $x = x$ for any x, while an **inequality** is *anti-reflexive* since $x \ngtr x$ for any x. A *non-reflexive* relation is one which is not reflexive for at least one element x.

REGULAR PLANE FIGURE. Synonym of **regular polygon.**

REGULAR POLYGON. Description of any **polygon** having angles of the same size, sides of the same length and consequently vertices on a circumscribing circle. A few are: equiangular trigon or equilateral triangle (3 sides, internal angles each $\pi/3$ radians or 60°), tetragon or square (4, $\pi/2$, 90), pentagon (5, $3\pi/5$, 108), hexagon (6, $2\pi/3$, 120), octagon (8, $3\pi/4$, 135), decagon (10, $4\pi/5$, 144), dodecagon (12, $5\pi/6$, 150). For any regular polygon of n sides or n vertices or n angles, the internal angle is $(n-2)\pi/n$ radians or $\{(n-2)/n\} \times 180°$.

REGULAR POLYHEDRON. A solid bounded by **regular polygons** with all vertices on a circumscribing sphere. *Platonic* convex solids: tetrahedron (4 faces, 4 vertices, 6 edges), hexahedron or cube (6, 8, 12), octahedron (8, 6, 12), dodecahedron (12, 20, 30), icosahedron (20, 12, 30) *Kepler-Poinsot* concave solids: small stellated dodecahedron (12, 20, 30), great dodecahedron (12, 12, 30), great icosahedron (20, 12, 30). All except the small stellated and great dodecahedra obey Euler's Law for polyhedra: $f + v = e + 2$.

REGULAR SOLID. Synonym of **regular polyhedron.**

RELATION. The relation, R, from set X to set Y, is any proper subset of $X \cdot Y$, i.e., the *Cartesian product*, (x, y), $x \epsilon X$, $y \epsilon Y$. This is the set of ordered pairs of elements obtained by selecting one element from each set. The relation is a set R of ordered pairs (x, y), x being related to y (written xRy) if (x, y) is a member of R. If a relation is seen as a mapping its graph can be represented in a schematic diagram. Set X is mapped into set Y, and the relation R is represented by the set of joins (called couples) of the elements of X and Y. The arrows indicate the sense of the relation. See **Equivalence Relation; Inverse Relation; Reflexive Relation; Symmetric Relation, Transitive Relation.**

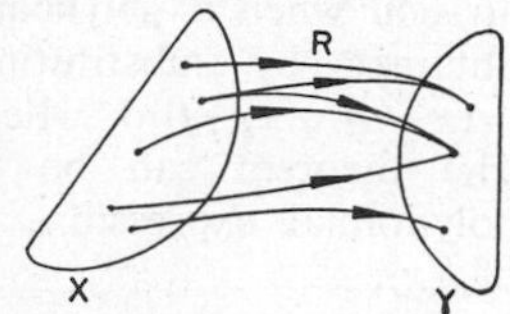

RELATIVE ACCELERATION, VELOCITY. The **acceleration** or **velocity** of a moving point with respect to another moving point. The acceleration or velocity computed with relation to a moving system of coordinates. Relative accelerations and velocities are treated as **vector differences.**

RELATIVE DENSITY. Synonym of **specific gravity.**

RELATIVE ERROR. For any set of measures of the same thing, the quotient error/mean of measures. It may be represented as a percentage error. See **Error.**

RELATIVE PROFIT, LOSS. The profit, loss expressed as a fraction of the cost price, the selling price or the turn-over according to circumstances. It may be expressed as a **percentage profit, loss.**

RELATIVE VELOCITY. See **Relative Acceleration, Velocity.**

RELATIVITY. The theory propounded by Albert Einstein (1879–1955) that all natural processes take place in a four-dimensional space–time **continuum.** This rests on the well-established facts that **absolute** speed or **velocity** cannot be ascertained, only its value relative to the observer; that observation depends itself on the emission or reflection of light from the object observed to the observer; and that nevertheless, when the speed of the emitted light is measured it gives a value independent both of the speed of the source of light and of the speed of the observer. It appears to be an absolute speed. Einstein devised a geometry of this space–time continuum which not only embodied these facts but led logically to a geometrical explanation of gravitational behaviour that did not demand a 'speed' for gravitational action. It was simply a property of space–time in the geometrical neighbourhood of a particle of matter.

REMAINDER IN DIVISION. If a dividend N is divided by a divisor D, $N/D = Q + R/D$, where Q is the quotient and R is the remainder. Hence, $R = N - Q\ D$. The remainder, in **division,** is the difference between the dividend and the product of the quotient and the divisor.

REMAINDER IN SUBTRACTION. If a subtrahend S is subtracted from a minuend M, $(M - S)$ is the difference or remainder. See **Subtraction.**

REMAINDER THEOREM. A method for determining the **remainder in division** when a **polynomial expression,** $f(x)$, is divided by $(x - a)$. It is obtained by substituting a for x in the polynomial. Thus, $f(x) = (x - a)\ q(x) + f(a)$, where $q(x)$ is the quotient and $f(a)$ the remainder. The theorem can be used for the determination of factors of a polynomial expression. If, for $f(x)$, $f(a) = 0$, then $(x - a)$ is a factor of $f(x)$.

REPEATED ROOTS. If $f(x)$ is divisible by $(x - a)^n$, then, in the solution of the equation $f(x) = 0$, a will be a root n times. In the sextic equation

$x^6 - 2x^5 - 8x^4 + 14x^3 + 11x^2 - 28x, + 12 = 0$ which is also $(x-1)^3(x+2)^2(x-3) = 0$, 3 is a simple root, -2 a double root, 1 a triple root. The curve representing this equation will pass through the point $\{3, f(3)\}$ once, the point $\{-2, f(-2)\}$ twice and the point $\{1, f(1)\}$ three times. An equation with a double root has this root in common with its first **derived equation.**

REPEATING DECIMAL. Synonym of **recurring decimal.**

REPRESENTATIVE FRACTION (R.F.). One way of stating the scale of a map or plan. Thus 1/10,560 means that one unit of length on the map is equivalent to 10,560 of the same units of length on the territory mapped or planned. R.F. 1/63,360 is equivalent to one mile to the inch.

RESECTION IN SURVEYING. The method of fixing a position by making observations on three fixed points.

RESIDUE CLASS. A term used in **modulo (modular) arithmetic.** The set of all numbers can be partitioned into a number of residue classes determined by the remainders after division by a given integer. The number 16, for example, is in residue class 2 when divided by integer 7. It is denoted $[16]_7$ or *residue class mod* 7, or Z_7.

RESISTING MOMENT, SHEAR. See **Bending Moment; Shear in Horizontal Beam.**

RESOLUTION OF FORCES. See **Resolution of Vectors.**

RESOLUTION OF VECTORS. The process of determining two vectors equivalent to a given one. A **vector quantity, R**, represented by $\overline{OC}$ may be resolved into *components* in any two directions $\overline{OX}$, $\overline{OY}$ by constructing a parallelogram with OC as diagonal and sides OA, OB lying along $\overline{OX}$ and $\overline{OY}$. Vectors $\mathbf{V}_1$ and $\mathbf{V}_2$ represented by $\overline{OA}$ and $\overline{OB}$ respectively, are the components of **R**. **R** is said to have been resolved into $\mathbf{V}_1$ and $\mathbf{V}_2$. If angle XOY is 90° and angle $XOC = \theta$, $\mathbf{V}_1 = \mathbf{R} \cos \theta$ and $\mathbf{V}_2 = \mathbf{R} \sin \theta$. This special case is employed in the resolution of forces in horizontal and vertical directions. See **Parallelogram of Vectors.**

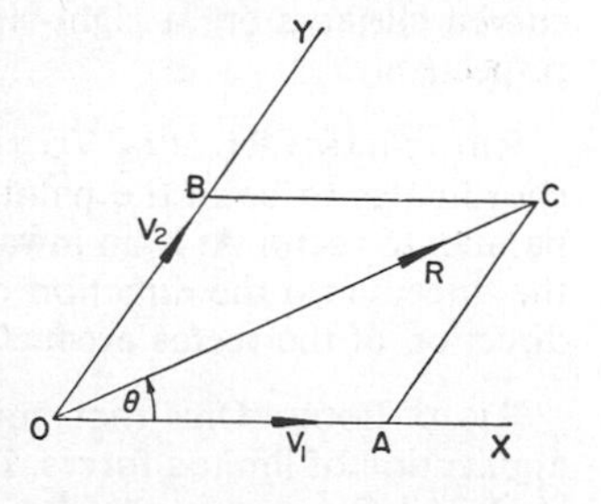

RESULTANT. The **vector quantity** which is equivalent in effect to two or more given vector quantities. The **parallelogram of vectors** is employed in its determination.

RETARDATION. A term used sometimes as a synonym of negative **acceleration.**

REVERSE ORDER. A **sequence** of n terms is put into reverse order when the nth, $(n-1)$th . . . terms are written as the first, second, . . . terms.

REVOLUTION. (1) A complete **rotation** about an axis or a point. (2) The **angle of rotation** required to complete a revolution—360° or 2π radians.

RHOMBOHEDRON. A **polyhedron** with six faces, each a **rhombus.**

RHOMBUS (RHOMB). A **quadrilateral** with four equal sides. Its opposite angles are equal, its diagonals bisect each other at right angles and its opposite sides are parallel.

RHUMB. A term in navigation. A change of one point on a mariner's compass. One sixteenth part of a reverse turn; $11\frac{1}{4}°$.

RHUMB-LINE. On the surface of a sphere a curve which cuts all the **meridians of longitude** at the same angle.

RHUMB-LINE NAVIGATION. Pursuing a course in one compass direction only.

RIDER. A subsidiary problem in mathematics.

RIGHT ANGLE. The concept of an upright **angle (plane angle)** between the vertical line and the horizontal plane at any point. One quarter of a revolution, 90° or $\frac{1}{2}\pi$ radians.

RIGHT LINE. Synonym of **straight line.**

RIGHT PRISM. See **Prism.**

RIGHT (RIGHT-ANGLED) TRIANGLE. A triangle containing a **right angle** and two **complementary angles.**

RIGHT-CIRCULAR CONE, CYLINDER. A **cone** or **cylinder** of which the curved surfaces are a right-circular **conical surface** or **cylindrical surface** respectively.

RIGHT-HAND RULE IN VECTOR MULTIPLICATION. Imagine the origin of coordinates to be in the palm of the right hand with the fingers directed parallel to vector **A**. If an inward movement of the wrist is needed to bring the fingers into the direction of the vector **B**, the thumb will indicate the direction of the **vector product,** $\mathbf{A} \times \mathbf{B} = |\mathbf{A}|\ |\mathbf{B}| \sin\theta\ \mathbf{n}$.

RIGID BODY. One that suffers no perceptible distortion during the application of limited forces. If three points distributed through the body, A, B and C, are separated by distances c, a, b, then the lengths, a, b and c are considered as unchanged during the application.

RIGID MOTION. A **transformation in geometry** in which a configuration is moved from one position to another in a plane or space without change of shape or size. An **isometry** such as **translation** or **rotation** or a combination of both such transformations.

RIGIDITY MODULUS. For an elastic body, the ratio of the unit **stress** to the unit **strain** associated with the stress, which acts in such a way that two contiguous parts of the body tend to slide relative to each other. If the stress T is applied over area $ABCD$, to produce a deformation θ in each angle, θ is the measure of the strain, and the ratio T/θ is the rigidity modulus.

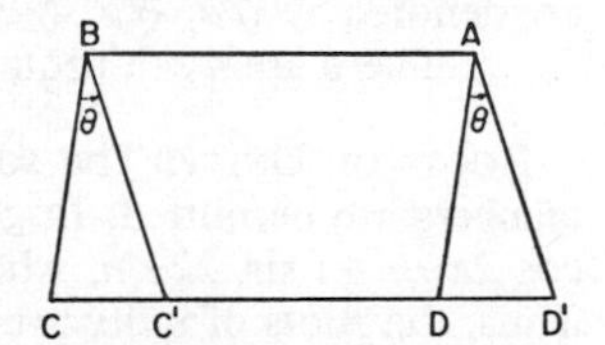

RING. A set of elements subject to two **binary operations** (addition and multiplication) and four axioms: (1) It is a **commutative (Abelian) group** with respect to addition; (2) The product of any two elements is unique; (3) The set obeys the **associative law** under multiplication; (4) The two operations are related by the **distributive law**. If the set obeys the **commutative law** under multiplication it is a *commutative ring*.

RISE. Any upward measurable change: as in altitude, atmospheric pressure, price, temperature, vital statistics, etc.

ROLLE'S THEOREM. If $f(x)$ is a single-valued continuous **function** for the interval $a \leqslant x \leqslant b$ and $f(x)=0$ when $x=a$ and $x=b$, then $f'(x)=0$ for at least one value of x distinct from and lying between a and b. If this is true for more than one value of x, there must be an odd number of such values.

ROLLING MOTION. This type of motion assumes two surfaces with a point or line of contact, and, one surface being considered fixed, a continuous change of this point or line of contact. Slipping is not involved under the conditions that the moving body rotates round its own **centroid** while this centroid revolves round the centre of **curvature** of the fixed surface at the point or line of contact. The **cycloid, epicycloid** and **hypocycloid** are loci which arise from rolling motion.

ROOT MEAN SQUARE. See **Deviation in Statistics.**

ROOT OF COMPLEX NUMBER. The **complex number**

$$r[\cos(2k\pi+\theta)+i\ \sin(2k\pi+\theta)]$$

has n nth roots given by

$$\sqrt[n]{r}[\cos\{(2k\pi+\theta)/n\}+i\ \sin\{(2k\pi+\theta)/n\}]$$

where k assumes the values 0, 1, 2, . . . $(n-1)$, and r is not a negative number. A special case of this is the **root of real number.**

ROOT OF EQUATION. In an equation $f(x)=0$, any value of x which satisfies it. The general linear equation $ax+b=0$ has one solution, $x=-b/a$. The general quadratic equation, $ax^2+bx+c=0$, has two roots, $(-b\pm\sqrt{b^2-4ac})/2a$. The cubic and quartic equations can be solved by formulae, but the quintic and higher powered equations have no explicit algebraic formulae for the solutions. See **Discriminant**; **Rule of Signs and Roots.**

Root of Real Number. The solution of the equation $x^n = a$. These roots are denoted by $\sqrt{a}$, $\sqrt[3]{a}$, $\sqrt[4]{a}, \ldots \sqrt[n]{a}$, where the order of the root is 2, 3, 4, . . . n. There are n nth roots of any non-zero number.

Roots of Unity. The solution of the equation $x^n = 1$ when **complex numbers** are permitted. In general, unity has n roots of order n, given by $\cos 2k\pi/n + i \sin 2k\pi/n$, where k assumes the values 0, 1, 2, . . . $(n-1)$. Thus, the roots of unity are:

square roots. 1; -1.

cube roots. 1; $\frac{1}{2}[-1+i\sqrt{3}]$; $\frac{1}{2}[-1-i\sqrt{3}]$, denoted, $1, \omega, \omega^2$.

fourth roots. 1; i; -1; $-i$.

fifth roots. 1; $\frac{1}{2}\sqrt{\frac{1}{2}(3-\sqrt{5})} + i\,\frac{1}{2}\sqrt{\frac{1}{2}(5+\sqrt{5})}$;

$$-\tfrac{1}{4}(1+\sqrt{5}) + i\,\tfrac{1}{2}\sqrt{\tfrac{1}{2}(5-\sqrt{5})}; \qquad -\tfrac{1}{4}(1+\sqrt{5}) - i\,\tfrac{1}{2}\sqrt{\tfrac{1}{2}(5-\sqrt{5})};$$

$$\tfrac{1}{2}\sqrt{\tfrac{1}{2}(3-\sqrt{5})} - i\,\tfrac{1}{2}\sqrt{\tfrac{1}{2}(5+\sqrt{5})}.$$

Other roots are found similarly. The values of the roots of unity can be represented in an **argand diagram** by points on the circumference of a unit circle, with centre at the origin. The point (1,0) is common to all orders: the points associated in one order are placed at equal intervals on the circumference.

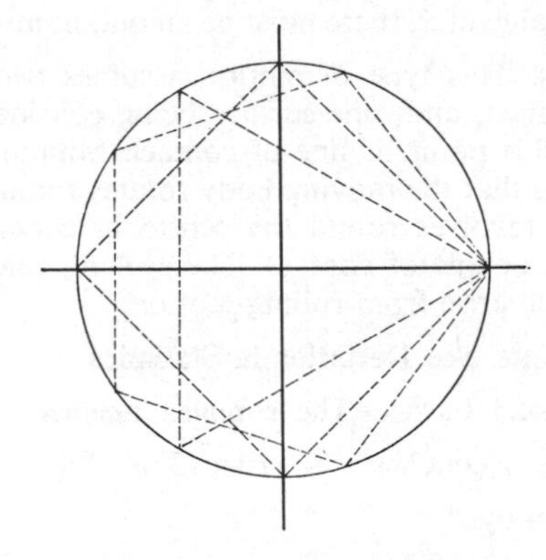

Rope Stretchers' Triangle. The **right triangle** with sides in the ratios 3:4:5 made by the segments of a rope of length 12 cubits, used by the Egyptians and Chinese as a surveying device. Thales (640-550 B.C.) first correctly surmised that the right angle opposite the longest side was not consequent upon a relationship between the numbers 3, 4 and 5, but upon the relationship between the squares: $9+16=25$. The Pythagoreans found many more integer triads which have the right angle property. See **Pythagoras's Theorem**; **Pythagorean Triangle**; **Pythagorean Triples.**

Rotation. This is a **rigid motion** where all points turn about a common centre in a plane or a common line of rotation in space. Every point turns through the same **angle of rotation**. The sense of rotation is positive if anti-clockwise. Half a **revolution**, positive or negative, gives the same result.

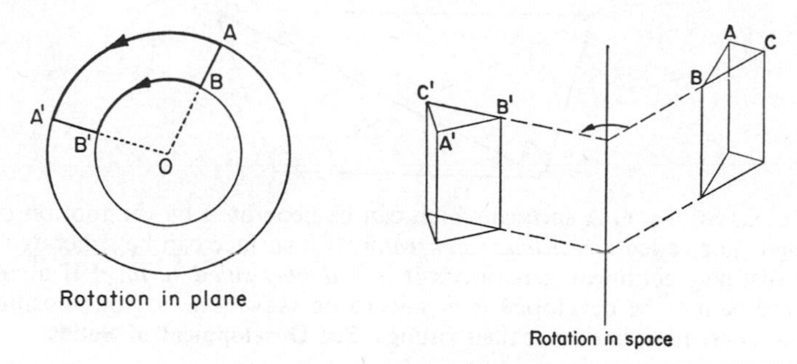

Rotation in plane

Rotation in space

ROTATION OF AXES. See **Transformation of Axes.**

ROTATION SYMMETRY. A rotation of a geometric figure which transforms the figure into itself. A rotation in a plane through 0°, 90°, 180°, 270°, 360° transforms a square into itself. See **Group of Symmetries; Order of Symmetry.**

ROULETTE. The locus of a point or the **envelope** of a line attached to a curve which rolls without slipping along a fixed curve; the point giving a *point-roulette*, the line, a *line-roulette*. The **cycloid, epicycloid, hypocycloid** are roulettes. **Epitrochoids** and **hypotrochoids** are roulettes formed by points on circles which roll on the outside or inside of fixed circles.

RULE OF SARRUS. A rule for writing down the terms of a **determinant.** To an array of n^2 elements, write below the first $n-1$ rows, as shown. A line along the principal diagonal gives the first positive term; $n-1$ lower parallel lines give the other positive terms. A set of n ascending lines give the negative terms.

$$\Delta = \begin{vmatrix} a_1 & b_1 & c_1 \\ a_2 & b_2 & c_2 \\ a_3 & b_3 & c_3 \end{vmatrix} = \begin{matrix} a_1b_2c_3 + a_2b_3c_1 + a_3b_2c_1 \\ -a_3b_2c_1 - a_1b_3c_2 - a_2b_1c_3 \end{matrix}$$

$$\begin{matrix} a_1 & b_1 & c_1 \\ a_2 & b_2 & c_2 \end{matrix}$$

RULE OF SIGNS. See **Algebraic Addition, Subtraction, Multiplication, Division.**

RULE OF SIGNS AND ROOTS. If, in a **polynomial expression** there are two positive or two negative terms in succession, there will be one negative root to the equation in which the expression is equated to zero. If a positive term succeeds a negative or vice-versa, there will be one positive root. For example: in the equation $x^3+4x^2-7x-10=0$, the first two terms indicate a negative root (-5), the middle two indicate a positive root (2), and the last two terms, another negative root (-1). The rule is attributed to Descartes.

RULE OF THREE PERPENDICULARS. A property of a **tetrahedron** every face of which is a right-angled triangle. In the diagram, if any three of the marked angles are right angles, the fourth is also a right angle.

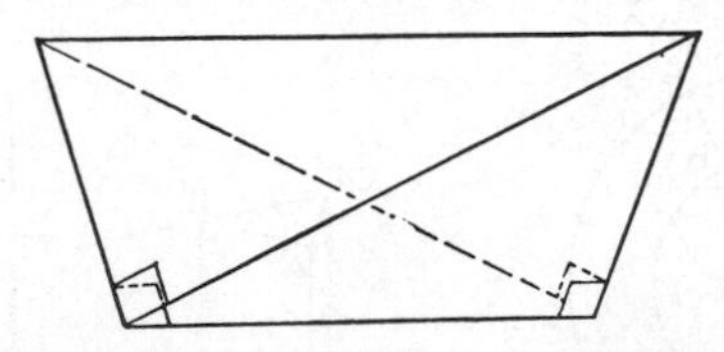

RULED SURFACE. A **surface** which can be generated by the motion of a straight line called a *rectilinear generator*. If a surface can be generated by two distinct rectilinear generators it is a *doubly-ruled surface*. If a ruled surface cannot be developed it is said to be *skew*. The various positions of the **generating line** are called rulings. See **Development of Solids**.

S

s. (1) **Second of time-interval**. (2) Distance in some forms of equations of motion. (3) Semi-Perimeter in **Heron's (Hero's) formula**.

SVP. Side vertical plane. See **First Angle, Third Angle Projections.**

SAMPLE. A finite part of a **population** chosen for the purpose of **inference in statistics.**

random sample. One in which every member of the population has an equal chance of being included in the sample.

stratified sample. One which is made up of random samples, taken from each of several sub-populations, called *strata*, which are sub-divisions of the population.

SAMPLING. The process in **statistics** of obtaining a **sample** of a population for the purpose of **inference in statistics.**

SCALAR MULTIPLICATION. The multiplication of a **vector quantity** by a **scalar quantity** which increases or diminishes the vector quantity so multiplied or reverses its direction.

SCALAR PRODUCT. A product of **vector quantities** which results in a **scalar quantity**: not to be confused with **scalar multiplication.** It is sometimes called the *inner product*, and because of its symbol, **A B**, the *dot product*, distinguishing it from **vector product** $\mathbf{A} \times \mathbf{B}$. If **A** is (p, q) and **B** (r, s), then $\mathbf{A}\ \mathbf{B} = pr + qs$. If $\mathbf{V}_1$ is (a_1, b_1, c_1) and $\mathbf{V}_2$, (a_2, b_2, c_2), then $\mathbf{V}_1\ \mathbf{V}_2 = a_1a_2 + b_1b_2 + c_1c_2$, a scalar quantity. If the vectors are given geometrically, $\mathbf{A}\ \mathbf{B} = |\mathbf{A}|\ |\mathbf{B}|\ \cos\theta$, θ being the angle between the vectors when they are placed with a common origin.

SCALAR QUANTITY. A single number possessing size but not involving the concept of direction.

SCALE. In general, a system of points placed at known intervals on a line for the purpose of measuring. It may be used for measuring concrete objects, or as the basis of comparison in map scales, **scaled drawings** or **comparative scales.** As a **number scale** it provides the basis for elementary computation with directed numbers. The **logarithmic scale** facilitates more accurate computation. The use of a scale is implicit in any **system of notation** where *scale of notation* can be interpreted as scale of measure in determining the relative sizes of numbers.

SCALE OF NOTATION. See **Scale; System of Notation.**

SCALE OF RELATION. See **Recurring Sequence, Series.**

SCALED DRAWING. A plan or map on which the length of every line segment bears a constant ratio to the length of the true line it represents. A scale of 1/10 implies that every area shown will be 1/100 of true area and every volume calculated will be 1/1 000 of the true volume.

SCALENE TRIANGLE. From the Greek *skolios*, crooked; on a plane or spherical surface, a triangle with no two sides of equal length.

SCATTER DIAGRAM (SCATTERGRAM). A diagram to show the frequency distribution of pairs of **variables.** The intersections of columns and rows form cells in which the frequencies of occurrence are indicated. A scattergram shows pictorially the **dispersion** of a set of data.

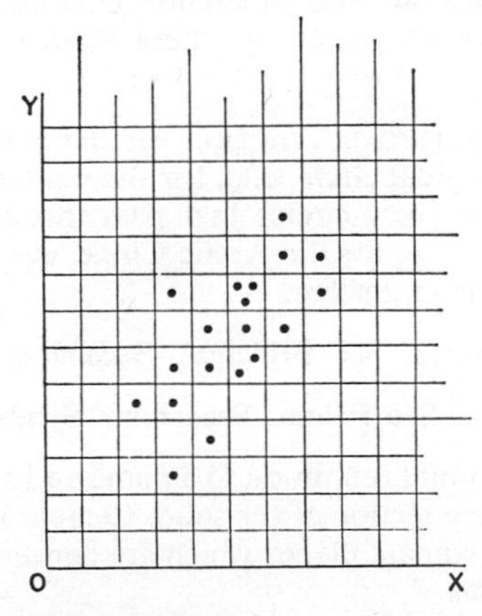

SCORE IN STATISTICS. See **T-score.**

SCREW. In **motion-geometry,** the combination of a **rotation** about a line and a **translation** along the line.

SEC. Abr. secant. See **Trigonometric Ratios.**

SECH. Abr. hyperbolic secant. See **Hyperbolic Functions.**

SECANT CONICAL PROJECTION. See **Map Projection.**

SECANT RATIO (sec). See **Trigonometric Ratios.**

SECANT, HYPERBOLIC (sech). See **Hyperbolic Functions.**

SECANT IN GEOMETRY. A straight line which has one or more points in common with a curve; the number depending upon the nature and position of the curve.

SECOND DERIVATIVE. See **Derivatives.**

SECOND DIFFERENTIAL COEFFICIENT. Synonym of **second derivative.**

SECOND MOMENT. Synonym of **moment of inertia.**

SECOND OF TIME-INTERVAL. The astronomical definition of time-interval is the fraction 1/31 556,925.974 7 of the **tropical year** for 1900, January 0 at 12 hr ephemeris time. For scientific purposes, an atomic constant based on a radiation frequency of the caesium atom is used (accurate to about 1 second in 3000 years). See **Minutes, Seconds; Sexagesimal.**

SECOND ORDER DIFFERENTIAL EQUATION. A **differential equation** involving **differential coefficients** of no higher order than the second. When y, dy/dx, d^2y/dx^2 occur to the first degree only it is called a linear differential equation. The general linear differential equation of the second order is $A\ d^2y/dx^2 + B\ dy/dx + C\ y + D = 0$, where A, B, C, D are constants or functions of x.

SECONDARIES TO SPHERICAL CIRCLES. All the **great circles** that pass through the poles of a great circle, e.g., for the equator, every **meridian of longitude.** Also all the great circles that pass through the near and far poles of a small circle, e.g., for the Arctic Circle, every meridian of longitude. See **Pole of Circle on Sphere.**

SECONDARY DIAGONALS. See **Principal, Secondary Diagonals.**

SECONDARY SYMBOL. See **Prime, Secondary Symbols.**

SECTION. The set of points common to a plane and a solid through which it passes is called a plane section of the solid. **Contour** lines are sections of a terrain made by approximate planes which are concentric shells round the earth. See **Cross section.**

SECTOR OF CIRCLE. Part of a circle bounded by two radii and an arc. See **Major, Minor Sector of Circle.**

SECTOR OF SPHERE. The solid generated by the rotation of a **sector of circle** about a diameter of the circle. If diameter DCD' passes through the sector ACB, there will be *one* base, a *cap* of the sphere; if it does *not* pass through the sector, there will be *two* bases, one plane circular, the other a *zone* of the sphere. The remaining surface in each case is a right-circular **conical surface.** See **Major, Minor Sector of Sphere; Spherical Cap; Zone.**

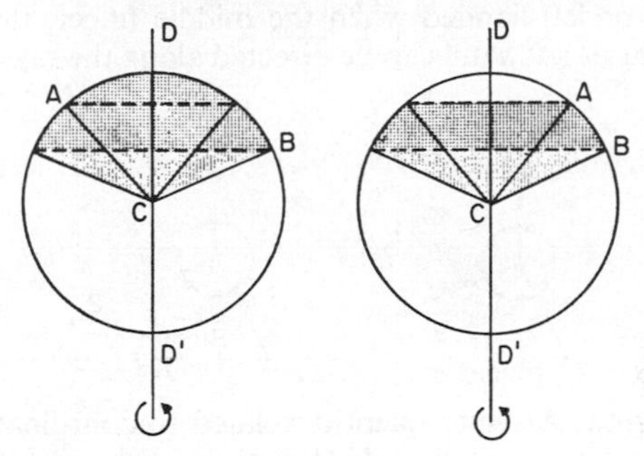

SEGMENT OF CIRCLE. Part of a circle bounded by a chord and an arc. See **Major, Minor Segment of Circle.**

SEGMENT OF CURVE. (1) Part of a curve between two points on it. (2) The area bounded by an arc and the chord joining its end points.

SEGMENT OF LINE. Part of a straight line between two points on it. The segment contains one end point only or both end points (an isolated segment contains both; the point of section of a line will lie in one or other of the two segments).

SEGMENT OF SPHERE. The solid bounded by two parallel planes, intersecting or tangential to the sphere and that portion of the spherical surface between them. See **Major, Minor Segment of Sphere.**

SELECTION. See **Combination.**

SEMICIRCLE. Half a circle, separated from its twin half by a diameter.

SEMICUBICAL PARABOLA. The curve which is the graph of either of the equations: $ay^2 = x^3$; $by^3 = x^2$.

SEMI-LATUS RECTUM. Half the **latus rectum** of a **conic section.** See **Focus-Directrix Definition of Conic.**

SEMITONE. Half a tone and one-twelfth part of an **octave.** In a major scale the difference in pitch between the third and fourth, and between the seventh and eighth notes are semitones. In a minor scale, the intervals between the second and third and the fifth and sixth notes are semitones.

SENARY. Associated with six, as in senary **system of notation.**

SENSE OF INEQUALITY. See **Inequality.**

SENSE OF LINE. Lines are positive or negative according to whether their sense of direction and that of an accepted positive direction are the same or opposite. For **directed line segments** and **rays** the sense is indicated by the order of their end points. Thus $\overline{AB} = -\overline{BA}$, or $\overrightarrow{AB} = -\overrightarrow{BA}$.

SENSE OF ORIENTATION. A description of the relative directions of the rays of a **trihedral angle** (for example, the axes of coordinates). The sense is right-handed or left-handed when the middle finger, thumb and fore-finger of the right or left hand can be directed along the rays (1), (2) and (3) respectively.

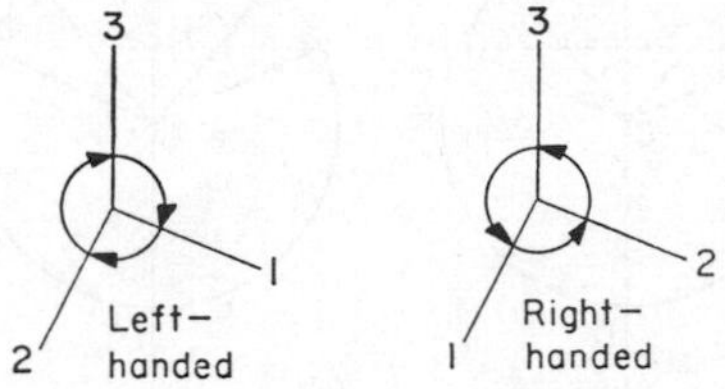

3 3

2 Left-handed 1 Right-handed

SENSE OF VECTOR. A **vector quantity** related to coordinate axes, $X'OX$, $Y'OY$, $Z'OZ$, and passing through O, may be directed from O (1) along any axis in a positive or negative sense of direction, (2) across any co-ordinate plane in a positive-positive, positive-negative, negative-positive or negative-negative sense of direction or (3) into any one of the eight **octants** about O. The coordinates (x,y,z) of any point on the line of the vector will show, by the eight combinations possible of + and − the octant through which it is directed.

SEPARATION (**1**). The partition of a set of elements into two classes by choosing an arbitrary separator such that all elements of one class are less than all elements of the other. There are two types of separation.

separation of first kind. The separator is some element of the set which can be assigned to one or other of the classes. For example, 2 separates the rational numbers into two classes defined as $A\{x \leqslant 2\}$ and $B\{x > 2\}$.

separation of second kind. The separator is not a member of the set. For example $\sqrt{2}$ separates the rational numbers into two classes defined as $A\{x \leqslant 0 \text{ and } x^2 < 2\}$ having no greatest member and $B\{x^2 > 2\}$ having no least number. This is an example of a **Dedekind cut.**

SEPARATION (**2**). The classification, by elements of set X, of the set $S = \{x, y \mid x \in X, y \in Y\}$ to reveal the set Y, where S is a set of **ordered pairs.** Written symbolically as $S \div X = Y$ it is the **inverse operation** of that which produces the **Cartesian product** $S = X \times Y$., e.g.

$$\{(a, p), (a, q), (a, r), (b, p), (b, q), (b, r)\} \div \{a, b\} = \{p, q, r\}.$$

SEPTENARY. Associated with seven as in septenary **system of notation.**

SEQUENCE. Any set of elements written consecutively.

alternating sequence. A sequence in which the elements are alternately positive and negative.

bounded sequence. An ordered sequence in which $|x_n| \leqslant k$ for each and every element of the sequence, i.e. for all n, k is the bound of the sequence.

monotone ascending (increasing) sequence. One in which $x_n \leqslant x_{n+1}$ for all n.

monotone non-ascending (non-increasing, descending, decreasing) sequence. One in which $x_n \geqslant x_{n+1}$ for all n.

null sequence. If ϵ is a given positive **rational number**, a sequence in which $|x_n| < \epsilon$ for all n, with, at most, a finite number of exceptions.

ordered sequence. One that has been written in **one-one correspondence** with the positive **integers**; the order being dependent upon some property of the elements or on some plan of procedure.

See **Bound of Set; Convergence, Divergence of Sequence.**

SERIES. The sum of an ordered **sequence**. A finite or infinite series has a finite or infinite number of terms.

alternating series. A series in which the terms are alternately positive and negative.

oscillating convergent series. One in which $|u_{n+1}| < |u_n|$ for all n and in which the terms are alternately greater and less than the limit; e.g., $1 - \frac{1}{2} + \frac{1}{3} - \frac{1}{4} + \frac{1}{5} \ldots$ oscillates and converges to the limiting value 0.694 14 . . . or $\log_e 2$.

oscillating divergent series. One in which $|u_{n+1}| > |u_n|$ for all n and for which the sign of the sum, $S_\infty = \infty$, is indefinite. For example, $1 - 2 + 3 - 4 + \ldots$

See **Convergence, Divergence of Series.**

SET. A collection or class of things called elements, the concept of which forms the basis of **set theory.**

SET THEORY. The **Boolean algebra** of sets or classes. The logical treatment of the relations between the **operations** on sets of things. Sets can be represented by letters, relations and operations by symbols. A **venn diagram** can be used as a geometric interpretation. A member of a set is an element, and if a is an element of b it is written $a \epsilon b$.

universal set (universe). The set of all those elements under consideration.

subset. If set A is included in set B, A is a subset of B, and written $A \subset B$.

unit set. A set containing only one element.

Union $A \cup B$

null set (empty set). The set which contains no element; symbol ϕ.

complement (inverse) of set. If A is a subset of a universal set, the subset containing all elements not in A is the complement (inverse) of A, written A'.

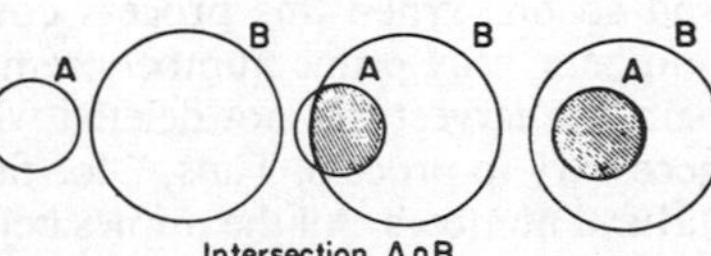

Intersection $A \cap B$

union (sum) of A and B. The smallest set containing all elements in either A or B, written $A \cup B$, read A cup B.

intersection (product) of A and B. The largest set containing elements in both A and B, written $A \cap B$, read A cap B.

overlapping sets. Sets having at least one common element.
disjoint sets. Sets having no common element.
The intersection and the union of sets obey the **associative law** and the **commutative law**. See **Algebra of Sets**; **Bound of Set**; **Cartesian Product**; **Countable Set**; **Countably Infinite Set**; **Denumerable Set**; **Difference of Two Sets**; **Finite Set**; **Partition of Set**; **Separation**; **Truth Set**.

SEXAGESIMAL. Associated with sixty, as in sexagesimal **system of notation.** This system was used by the Babylonians and formed the basis of the measurement of **angle (plane angle)** and the divisions of the **mean solar day**.

SHARING. The process of partition in **division.**

SHEAF OF PLANES. The set of all planes which pass through a given point called the *centre* of the sheaf. See **Pencil of Planes.**

SHEAR. See **Bending Moment, Shear in Horizontal Beam.**

SHEAR TRANSFORMATION. A **translation** in a plane or space in which one coordinate axis or coordinate plane remains unchanged, and movement of any point is (1) parallel to the fixed axis or plane and (2) proportional to its distance from the axis or plane. This shearing is of the form $\{x' = x, y' = y + kx\}$ or $\{x' = x + ky, y' = y\}$. It is one of the **affinities.**

SHEAR TRANSLATION. See **Shear Transformation.**

SHORT RADIUS. Synonym of radius of **inscribed circle of polygon.**

SIDEREAL DAY. The time interval between two successive transits of the first point of the constellation Aries: 23 hours 56 minutes 4 seconds in contrast with **mean solar day** of 24 hours.

SIDEREAL YEAR. The period of rotation of the earth about the sun determined by its rotation relative to the stars: 365.256 36 **mean solar days or** 365 days 6 hours 9 minutes 9.5 seconds. $366\frac{1}{4}$ sidereal days are approximately equal to $365\frac{1}{4}$ mean solar days. See **Anomalistic Year**; **Tropical Year.**

SIEVE OF ERATOSTHENES. The process used by Eratosthenes in the third century B.C. to find the **prime numbers**. From a list of the **natural numbers** he crossed out all the second numbers but 2, all the third numbers but 3, and so on. When this process could not be repeated for a given set of numbers, only prime numbers remained. In a sequence of n natural numbers, the largest one not deleted which is less than $\sqrt{n}$ is the last one it is necessary to process. Thus, 7 for finding all prime numbers in the first 100 natural numbers. All the primes below 10^9 have been found by this process.

SIGN OF AGGREGATION. See **Aggregation.**

SIGN OF OPERATION. A symbol used in association with an **operator** to denote an **operation.**

SIGNED MINOR. Synonym of cofactor of a **determinant.**

SIGNIFICANCE TESTS. In **statistical analysis** it is important to be able tc decide whether any **deviation** from the normal shown by a **sample** of a **population** can be accounted for by accidental variations or must be ascribed to some systematic influence at work. The criteria used, which enable a **probability** of randomness to be calculated for the sample, are known as significance tests.

SIGNIFICANT FIGURES. The specified number of figures in any number expressed as an **integer** and/or a decimal **fraction**, which, used with their place value, is an accepted **approximation** for the number. Thus, 92,897,000 miles is 92,800,000 miles to three significant figures; 3.141 5926 is 3.14 to three significant figures; 0.090,909,09 is 0.0909 to three significant figures. One does not 'correct' to significant figures: 0.0909 is 0.091 correct to three decimal places.

SIMILAR FIGURES. Plane or solid geometrical figures which can be made to correspond in such a way that the **ratio** of distances between pairs of points of one figure and distances between corresponding pairs of points in the other figure are in constant ratio.

SIMILAR FRACTIONS. **Fractions** having a common denominator. Common fractions such as 3/15, 8/15 or decimal fractions such as 0.13, 0.85 (denominator 100).

SIMILAR (LIKE) TERMS. In general, terms in an expression which differ in their numerical **coefficients** only. The sum of similar terms can always be expressed as a single term, e.g., $7ab^2c^3 + 3ab^2c^3 = 10ab^2c^3$. If from the context of a problem certain letters are known to be literal coefficients, then some terms may be referred to as similar (like) terms because they differ only in the numerical and literal coefficients, e.g., $3x^2 + 5x + a + bx^2 + cx + 5 = (3 + b)x^2 + (5 + c)x + (a + 5)$. See **Collection of Terms.**

SIMILAR POLYGONS. **Polygons** which are **similar figures.** Corresponding angles are equal and corresponding sides are in **proportion.**

SIMILAR SOLIDS. Solids bounded by **similar surfaces.**

SIMILAR SURFACES. Surfaces that can be made to correspond in such a way that distances between pairs of points on one surface and distances between corresponding pairs of points on the other surface are in constant **ratio.**

SIMILAR TRIANGLES. **Triangles** which are **similar figures.** Corresponding angles are equal and corresponding sides are in **proportion.**

SIMILITUDE. If two **similar figures** are drawn so that a pair of corresponding sides are parallel, the common point of the joins of their corresponding ends will be a *centre of similitude*. The lines drawn through this point and any point on one figure will pass through the corresponding point on the other. Such **radially related figures** have linear dimensions in the **ratio** of two corresponding **rays.** Thus figures α and β have similitude because segments 1 and 2 are parallel and fix centre of similitude C. See **Transformation of Similitude.**

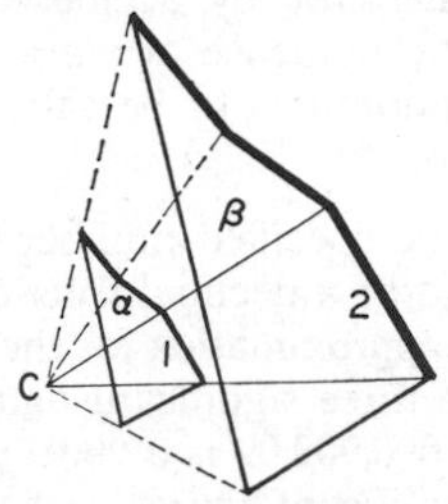

Simple Closed Curve. A **closed curve** which does not intersect itself.

Simple Expression. Synonym of *monomial expression.* See **Multinomial Expression.**

Simple Harmonic Motion (S.H.M.). Motion in a straight line with the condition that **acceleration** is proportional to the distance from a fixed point in the line and is directed towards this point. This is the type of motion associated with springs, tuning forks and the simple pendulum. This type of motion is referred to as an **oscillation.** It is associated with the uniform movement of a point on the circumference of a circle. If P is the projection of P', and P moves backwards and forwards along a diameter while P' moves uniformly round the circumference with **angular velocity** ω, then the displacement of P from centre O is $r \cos \theta$, i.e. $r \cos \omega t$. If $OP = x$, $\dot{x} = -r\omega \sin \omega t$, and $\ddot{x} = -r\omega^2 \cos \omega t = -\omega^2 x$. Angle POP' is called the *phase* of the motion; AA', the *range*; OA, the **amplitude**. The *period* is that time taken by P' to complete a circuit or by P to travel a distance equal to two diameters. It is $2\pi/\omega$. The *frequency* of the oscillation is $\omega/2\pi$. See **Pendulum, Simple; Phase in Simple Harmonic Motion.**

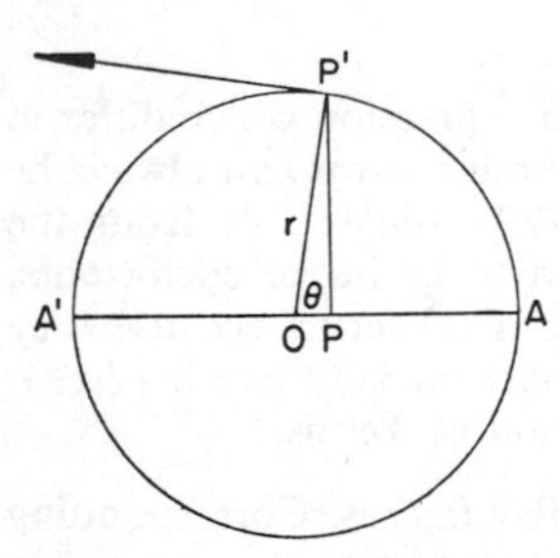

Simple Interest. The type of **interest** in which the principal remains unchanged throughout the period of the loan. If P is the principal, r the rate per cent of interest and n the number of payments, the simple interest is $Prn/100$.

Simple Pendulum. See **Pendulum, Simple.**

Simple Practice. See **Practice.**

Simple Proportion. The equality of two **ratios** gives four quantities in simple **proportion.** If any three of the quantities are known the fourth can be found. If $a/b = c/d$, $a = b(c/d)$; $b = a(d/c)$; $c = d(a/b)$; $d = c(b/a)$.

SIMPLEX. A term used in **topology**. An n-simplex is the simplest geometric figure in an n-dimensional space. For example, point, line, triangle, tetrahedron, in zero, one, two, three dimensional space.

SIMPSON'S RULE. A rule for determining the area of a shape with irregular boundaries. It assumes that the curved outline can be replaced by consecutive parabolic arcs. The area is divided into a number of strips of equal width as shown. The area of portion $AA'N'N$ is then $\frac{1}{3}d(y_1+4y_2+2y_3+4y_4+\ldots+2y_{n-2}+4y_{n-1}+y_n)$. This is a closer approximation to the truth than the one obtained by using the **trapezoidal rule.**

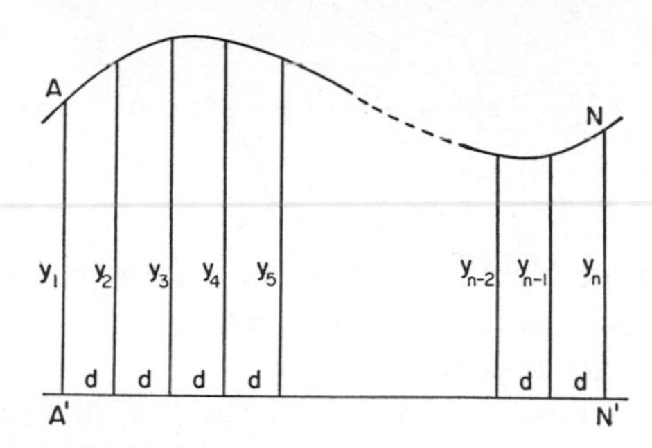

SIMSON'S LINE. See **Pedal Triangle.**

SIMULTANEOUS EQUATIONS. Two or more equations that are true at one and the same time and are therefore satisfied by the same values of the unknown quantities involved. See **Cramer's Rule.**

SIN. Abr. sine. See **Trigonometric Ratios.**

SINH. Abr. hyperbolic sine. See **Hyperbolic Functions.**

SINE CURVE. The graph of the equation $y=\sin x$. It is repetitive every 2π radians (360°) and $-1\leqslant y\leqslant +1$. See **Circular Functions.**

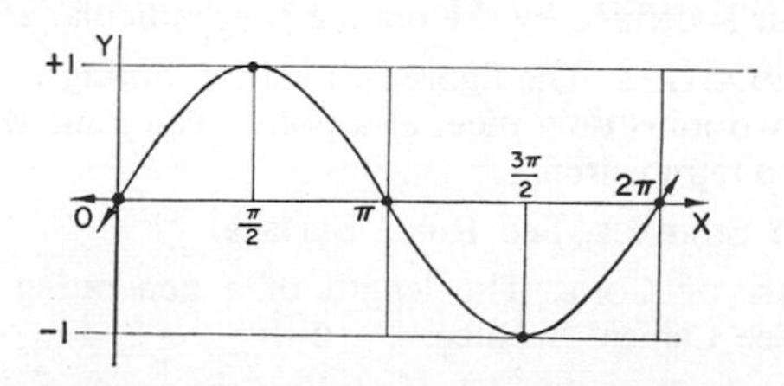

SINE FORMULAE (SINE RULE). In a triangle $AcBaCbA$, $a/\sin A=b/\sin B=c/\sin C$. The formulae can be derived by alternative methods.

(1) The perpendicular p_1, drawn from A to a is both $c\sin B$ and $b\sin C$, hence $c/\sin C=b/\sin B$. The perpendicular p_2, drawn from B to b is both $c\sin A$ and $a\sin C$, hence $c/\sin C=a/\sin A$.

(2) If BA' is a diameter of the circumcircle ABC, $A'C$ will be perpendicular to BC and angle A' = angle A. Hence, $\sin A = \sin A' = a/2R$, and $2R = a/\sin A$. Similarly, $2R = b/\sin B$, etc.

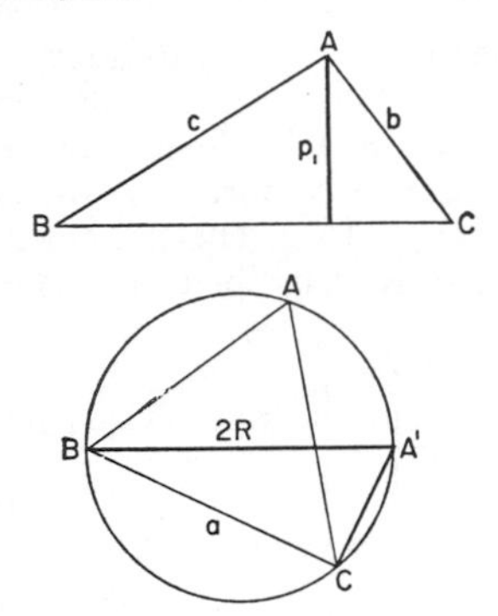

SINE, HYPERBOLIC (sinh). $\frac{1}{2}(e^{x} - e^{-x})$. See **Hyperbolic Functions.**

SINE RATIO (sin). See **Trigonometric Ratios.**

SINGLE ROOT. See **Repeated Roots.**

SINGLE-VALUED FUNCTION. A function $y = f(x)$, in which y has one value only for each value assigned to x. Synonymous with **function** when a **multi-valued function** is defined as a **relation.** For each value of x there is a unique value of y.

SINGULAR POINT. A point on a curve where the **gradient of curve** dy/dx is indeterminate.

SINUSOIDAL. Shaped like a **sine curve,** $y = \sin x$.

SKEW CURVE. See **Twisted (Skew) Curve.**

SKEW DISTRIBUTION. See **Distribution.**

SKEW LINES. Non-intersecting, non-parallel lines in space. The distance between any pair is defined by the unique perpendicular to both.

SKEW QUADRILATERAL. The figure formed by joining four non-**coplanar points** so that two joins only meet at a point. The joins will form four of the six edges of a **tetrahedron.**

SKEW RULED SURFACE. See **Ruled Surface.**

SLANT HEIGHT OF CONE. The length of a **generating line** of a right circular **cone.** See **Conical Surface.**

SLANT HEIGHT OF PYRAMID. The length of the median through the apex of any lateral face of a regular **pyramid.** See **Median of Triangle.**

SLANT HEIGHT ON PLANE. If AB is a horizontal line on an **inclined plane,** the slant height of a point P on the plane is the perpendicular distance from P to AB.

SLIDE RULE. A mechanical device consisting of two **logarithmic scales** which can slide relative to each other. The multiplication or division of numbers is achieved by the addition or subtraction of the logarithms of the numbers which are represented as lengths on the scale.

SLOPE ANGLE. Synonym of **angle of inclination.**

SLOPE OF CURVE. Synonym of **gradient of curve.**

SLOPE OF LINE. Synonym of **gradient of line.**

SMALL CIRCLE. The section of the surface of a sphere made by a plane which does not pass through the centre of the sphere. The centre of a small circle is defined as the nearer pole and its radius as the arc of a great circle (and its length) from that pole to the small circle. See **Pole of Circle on Sphere.**

SMOOTH SURFACE. A surface is said to be smooth when the reaction between it and a body moving on it in any way is always in the direction of the **normal** to the surface.

SOLAR DAY. See **Apparent Solar Noon; Apparent Solar Time.**

SOLAR TIME. See **Apparent Solar Time; Mean Solar Day.**

SOLID ANGLE. The geometrical configuration formed by all the **half-lines** which share a common vertex and pass through a **simple closed curve**. The solid angle subtended at a point P, by a portion of a surface S, is measured by the area u. This is that portion of the surface of a sphere having centre P, and radius 1, lying within the solid angle. The unit of solid angle is the *steradian*, and the largest solid angle contains 4π steradians. If the steradian is defined by the solid angle at the vertex of a **right circular cone**, the plane angle between opposite generating lines of the cone is about 65° 32′. See **Angle (Plane Angle).**

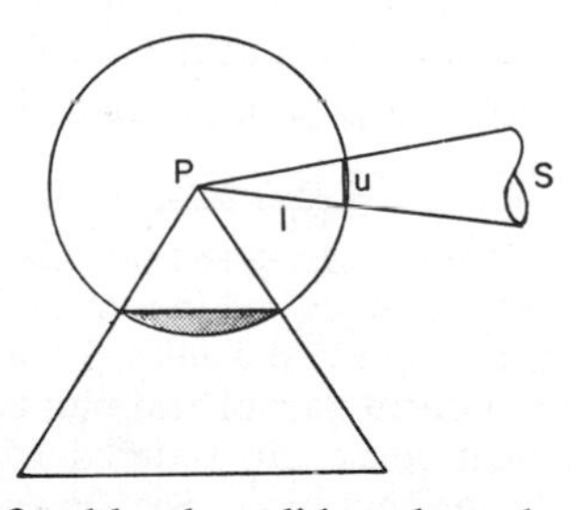

SOLID BODY. The concept of one state of matter when we distinguish it from liquid body (shape of containing vessel and with free surface part of a geoidic shell) and gaseous body which must be wholly contained to prevent its dispersion throughout the atmosphere.

SOLID GEOMETRY. The study of geometrical **configurations** in space. See **Geometry.**

SOLID IN GEOMETRY. A body occupying a portion of three-dimensional space and therefore bounded by a closed surface which may be curved (e.g., sphere), curved and planar (e.g., cylinder) or planar (e.g., cube). Only its spatial relations are studied.

SOLID IN MECHANICS. Synonym of *body*. A piece of matter which occupies a portion of three-dimensional space. **Forces** can be recognized by the effects they have on the state of rest or uniform motion in a straight line of a body. Sometimes a body is referred to as a **mass**, when mass is loosely defined as the quantity of matter in a body. See **Rigid Body**.

SOLID OF REVOLUTION. When a plane surface is rotated through a complete revolution about a line, the *axis of revolution*, in the plane, a solid of revolution is generated. The perimeter generates a *surface of revolution*. The integral calculus is used to determine the *area of revolution* and the *volume of revolution*. If the curve $y=f(x)$ between $x=a$ and $x=b$ is rotated about the x-axis, a surface of revolution is generated, the area of which is given by $A=\int_a^b 2\pi y\sqrt{1+(dy/dx)^2}dx$. The volume bounded by this surface of revolution and the circles generated by the ordinates $f(a)$ and $f(b)$ is given by $V=\int_a^b \pi y^2 dx$. See **Pappus's Theorems**.

SOLIDUS. The oblique line used sometimes in a **common fraction** to separate the numerator and the denominator, thus, N/D.

SOLUTION OF EQUATION. See **Root of Equation**.

SOLUTION OF QUADRATIC EQUATION. See **Completing the Square**; **Quadratic Equation, Formula**.

SOLUTION OF SIMULTANEOUS EQUATIONS. See **Cramer's Rule**.

SPACE CURVE. See **Curve**.

SPACE IN ELEMENTARY GEOMETRY. The space of the physical environment as conceived in terms of direction and distance. Points can be represented by a **real number** (x), a linear coordinate in one-dimensional space; an ordered pair of real numbers (x, y), planar coordinates in two-dimensional space; an ordered triple (x, y, z), spatial coordinates in three-dimensional space. See **Dimension of Space**.

SPACE IN MODERN GEOMETRY. A set of elements or points which satisfy a set of **postulates**. Such points can be represented in a coordinate system by 1, 2, 3, . . . n **real numbers** according to whether the set of points belong to a 1, 2, 3, . . . n-dimensional space. The **space in elementary geometry** is one of such spaces. See **Dimension of Space**; **Euclidean Space**; **Non-Euclidean Space**.

SPECIFIC GRAVITY. Synonym of *relative density*. The ratio of the **density of substance** to that of some specified substance; usually in the case of solids and liquids, water at 4°C, and in the case of gases, air, hydrogen or oxygen at 0°C with barometric pressure 760 mm.

SPECIFIC HEAT. The number of **calories** needed to raise the temperature of 1 gramme of a substance 1°C, or the number of **British thermal units** needed to raise the temperature of 1 lb of the substance 1°F is called the *thermal capacity* of the substance. The comparison of the thermal capacity of any substance with the thermal capacity of water gives the ratio known as the specific heat of the substance. (The relation between thermal capacity and specific heat is analogous to that between density and specific gravity.)

SPEED. The rate of change of displacement with respect to time. The direction of motion is not specified as in **velocity**. If the distance moved by a point is s the speed of the point is the derivative ds/dt or $\dot{s}$.

average speed. The distance moved by a point in an interval of time divided by the interval of time, written $\Delta s/\Delta t$ or $\delta s/\delta t$.

instantaneous speed. The limit of the average speed of the point as the time interval, Δt, approaches zero.
See **Equations of Motion.**

SPHERE. A solid bounded by a **spherical surface** and generated by a plane semicircle rotating about its diameter. Its volume is $4\pi r^3/3$ or $\pi d^3/6$, where r is the radius, d the diameter of the semicircle. If the centre is the origin of **rectangular coordinates** then the solid is the set of points which satisfy the relation $x^2+y^2+z^2 \leqslant r^2$.

SPHERE, CELESTIAL. See **Celestial Sphere.**

SPHERICAL ANGLE. The geometric configuration on a **spherical surface** consisting of two arcs of **great circles** and their point of intersection called the vertex of the angle. Its measure is that of the **dihedral angle** between the planes of the great circles. See **Spherical Triangle.**

SPHERICAL CAP, ZONE. The portion of the surface of a **sphere** lying between two parallel plane **sections** is a *zone*. A *cap* is a zone in which one of the planes is tangential to the sphere. On the earth, the region between two **parallels of latitude** is a zone. The Arctic Circle is the southern boundary of a cap. The area of a zone or cap is equal to that of the belt on a cylinder perpendicular to the planes and lying between them, the diameters of cylinder and sphere being equal. The area of the cap shown is πdh_1; that of the zone πdh_2. See **Sector of Sphere.**

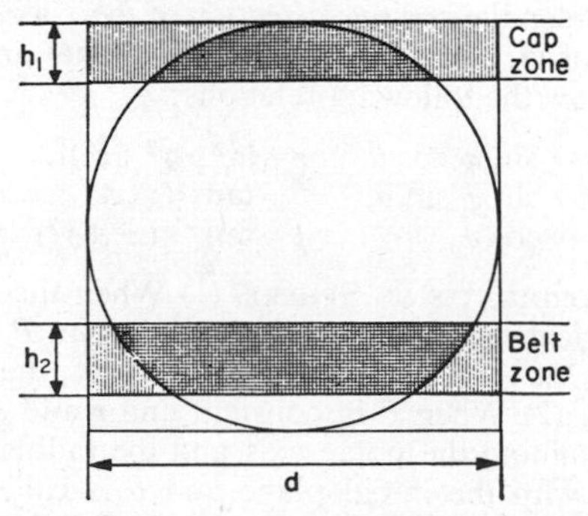

SPHERICAL COORDINATES in ASTRONOMY. The system of **spherical coordinates in geodesy** extended to cover the concept of their projection from the centre of the earth on to the **celestial sphere.** The celestial north pole is near the North Star; the celestial south pole is near the Southern Cross constellation. The *celestial equator* is the intersection of the earth's **equatorial plane** and the celestial sphere.

SPHERICAL COORDINATES IN GEODESY. The **spherical coordinates in space** of $P(r, \theta, \phi)$, when the given sphere is the earth. The pole, O, is at the centre of the earth, the initial plane contains the axis of the earth and passes through Greenwich, London. The segment OP is a radius of the earth and is assumed to be constant. The initial plane is called the *zero meridian plane* and θ, called the *longitude* or *azimuth* of P, is the angle between the zero meridian plane and the projection of OP on the **equatorial plane.** The complement of ϕ, called the *latitude* of P, is the angle between OP and the equatorial plane. The point of zero longitude and zero latitude lies in the Gulf of Guinea.

SPHERICAL COORDINATES IN SPACE. Synonym of polar coordinates in space. The position of a point P in space can be described in terms of its distance from a fixed point O, the angle ϕ between *terminal line* OP and *initial line* OZ and the **dihedral angle** θ between initial plane through OZ and plane containing OP and OZ. The point O is called the *pole*, r the length of OP (the *radius vector*), OZ the *polar axis*, θ the *longitude* or *azimuth* and ϕ the *colatitude* of the system, and the coordinates of P are written (r, θ, ϕ). The spherical (polar) coordinate system is related to the rectangular coordinate system by associating the pole with the origin and the polar axis with the z-axis. The initial plane coincides with the zx-plane and θ is the angle between the positive direction of the x-axis and the projection on the xy-plane of OP. The **rectangular coordinates** and the spherical coordinates of P show the following relations:

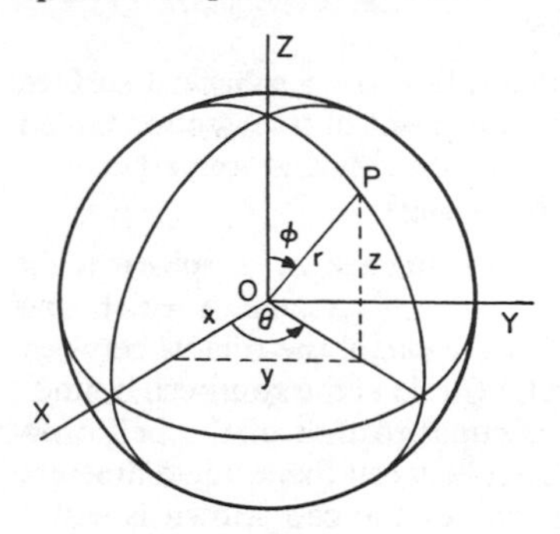

$$x = r \sin\phi \cos\theta, \quad r = (x^2 + y^2 + z^2)^{\frac{1}{2}},$$
$$y = r \sin\phi \sin\theta, \quad \theta = \tan^{-1}(y/x),$$
$$z = r \cos\phi, \quad \phi = \tan^{-1}\{(x^2 + y^2)^{\frac{1}{2}}/z\}.$$

SPHERICAL COORDINATES ON SPHERE. (1) When the radius vector, r, of (r, θ, ϕ), the **spherical coordinates in space** of a point P, is constant while θ varies from 0° to 360° and ϕ varies from 0° to 180°, the point P generates a **spherical surface.** (2) When θ is constant and r and ϕ vary, P generates a half plane containing the polar axis and the radius vector. It makes a **dihedral angle** θ with the initial plane and θ is called the *longitude* or *azimuth* of P. (3) When ϕ is constant and r and θ vary, P generates the surface of a **nappe** of a cone, and ϕ is called the *colatitude* of P. For a given sphere, a *meridian* is the intersection of the plane (θ constant) and the surface of the sphere; a *circle of latitude* is the intersection of the cone (ϕ constant) and the surface of the sphere.

SPHERICAL DEGREE. The amount of area on the surface of a sphere equal to 1/720 of the whole **spherical surface.** It is defined by the **birectangular spherical triangle** whose third angle is one (ordinary) degree.

SPHERICAL DISTANCE. Synonym of **geodesic.**

SPHERICAL EXCESS. For a **spherical polygon,** the difference between the sum of the angles of the polygon (measured in degrees) and $(n-2)$ 180°. For a **spherical triangle,** $n=3$, the difference between the sum of its angles and 180°.

SPHERICAL HELIX. Synonym of **loxodromic spiral** (loxodrome), **rhumb line.**

SPHERICAL POLYGON. Part of a **spherical surface** bounded by arcs of **great circles.** It subtends a central **polyhedral angle** at the centre of the sphere. See **Spherical Angle; Spherical Excess; Spherical Triangle.**

SPHERICAL RADIUS OF SMALL CIRCLE. See **Small Circle.**

SPHERICAL SECTOR. See **Sector of Sphere.**

SPHERICAL SEGMENT. See **Segment of Sphere.**

SPHERICAL SURFACE. The set of points equidistant from one point called the centre. It is generated by the arc of a semicircle rotating about its diameter. Its area is $4\pi r^2$ or πd^2, where r and d are the radius and diameter of the semicircle. It is thus four times as extensive as the area of a **great circle** section. If the centre is the origin of **rectangular coordinates,** then the surface is the set of points which satisfy the equation $x^2+y^2+z^2=r^2$. See **Sphere.**

SPHERICAL TRIANGLE. Part of a **spherical surface** bounded by arcs of three **great circles.** The sum of its angles ranges from 180° to 540°. See **Spherical Angle; Spherical Excess; Spherical Polygon.**

SPHERICAL TRIGONOMETRY. The study and measurement of triangles on a sphere in contrast to **trigonometry** of **plane figures.**

SPHERICAL WEDGE. Any one of four parts of a **sphere** formed when two **great circle** planes have a common diameter line of intersection. The curved portion of the surface, referred to as the base is a *lune* of the **spherical surface.** If r is the radius and θ the **dihedral angle** of the planes, the volume of the wedge is $(\theta/270)\pi r^3$

SPHEROID. Synonym of **ellipsoid of revolution.**

SPINODE. Synonym of **cusp.**

SPIRAL. A curve described by the rotation of a point about a fixed point in such a way that its distance (r, the *radius vector*) from the fixed point has a specific relation to the angle of rotation (θ, the *vectorial angle*).

Archimedes' spiral. $r = a\theta$.
logarithmic (equiangular, logistic). $\log r = a\theta$.
parabolic (Fermat's). $r^2 = a\theta$.
hyperbolic (reciprocal). $r\theta = a$.

See **Helix.**

SQUARE IN ALGEBRA. The second power, represented by the exponent 2, as x^2, the square of x, standing for the product $x\ x$.

SQUARE IN GEOMETRY. A **quadrilateral** with four equal sides and four right angles. Its opposite sides are parallel; its diagonals are equal in length and bisect each other at right angles.

SQUARE INCH (in²). A unit of **area** equal to that of a square with a one inch side (one inch square).

SQUARE MEASURE. The measurement of **area.**

SQUARE NUMBERS. 1, 4, 9, 16, etc. Numbers which may be arranged in a square (geometric) **array**. Any square number, n^2, is the sum of the first n odd numbers.

$1 = 1$	first odd number,	1^2
$4 = 1 + 3$	first two odd numbers,	2^2
$9 = 1 + 3 + 5$	first three,	3^2
$16 = 1 + 3 + 5 + 7$	first four,	4^2

See **Gnomonic Numbers; Polygonal Numbers.**

SQUARE OF BINOMIAL. An expression of the form $(a \pm b)^2$ expanded to give $a^2 \pm 2ab + b^2$, examples of a *perfect trinomial square.* The terms $(\pm 2ab + b^2)$ are referred to as a **gnomon** and can represent a **gnomonic number.**

SQUARE ROOT. The square root of a number is that number which when multiplied by itself produces the given number. The solution of the equation $x^2 - n = 0$, where n is the given number. If n is positive and real x has two real values, equal in magnitude, opposite in sign. If n is negative, x has two values which are **imaginary numbers.**

SQUARING THE CIRCLE. See **Quadrature in Geometry.**

STABLE EQUILIBRIUM. See **Equilibrium.**

STADE. A Greek unit of length equal to 125 paces or 625 feet. A foot was a length of approximately 12.13 British inches; hence the stade was approximately .1196 mile.

STANDARD DEVIATION (S.D., σ). The square root of the mean value of the squares of individual deviations, calculated from the formula $\sigma=\sqrt{\{(\Sigma x^2)/n\}}$, where n is the number of cases under consideration and Σx^2 the sum of the squares of the n deviations. See **Deviation in Statistics.**

STANDARD FORM. An accepted way of writing an equation, expression, formula, etc. Examples: equation of circle, $x^2+y^2=a^2$; binomial expression, $a^2+2ab+b^2$; formula for solution of quadratic equation, $x=\{-b\pm\sqrt{(b^2-4ac)}\}/2a$; multiples of 10, $215{,}000=2.15\times 10^5$.

STANDARD TIME. Synonym of *zone time*. The time based on the **mean solar day** of some standard **meridian of longitude** is chosen as the civil time for the zone the meridian crosses. The extreme eastern and western solar times in the zone may differ by more than an hour. The times in two contiguous zones differ by one hour, the more easterly one being an hour in advance.

STANDARD TRIANGLE. The standard nomenclature due to Euler, *AcBaCbA* in which vertices *A, B, C* are opposite sides *a, b, c*, the six letters being written anti-clockwise around the triangle. The triangle is referred to equally well as $\triangle ABC$ or as $\triangle abc$. The interior angles (internal angles) are $\angle A$, $\angle B$, $\angle C$.

STANDING (STATIONARY) Waves. A phenomenon which occurs when a travelling wave (e.g., light wave or sound wave) is reflected to travel the same path. The incident and reflected travelling waves reinforce one another at certain points and neutralize one another at others. The resulting **oscillation** is a *standing wave*. The points where there are zero displacement and zero velocity in the medium are called *nodes*; the points where there is maximum displacement and velocity are called *anti-nodes*.

STATICS. See **Dynamics; Mechanics.**

STATIONARY POINTS, VALUES. See **Maximum, Minimum Values.**

STATIONARY TANGENT. A tangent through a stationary point. See **Maximum, Minimum Values.**

STATISTICAL ANALYSIS. A term often used synonymously for **statistics.** The use of the various principles of statistics.

STATISTICS. The study of the methods of collecting and analysing data. The subject has various aspects: classification of data; inference involving probability; methods of collecting data for inference (**sampling**). See **Inference in Statistics; Probability in Statistics; Significant Tests.**

STATUTE MILE. British Imperial and U.S. standard unit of length; equal to 8 furlongs, 1 760 **yards.** Its metric equivalent is 1.609 3 Km. (The popular conversion is 5 miles ≡ 8 Km.) See **Geographical Mile; Nautical Mile.**

STERADIAN. The unit of measure of a **solid angle.**

STEREOGRAPHIC ZENITHAL PROJECTION. See **Map Projection.**

STIRLING'S THEOREM. For large n, approximately, $n! = n^n e^{-n}\sqrt{(2\pi n)}$. (For 8! this gives a result 1 per cent short; for 20! this gives a result less than $\frac{1}{2}$ per cent short.) The theorem is necessary for deriving Gauss's Normal Law. See **Factorial *n* (*n*! $\lfloor n$); Gauss-Laplace Law.**

STRAIGHT LINE. The line, in **Euclidean geometry,** which passes through two points in such a way that the length of the segment between the points is a minimum. Both words have significance derived from the phrase 'a *stretched linen* thread'. It is an ideal concept closely approximated to in the **plumb-line,** a ray of light and the edges of crystals. In general the term *line* is synonymous with *straight line*. See **Rope Stretchers' Triangle.**

STRAIN. A measure of deformation in a non-rigid body associated with external forces and the **stress** involved. Strain may be *linear*, *superficial* or *volumetric* and is computed as deformation per unit of length, area or volume. See **Bulk Modulus; Young's Modulus.**

STRATA IN STATISTICS. See **Sample.**

STRATIFIED SAMPLE. See **Sample.**

STRESS. Internal forces of a body subjected to external forces. Stress is always associated with an accompanying deformation of the body, called the **strain.** Stress does not occur in a theoretical ideal solid, a rigid body. Stresses may be *tensile*, *compressive* or *shear*, and are computed as force per unit area of cross-section. See **Bulk Modulus; Young's Modulus.**

STRIKE. See **Dip.**

STROPHOID. Let A and B be two fixed points, and a variable line through A cut a fixed curve C at Q. If P and P' are points on the line AQ such that $PQ = QP' = QB$, the locus of P and P' is called the *strophoid* of curve C with respect to the pole A and the fixed point B. The strophoid has various forms depending on the form of the fixed curve.

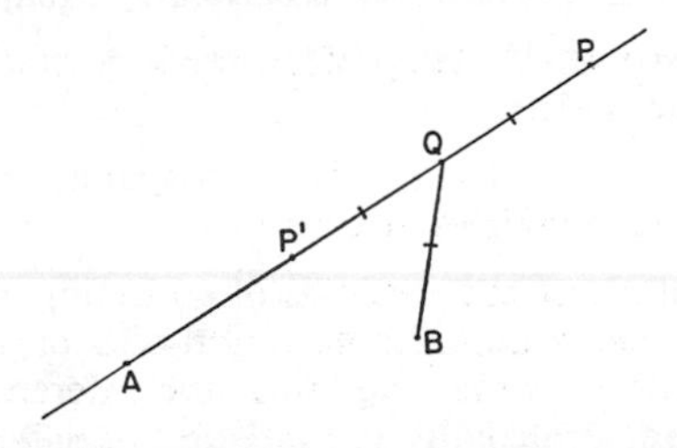

STRUCTURE. Any arrangement of **struts and ties** articulated so as to form a framework for some special purpose such as supporting a roof or carrying a roadway. See **Funicular Polygon.**

STRUT AND TIE. A rod in a **structure** may be a strut exerting forces outwards along its length and thus preventing deformation, or a tie exerting forces inwards. A *thin* rod or a wire can act only as a tie.

SUBFIELD. Any subset of a field which is itself a field. Any subset of a field which has no proper subset is called a **prime field**. See **Field Theory.**

SUBGROUP. Any subset of a group which itself is a group. See **Group Theory.**

SUBNORMAL, SUBTANGENT. For any point on a curve, the **orthogonal projections** of the normal and the tangent on to the x-axis between the limits shown.

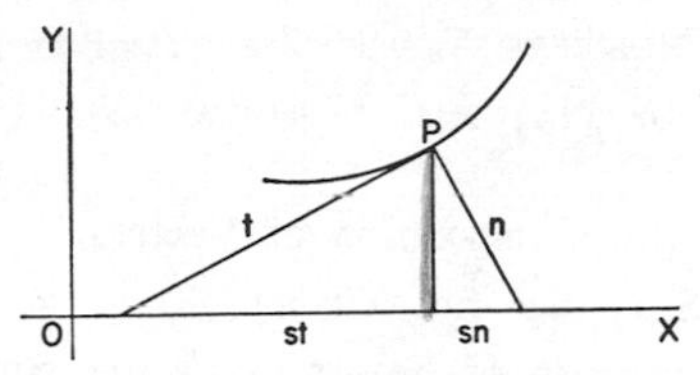

SUBSCRIPT, SUPERSCRIPT. Any symbol placed below or above and normally to the right of another symbol to distinguish it from the same symbol with a different meaning. Examples are: (1) v_0 and v_t are velocities at 0 seconds and t seconds; (2) y' and y'' are first and second **derivatives** of y with respect to x with special cases, $\dot{x}$ and $\ddot{x}$ for dx/dt **(velocity)** and d^2x/dt^2 **(acceleration)**; (3) D_1, $D_2 \ldots D_n$ are the first, second . . . nth derivatives of a function; (4) ${}_nP_r$ or ${}^nP_r$ is the number of **permutations** of n things taken r at a time. See **Primary, Secondary Symbols.**

SUBSET. See **Set Theory.**

SUBTANGENT. See **Subnormal, Subtangent.**

SUBTEND. (1) If two points A, B on a curve are joined by line segments to a point P in the plane of the curve, then arc AB or chord AB is said to subtend the angle APB at P. (2) If a **solid angle** is formed at a point P by the **half-lines** through P and a **simple closed curve** C, then the curve C or any surface segment bounded by C is said to subtend the solid angle at P.

SUBTRACTION. A general term for the **binary operation** equivalent to determining the remainder when a part of a whole is removed; the reverse operation of **addition.** It can also be interpreted as the process of determining the difference in size of two groups. It is expressed in the form $a - b$, where a is the *minuend* and b the *subtrahend.* Fundamentally, the operation is an abstraction from concrete experience with real objects and is expressed in terms of **natural numbers,** for which it cannot occur when a is less than b. The concept of subtraction is developed by extending the number concept to embrace rational numbers (fractions). The concept is further developed by the **algebraic subtraction** of **directed numbers,** and further still by its application to the field of **complex numbers.**

The term subtraction can be applied to fields outside traditional arithmetic. A set of things may be separated in a subtractive sense into two sets, as a reverse process to union (or sum) in **set theory.**

SUCCESSIVE ADDITION. Synonym of extended count, one aspect of **multiplication.**

SUCCESSIVE SUBTRACTION. Synonym of quotition (grouping), one aspect of **division.**

SUM. The result of addition; synonym of total. See **Aggregate.**

SUM OF CUBES OF NATURAL NUMBERS. $S_n = 1^3 + 2^3 + \ldots + n^3 = \{\frac{1}{2}n(n+1)\}^2$. This is the square of the **sum of natural numbers.**

SUM OF NATURAL NUMBERS. $S_n = 1 + 2 + \ldots + n = \frac{1}{2}n(n+1)$.

SUM OF SQUARES OF NATURAL NUMBERS. $S_n = 1^2 + 2^2 + \ldots + n^2 = \frac{1}{6}n(n+1)(2n+1)$.

SUM OF VECTORS. See **Composition of Vectors.**

SUM (UNION) OF SETS. See **Set Theory.**

SUMMATION. The process of determining the sum of a **series.** See **Convergence, Divergence of Series.**

SUMMATION SIGN. The Greek capital letter *sigma*, Σ, is used to denote the sum of a series. Thus, the sum of the series of n terms which has a general term a_r is written symbolically and in expanded form:

$$\sum_{r=1}^{n} a_r = a_1 + a_2 + a_3 + \ldots + a_r + \ldots + a_n.$$

The expanded form is sometimes described as S_n.

sup. Abr. least upper bound. See **Bound of Set.**

SUPERFICIAL. See **Coefficient of Superficial Expansion.**

SUPERPOSABLE. Synonym of congruent. See **Superposition.**

SUPERPOSE. To place one figure on another so that corresponding elements coincide. See **Congruence (Congruency) in Geometry.**

SUPERPOSITION. The axiom of superposition states that any figure can be moved in space without distortion of shape or size. This is a necessary condition for *superposable* figures being **congruent.**

SUPPLEMENTAL CHORDS. In a circle or an ellipse, chords joining ends of any diameter to a point on the perimeter. Two diameters parallel to the chords are a pair of **conjugate diameters.**

SUPPLEMENTARY ANGLES. Any pair, the sum of which is two right angles (180° or π radians). One angle is the supplement of the other.

SURD. A numerical expression containing an **irrational number.** Sometimes used as a synonym of irrational number. Surds are quadratic, cubic, quartic, etc., as the index of the radical is 2 (understood), 3, 4, etc. (e.g., $\sqrt{3}$, $\sqrt[3]{4}$, $\sqrt[4]{5}$, etc.).

mixed surd. One containing a rational factor (e.g., $7\sqrt{2}$) or a rational term (e.g., $7 - \sqrt{2}$).

pure surd. One in which each term is a surd (e.g., $3\sqrt{5}+\sqrt{2}$).
entire surd. One containing no rational factor or term.

SURFACE. A set of points forming a space which has only two dimensions. The surface may be planar or curved. When the curved surface is closed it forms the boundary of a **solid.** See **Development of Solids; Ruled Surface.**

SURFACE OF REVOLUTION. See **Solid of Revolution.**

SURVEY. The process of mapping a given region of land. The techniques include **triangulation, traversing** and **levelling.**

SYMBOL. Any letter or device used to represent a quantity, relation or operation.

SYMMETRIC DIFFERENCE OF TWO SETS. See **Difference of Two Sets.**

SYMMETRIC DISTRIBUTION. See **Distribution.**

SYMMETRIC EQUATION. An equation in two **variables,** x and y, which is unaltered when x and y are interchanged; e.g., $xy=4$; $x^2+y^2=9$. The graph of any symmetric equation has **axial symmetry** with respect to the line $y=x$.

SYMMETRIC EXPRESSION. An expression of two or more **variables,** in which the exchange of one variable for another does not alter the total value; e.g., $\tan(A+B)$ is symmetric, $\tan(A-B)$ is not.

SYMMETRIC FUNCTION. An expression in two or more **variables,** which remains unaltered in value when any two variables are interchanged; e.g., $x^2+y^2+z^2+xyz$. If the function remains unaltered under **cyclic order** changes, the function is **cyclosymmetric.**

SYMMETRIC FUNCTIONS OF ROOTS. If a **polynomial expression** of the nth degree is factorized as

$$f(x)=a(x-\alpha_1)(x-\alpha_2)(x-\alpha_3)\ldots(x-\alpha_n),$$

then $\alpha_1, \alpha_2, \alpha_3, \ldots \alpha_n$ are the roots of the equation $f(x)=0$. If the equation $f(x)=0$ is

$$a\ x^n+b\ x^{n-1}+c\ x^{n-2}+d\ x^{n-3}+\ldots=0,$$

the sum of the roots taken one at a time is $-b/a$; taken two at a time, c/a; taken three at a time, $-d/a$; etc. The sums are alternately negative and positive. *Three at a time* means that the terms are *products* of three roots. Thus, for a quadratic,

$$\alpha_1+\alpha_2=-b/a, \qquad \alpha_1\alpha_2=c/a,$$

and a cubic,

$$\alpha_1+\alpha_2+\alpha_3=-b/a,\ \alpha_1\alpha_2+\alpha_2\alpha_3+\alpha_3\alpha_1=c/a,\ \alpha_1\alpha_2\alpha_3=-d/a.$$

These are all **symmetric functions** of the roots.

SYMMETRIC GROUP. See **Group Theory.**

SYMMETRIC PAIR OF EQUATIONS. A pair of equations which remains unaltered as a pair when the variables are interchanged. For example, the equations of a pair of circles:

$$x^2+y^2+4x+3y+2=0 \text{ and } x^2+y^2+3x+4y+2=0.$$

SYMMETRIC RELATION. A **relation** which is identical to its own **inverse relation**. Thus, for a given relation R, $xRy \Leftrightarrow xR^{-1}y$. Examples of symmetric relations are parallelism and perpendicularity: $a//b \Leftrightarrow b//a$; $a \perp b \Leftrightarrow b \perp a$. ($A$ is married to B) $\Leftrightarrow$ (B is married to A). A relation which is not symmetric is **asymmetric**.

SYMMETRICAL, SYMMETRIC. Having **symmetry**.

SYMMETRY. Two points, P and P', are symmetrical with respect to a plane p, a line l (*axial symmetry*) or a point O (*central symmetry*). P and P' correspond under **reflection** in the plane p, in the line l or in the point O, respectively. See **Congruent Figures in Space**; **Group of Symmetries**; **Rotational Symmetry**.

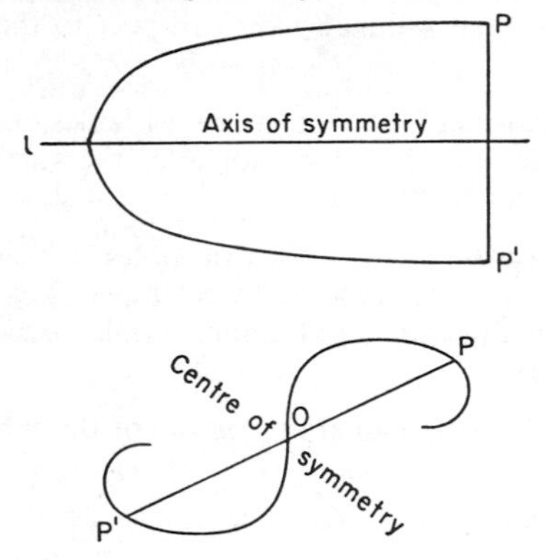

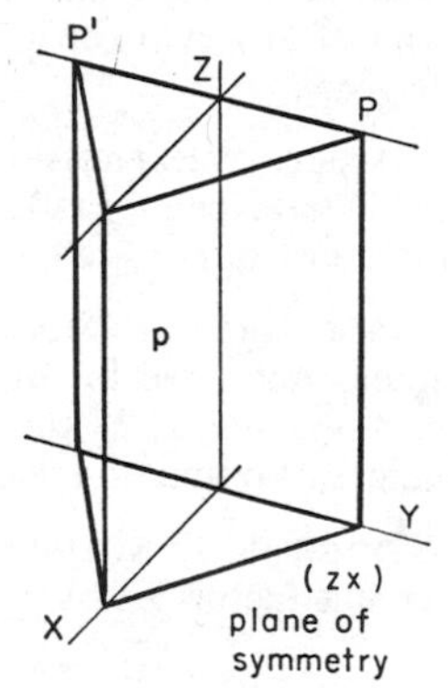

SYSTEM. A set of elements which have a common property; e.g., system of positive integers; system of rays from a common centre. Also any set of elements, their properties, relationships and principles involved, which form an integrated study; e.g., a **system of coordinates**; a **system of notation**; **group theory**.

SYSTEM OF COORDINATES. A method of determining the position of any point by using its coordinates and a **frame of reference**. See **Coordinates of Point**.

SYSTEM OF NOTATION. A method of denoting real numbers by a **scale** of notation together with the rules of computation in that system. The *common scale of notation* is the **denary number system** in which each digit used has *place value* which is a power of 10. Ten is called the *base* or *radix* of the system. Other systems of notation are the **binary number system** (base, 2), *ternary* (3), *quaternary* (4), *quinary* (5), *senary* (6), *septenary* (7), *octonary* (*octonal*) (8), *nonary* (9), *undenary* (11), *duodenary* (12). See **Base (Radix) in Number System.**

SYSTEMATIC ERROR. An error in data due to the method of collecting, often attributable to bias, or an error due to incorrect treatment of the data. See **Statistics.**

T

T.B.M. Temporary Bench Mark. See **Bench Marks.**

T SCORE. A standardized score which has a mean of 50 and a **standard deviation** of 10. See **Mean in Statistics.**

TACNODE. Synonym of double **cusp** or point of osculation.

TAN. Abr. tangent. See **Trigonometric Ratios.**

TANH. Abr. hyperbolic tangent. See **Hyperbolic Functions.**

TANGENT CURVES. Two curves which touch at a point P, called the *point of contact*, such that they have the same **tangent line** at P.

TANGENT GRAPH. The graph of the equation $y = \tan x$. It is repetitive every π radians (180°). The lines $x = (2n+1)\pi/2$ are **asymptotes.** See **Circular Functions.**

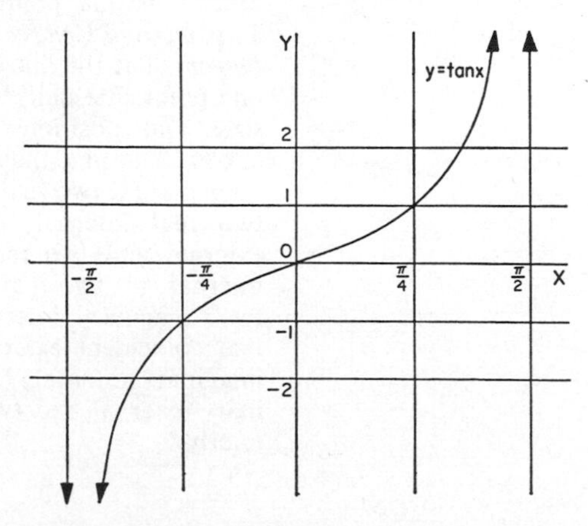

TANGENT, HYPERBOLIC (tanh). The ratio $\sinh x/\cosh x$, which is equal to the **exponential function** $(e^x - e^{-x})/(e^x + e^{-x})$. See **Hyperbolic Functions.**

TANGENT LAW. The relation between the ratio of the difference and the sum of two sides of a triangle, expressed trigonometrically as follows: $(a-b)/(a+b)=\{\tan \frac{1}{2}(A-B)\}/\{\tan \frac{1}{2}(A+B)\}$. It is of great use in logarithmic calculations.

TANGENT LINE. If a fixed point P and a variable point P' lie on the curve of the function $f(x)$, the limiting position of the secant line through P and P' as P' approaches P is the tangent line at the point of contact P. The gradient of the tangent at P and the gradient of the curve at P are both equal to the value of the first **derivative** of the function $f(x)$ for the value of x at P.

TANGENT PLANE. A plane which touches a surface at a point P, called a *point of contact*, such that any line in the plane passing through P is a **tangent line** to the surface at P (and perpendicular to the normal at P). In the cases of conical and cylindrical surfaces the tangent plane touches the surface along an **element of contact**, a straight line.

TANGENT RATIO (tan). See **Trigonometric Ratios.**

TANGENTIAL POLAR EQUATION. If (r_1, a) are the polar coordinates of P and (r_2, β) those of Q on **conic section** with polar equation $l/r=1-e\cos\theta$, then the polar equation of the secant through P and Q is $l/r=\sec \frac{1}{2}(\beta-a).\cos \frac{1}{2}(2\theta-a-\beta) \; -e\cos\theta$. In the limit as $Q\rightarrow P$, $a=\beta$ and the equation becomes $l/r=\cos(\theta-a)-e\cos\theta$. This is the polar equation of the tangent at the point P. See **Pedal Equation; Polar Coordinates in Plane; Polar Equation of Conic Section.**

TANGENTS TO TWO CIRCLES. In general there are four common **tangent lines** to two circles. Two *external* (*direct*) *tangents* cut the line of centres at one point, externally. Two *internal* (*indirect, transverse*) *tangents* cut the line of centres at one point internally. The relative sizes and positions of the two circles determine the nature of the tangents: (1) two real external and two real internal; (2) two real external and two real coincident internal; (3) two real external and two imaginary internal; (4) two real coincident external and two imaginary internal; (5) two imaginary external and two imaginary internal.

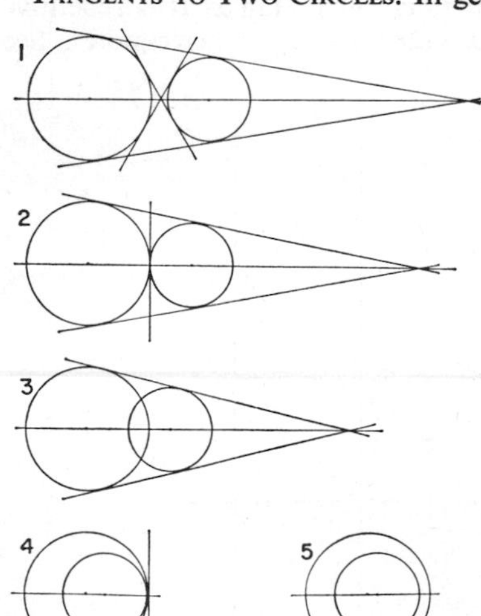

TAXES. Outgoings of cash payments directly to a central government in contrast with **rates** which go to local authorities. Taxes fall into several classes: *direct taxes* levied on income, property, etc.; *excise duties* levied on stamps, postal orders, cheques, legal agreements, etc.; and *indirect taxes* levied on personal spending, customs duties, import charges, etc.

TANGRAM. A mathematical game reputed to have been invented by a legendary Chinese, called Tan, about 4000 years ago. Such a person probably never existed and some students of the game have suggested the name is derived from Cantonese for Chinese, *t'ang*, or from the Chinese *t'an*, to extend. Whatever its origin the puzzle has remained unaltered throughout its long history. A square is cut into seven pieces and then rearranged without overlapping to form realistic designs. The number of possible combinations is infinite; some are intriguing. The figures shown are different arrangements of the seven shapes in the square.

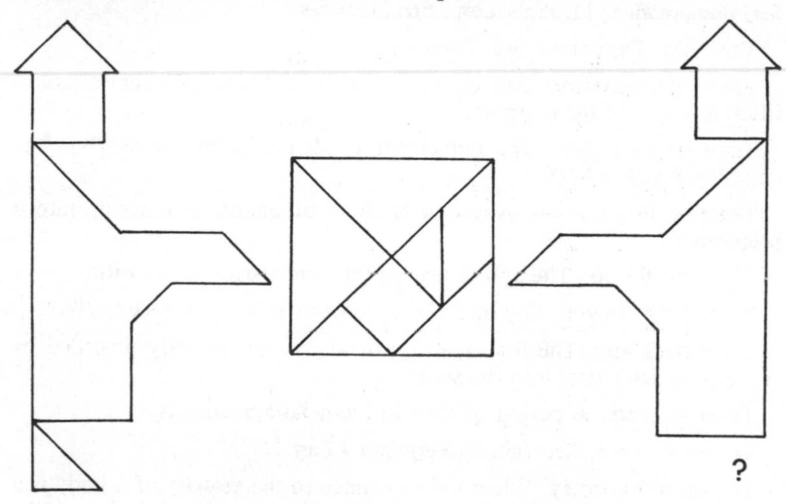

TAYLOR'S THEOREM. For a function of one variable x and a fixed value a for x,

$$f(x)=f(a)+f'(a)\ (x-a)+f''(a)\ (x-a)^2/2! \\ +f'''(a)\ (x-a)^3/3!+\ \dots \\ +f^{(n-1)}(a)\ (x-a)^{n-1}/(n-1)!+R_n.$$

If $a=0$, this theorem is known as *Maclaurin's theorem*:

$$f(x)=f(0)+f'(0)\ x+f''(0)\ x^2/2!+f'''(0)\ x^3/3!+ \\ \dots\ +f^{(n-1)}(0)\ x^{n-1}/(n-1)!+R_n.$$

Throughout, $f'(a)$, $f''(a)$, etc., are the evaluations of first, second, etc., **derivatives** of $f(x)$ for values $x=a$, and R is the remainder after n terms.

TEMPERATURE. The **absolute** measure of the total **kinetic energy** of the molecules of a body. Zero is probably -273.18 on the **celsius scale.**

TEMPERATURE CONVERSION. Any change of 5° celsius is the same as one of 9° Fahrenheit. A **temperature** of −40° is the same in both scales. Hence:

$$x^\circ\,C = \{(x+40)(9/5) - 40\}^\circ\,F,$$
$$y^\circ\,F = \{(y+40)(5/9) - 40\}^\circ\,C.$$

An alternative method is:

$$x^\circ\,C = \{9x/5 + 32\}^\circ\,F,$$
$$y^\circ\,F = \{(y-32)5/9\}^\circ\,C.$$

TENSILE. Appertaining to **tension** as in tensile **stress.**

TENSION. The internal **forces** brought into play to resist the external forces tending to increase the length of a body (springs, tie-rods, etc.). See **Compression; Hooke's Law; Strain; Stress.**

TERCIMAL FRACTION. See **Fraction.**

TERM OF EXPRESSION. For any expression in the form of a sum of quantities, any one of the quantities.

TERM OF FRACTION. The numerator or denominator: N or D in the common **fraction** N/D.

TERM OF PROPORTION. Any one of the four quantities entering into a **proportion.**

TERM OF RATIO. The antecedent or the consequent of the **ratio.**

TERM OF SEQUENCE. Any one of the elements of the **sequence.**

TERM OF SERIES. The first quantity or any other quantity preceded by + or − which enters into the **series.**

TERM OF TIME. A period of time in financial agreements.

TERMINAL LINE. See **Initial, Terminal Lines.**

TERMINAL VELOCITY. When the resistance to the **velocity** of a body is a function of the velocity, a terminal velocity will be reached and the **acceleration** will be zero. Air resistance to a falling body mounts with speed until it balances the **force of gravity**. The body then continues to fall with constant, terminal velocity.

TERMINATING CONTINUED FRACTION. See **Continued Fraction.**

TERMINATING DECIMAL FRACTION. A decimal **fraction** in which the number of digits is finite.

TERNARY FRACTION. See **Fraction.**

TERNARY NUMBER SYSTEM. One which uses the base 3 combined with place value notation. Thus, the denary numbers, 1, 2, 3, 4, etc., are written 1, 2, 10, 11, etc., in the ternary system. The units, tens, hundreds, etc., places are replaced by units, threes, three-squared, etc., places. See **Base (Radix) in Number System.**

TERNARY OPERATION. See **Operation in Set.**

TESSELLATION. Originally a term used to describe the pattern formed by covering a plane surface with congruent squares, from *tessara*, Latin from Greek, four, the cubical tiles used in mosaics. The term is often applied to patterns using congruent **equilateral triangles** or congruent regular **hexagons.** Sometimes the term is used synonymously with *parquet* to describe all possible patterns formed by covering a plane with shapes in some ordered sequence. Occasionally the concept is extended to embrace the **polyhedra** where the surface covered is a continuous surface. Plane tessellations can be classified into three types.

regular. One kind of regular polygon is used. Only three patterns are possible, using triangles, squares or hexagons.

homogeneous (or semi-regular). Regular polygons of any kind are used, but all common vertices (called *nodal points*) must be congruent. If (n_1, n_2, . . . n_k) sided polygons share a nodal point of *order k*, then $k-2$ $(1/n_1+1/n_2+\ .\ .\ .\ 1/n_k)=2$. This relation permits 17 arithmetic possibilities, but ten only can cover a plane entirely, one being used in two possible arrangements. The eleven possible patterns are classified by their order: ternary, $(k, 3)$, (3, 12, 12), (4, 6, 12), (4, 8, 8), (6, 6, 6); quaternary, $(k, 4)$ (3, 6, 3, 6), (3, 4, 6, 4), (4, 4, 4, 4); quinary, $(k, 5)$, (3, 3, 3, 4, 4), (3, 3, 4, 3, 4), (3, 3, 3, 3, 6); senary, $(k, 6)$, (3, 3, 3, 3, 3, 3).

non-homogeneous. These are infinite in variety and include patterns using only one irregular shape (with straight or curved edges), various sizes of one shape, and various non-homogeneous combinations of the homogeneous patterns.

TETRAGON. A **polygon** with four sides. A **quadrilateral.**

TETRAHEDRON. A **polyhedron** with four faces. It is one of the **regular solids** when the faces are regular **trigons** (equilateral triangles).

TETROMINO. A **polyomino** made of four adjacent squares.

THEOREM. The first word in all of Euclid's **enunciations,** from the Greek *theoreo*, I look at. Now synonym of the whole enunciation. It may be a general conclusion based on certain assumptions, which are to be proved; or it may be a **conclusion, proof** of which has already been accepted. See **Conjecture; Proposition.**

THEOREMS. Listed separately are the theorems associated with the following names: **Apollonius**; **De Moivre**; **Desargues**; **Lagrange**; **Maclaurin**; **Pappus**; **Ptolemy**; **Pythagoras**; **Rolle**; **Taylor**; **Torricelli.** See **Binomial Theorem**; **Mean Value Theorem**; **Multinomial Theorem.**

THEORY. The principles involved in the development of some central concept. The **assumptions, axioms, conjectures, postulates, propositions, theorems,** rules of procedure and **proofs,** which unify ideas, observations and experiments into an abstract logical system.

THEORY OF EQUATIONS. The study of **equations**: the methods of solution; the existence or non-existence of roots; the relations between the roots; the relations between the roots and the coefficients. See **Root of Equation.**

THEORY OF GROUPS. See **Group Theory.**

THEORY OF INDICES. The **power of quantity** is usually expressed by adding a superscript after the quantity. Thus, a^3 represents $a \times a \times a$. The laws for the multiplication and division of quantities which are powers of the same quantity are expressed in the identities:

$$a^m \times a^n = a^{m+n};$$
$$a^m \div a^n = a^{m-n};$$
$$(a^m)^n = a^{mn};$$
$$\sqrt[n]{a^m} = a^{m/n}.$$

From these are derived: $a^{-m} = 1/a^m$; $a^1 = a$; $a^0 = 1$.

THEORY OF LOGARITHMS. If $n = b^l$ is a relation between three numbers, then we say that l is the logarithm of the number n to the base b. Since the logarithm is thus an **index**, the laws for the operations with logarithms are essentially the same as those for operations with indices. The index laws expressed in logarithmic form are:

$$\log m\ n = \log m + \log n; \ \log m/n = \log m - \log n;$$
$$\log m^n = n \log m; \ \log \sqrt[n]{m} = (1/n) \log m.$$

The integral part of a logarithm, called its **characteristic**, is zero for log 1, positive for the logarithms of all numbers greater than 1 and negative for all fractions less than 1. There are no logarithms for negative numbers. The fractional part of a logarithm is called its **mantissa.**

THEORY OF NUMBERS. See **Number Theory.**

THEORY OF PARTITIONS. A **partition** of a number is an expression of the form $a_1 + a_2 + a_3 + \ldots + a_k$ where each term is an integer. Thus, partitions of 5 are 5, 4+1, 3+2, 3+1+1, 2+2+1, 2+1+1+1 and 1+1+1+1+1. The number of the partitions of a number is denoted $p(n)$; as above, $p(5) = 7$.

THEORY OF SETS. See **Set Theory.**

THIRD PROPORTIONAL. If a, b, c are numbers or quantities related by the proportion $a/b = b/c$, then c is the third proportional and equals b^2/a.

THREE DIMENSIONAL GEOMETRY. Synonym of **solid geometry.** See **Dimension of Ordinary Space.**

THREE DIMENSIONAL SPACE. See **Dimension of Space.**

TIE ROD. See **Strut and Tie.**

TIME. The concept of continuous existence, duration, succession, as implied by a sequence of events. It is the fourth dimension of mechanics, entering into the space–time continuum of the theory of **relativity.** See **Anomalistic Year; Apparent Solar Time; Calendar; Greenwich Mean Time; Mean Solar Day; Standard Time.**

TON. British Imperial unit of bulk weight, the same as 20 hundredweights or 2 240 lb **avoirdupois.** It is equivalent to 1.120 U.S.A. short tons of 2 000 lb and to 1.016 metric tons **(tonnes).**

TONE. One-sixth part of an **octave.** Any sound considered in terms of its pitch. See **Pitch in Music.**

TONNE. The weight of 1 000 kilogrammes. The counterpart of the **ton** but equal to 0.984 2 British tons. See **Metric System.**

TOPOLOGICAL PROPERTY. Any property of a geometric figure which remains invariant under a **topological transformation.**

TOPOLOGICAL TRANSFORMATION (HOMEOMORPHISM). The **transformation** between two geometrical figures A and B such that there is a continuous **one-one correspondence** in both directions between each point of A and a matching point of B. If such a transformation is possible the figures A and B are *topologically equivalent*. All convex **polyhedra** are topologically equivalent. See **Analysis Situs; Genus; Mapping; Topology.**

TOPOLOGICALLY EQUIVALENT. See **Topological Transformation (Homeomorphism).**

TOPOLOGY. A branch of geometry which studies the **topological properties** of figures. The study of those properties of figures which are unaffected by distortion (e.g., pulling, bending, stretching, etc.) without tearing. See **Analysis Situs.**

TORQUE. The turning effect of a **force** acting on a body is measured by the **moment of force** about an axis of rotation. It is equivalent to the product of the **moment of inertia** and the **angular acceleration.**

TORRICELLI'S THEOREM. The velocity, v, of discharge of water from a small orifice at a depth h below the free surface is given by $v=\sqrt{2gh}$, g being the acceleration due to gravity.

TORUS. Synonym of anchor ring. The solid generated by the revolution of a closed plane surface about an axis in its plane, the axis *not* intersecting the surface. If the two circles shown in the diagram touch one another, the torus is called a solid ring. If r is the radius of cross-section and ρ is the radius of the torus, the volume of the solid will be $2\pi^2\rho r^2$ and the area of its surface, $4\pi^2\rho r$. See **Pappus's Theorems; Solid of Revolution.**

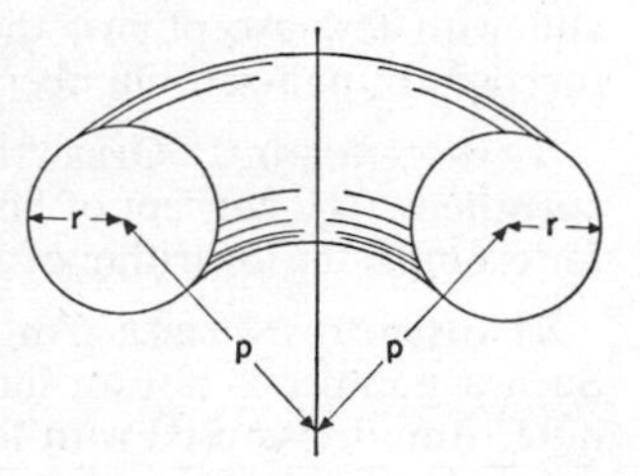

TOTAL. See **Aggregate**; **Sum.**

TOTAL CURVATURE. See **Curvature.**

TOTAL DERIVATIVE. The total derivative of a function of several variables, $V=F(x,y,z,\ldots)$, where $x,y,z,\ldots$ are functions of t, is

$$\frac{dV}{dt}=\frac{\delta V}{\delta x}\frac{dx}{dt}+\frac{\delta V}{\delta y}\frac{dy}{dt}+\frac{\delta V}{\delta z}\frac{dz}{dt}+\ldots$$

TOWER OF HANOI. One of the ancient mathematical games. It consists of a horizontal board with three vertical pegs; and on one peg a set of pierced discs placed in order of size with the largest at the bottom. The game involves moving one disc at a time on to another peg without placing it on a smaller disc, until all discs are transferred to another peg. If there are n discs, 2^n-1 transfers are required. The original tower is in the great temple of Benares, India, and it contains sixty-four gold discs. The priests are reputed to transfer them while keeping watch. If one disc is transferred each second of the day, about 5.82×10^{11} years would be needed to complete the game.

TOWER OF SETS. Synonym of **nested sets.**

TRACE OF LINE. (1) The intersection, called a *piercing point*, of a line with a **coordinate plane.** (2) The **orthogonal** projection of a line on a coordinate plane.

TRACE OF MATRIX. The sum of the elements of a principal diagonal of a **matrix.**

TRACE OF SURFACE. A curve of intersection of a surface with a **coordinate plane.**

TRACING. See **Curve Tracing.**

TRAJECTORY. The path described by a body such as a projectile or a planet. The trajectories of heavenly bodies are referred to as **orbits.** See **Parabolic Motion.**

TRANSCENDENTAL FUNCTIONS. Functions which are not algebraic. See **Circular, Exponential, Logarithmic Functions.**

TRANSCENDENTAL NUMBERS. **Numbers** which are not **algebraic numbers,** such, for example, as e, π, $2^{\sqrt{3}}$ (any irrational power of a rational number); and with few exceptions, the **hyperbolic, logarithmic** and **trigonometric functions** of non-zero numbers.

TRANSCENDENTAL OPERATIONS. Operations which are not **algebraic operations.** The concept of limit is necessary for their complete definition, for example the logarithm of a number.

TRANSFINITE NUMBER. An infinite **cardinal number** or **ordinal number.** Such a number is not an integer, and the rules of operation applicable differ from those used with finite numbers. If A is the transfinite number associated with the simplest of all the infinite classes—that of all natural numbers—and n is a finite number, then

$$A+n=A+A=n\ A=A^n=A.$$

The significance of transfinite numbers was discussed by Galileo in 1638 in his *Dialogues*. Interest in them was reawakened by Bolzano in 1851. It remained for Cantor in 1873 to investigate and establish degrees of infinitude which led to the theory of aggregates. He showed that the class of all real numbers was more numerous than the class of all natural numbers and the transfinite number C is given to it. It is not known whether C is the next transfinite number after A, nor whether the class of all transfinite numbers is itself finite, or of some finite or infinite order of infinitude. Further rules of operation are: $2^A=A^A=C$; $n\ C=C^n=C^A=C$; $2^C=C^C$, the latter is a transfinite number of higher order.

TRANSFORMATION. In general a **correspondence, mapping, function.**

TRANSFORMATION IN GEOMETRY. A transformation of a plane (or space) into itself is a **correspondence** between any point P in the plane (or space) and another point P' in the plane (or space). **Translations** and **rotations** are examples of such transformations which arise in the study of **congruent figures.**

TRANSFORMATION IN TOPOLOGY. See **Topological Transformation.**

TRANSFORMATION, LINEAR. See **Linear Transformation.**

TRANSFORMATION OF AXES. Axes are said to be *translated* when a new origin is chosen and they are drawn through it in directions parallel to those they had through the old one. They are said to be *rotated* when they are turned through the same angle. If (x, y) are the coordinates of a point in a given system of coordinates and (X, Y) are the coordinates of the same point in a transformed system, the equations of curves are altered by making the following substitutions: (1) if the y-axis is translated a distance h and the x-axis a distance k, $x=X+h$, $y=Y+k$; (2) if both axes are rotated through an angle θ, $x=X\cos\theta-Y\sin\theta$, $y=X\sin\theta+Y\cos\theta$.

TRANSFORMATION OF FORMULAE. This is sometimes referred to as *changing the subject*. Thus, the formula for simple interest can be transformed to give an expression for the number of units of time. Given $I=PRN/100$, $N=100\ I/PR$ is obtained by multiplying throughout by $100/PR$. See **Cosine Formulae.**

TRANSFORMATION OF SIMILITUDE. Synonym of *homothetic transformation.* In **rectangular coordinates,** the transformations $x'=kx$ and $y'=ky$. The distance between any pair of points is multiplied by k, the ratio of **similitude.** If $k<1$ the transformation shrinks the plane. If a figure is subjected to a homothetic transformation followed by a **translation** or a **reflection,** two types of similitude can occur: *direct* if the sense of rotation is maintained, *indirect* if it is reversed. See **Radially Related Figures.**

TRANSIT. The passage of a heavenly body across the **meridian of celestial sphere** of a place. Also the passage of a planet across the disc of the sun.

TRANSITIVE RELATION. A **relation** having the property: if A has a relation to B and B has the same relation to C then A has the same relation also to C. For example; equality in numbers, $x=y$, $y=z$ then $x=z$. Relations which do not have this property are *non-transitive* or *intransitive*; e.g., A is father of B, B is father of C then A is not father of C.

TRANSITIVITY. See **Order Relations; Transitive Relation.**

TRANSLATION. A change of position without change of direction. A **rigid motion** in which each point is moved in the same direction by the same amount. The directed line segment AB represents a translation. When applied to P it gives P'; to XY, $X'Y'$; to α, α'.

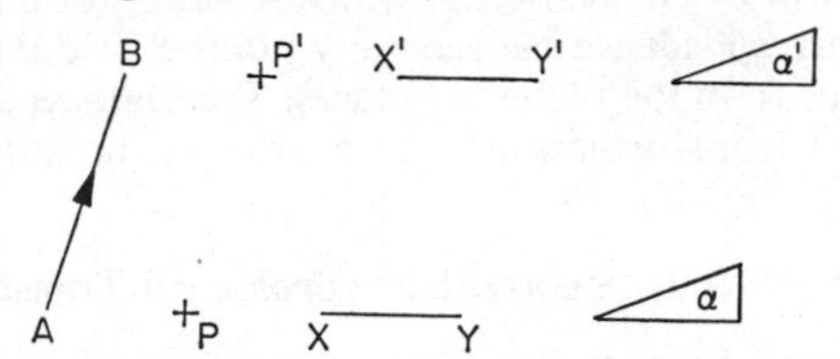

TRANSLATION OF AXES. See **Transformation of Axes.**

TRANSPOSE OF MATRIX. The **matrix** obtained by interchanging rows and columns in a given matrix.

TRANSPOSE OF TERM. To move a term from one side of an equation or an inequality to the other, involving a change of sign. This process is strictly one of applying to both sides of an equation or an inequality one operation which cancels the term on its original side. Thus:

$$\begin{aligned} ax+b \quad &>=< c, \\ ax+b-b &>=< c-b, \\ ax \quad &>=< c-b. \end{aligned}$$

TRANSPOSITION. (1) The process of transposing a term in an equation or an inequality. See **Transpose of Term.** (2) The interchange of two things. A **cyclic permutation** of degree two.

TRANSVERSAL. A straight line which crosses **coplanar lines**. The eight angles formed by its crossing two lines are paired and defined as follows:
alternate angles—(c, e), (d, f);
corresponding angles—(a, e), (b, f), (c, g), (d, h);
vertically opposite angles—(a, c), (b, d), (e, g), (f, h).

The alternate and corresponding angles are also pairs of equal angles when the transversal crosses two parallel lines. See **Adjacent Angles; Vertically Opposite Angles.**

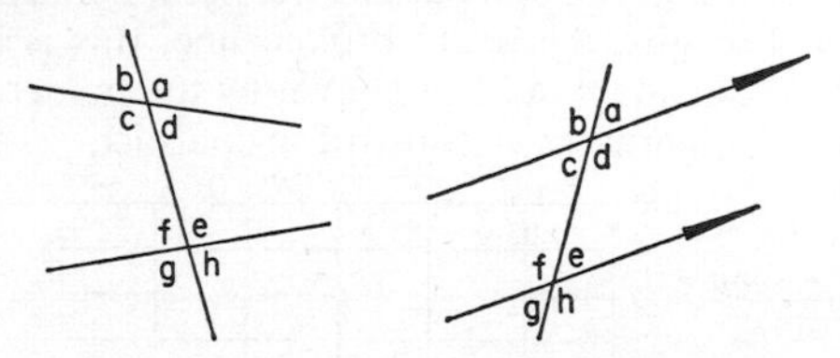

TRANSVERSE AXIS OF HYPERBOLA. See **Axes of Hyperbola.**

TRANSVERSE AXIS OF HYPERBOLOID. See **Axes of Hyperboloid.**

TRANSVERSE COMMON TANGENTS. See **Tangents to Two Circles.**

TRAPEZIUM. A **quadrilateral** with one pair of parallel sides. If the non-parallel sides are equal it is described as *isosceles.* The definitions of trapezium and **trapezoid** are often interchanged.

TRAPEZOID. A **quadrilateral** having no sides parallel. The definitions of trapezoid and **trapezium** are often interchanged.

TRAPEZOIDAL RULE. A rule for determining the area of a shape with irregular boundaries. It assumes that the curved outline may be replaced by consecutive line segments. The area is divided into a number of strips of equal width, as shown. The area of portion $AA'N'N$ is $\frac{1}{2}d(y_1 + 2y_2 + 2y_3 + \ldots + 2y_{n-1} + y_n)$. See **Simpson's Rule.**

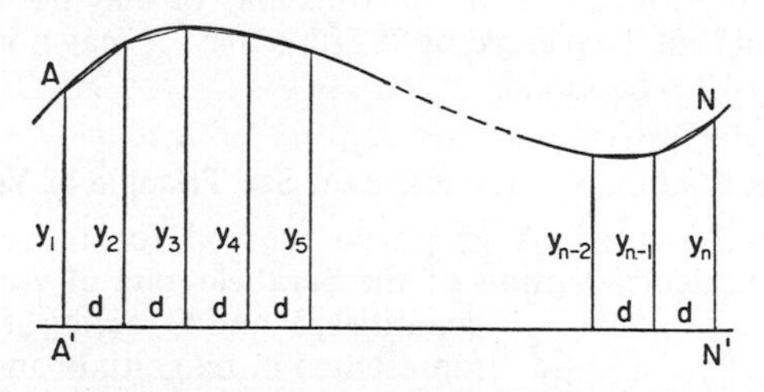

TRAVERSING. A method of surveying a region by surveying chosen paths formed by line segments whose lengths and relative directions are measured. A *closed traverse* bounds an area (*ABCDEA*); an *open traverse* does not (*APQD*). A *compound traverse* is a combination of closed and open traverses. Traversing is used in conjunction with **offsets.** See **Triangulation.**

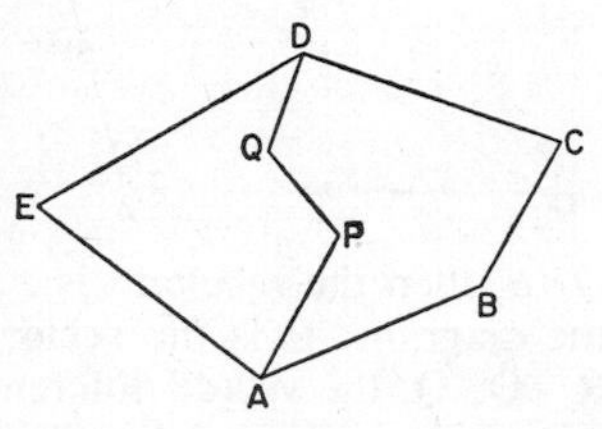

TREFOIL. A **multifoil** bounded by three congruent arcs of a circle.

TREND. A statistical term for a general tendency in data. If seven 11 a.m. daily temperature readings for one patient were graphed by isolated points then crudely joined to give the usual broken line, this is of only limited value. The trend of the data would be shown by the bolder line, called the **trend line**. Normal temperature is shown for contrast.

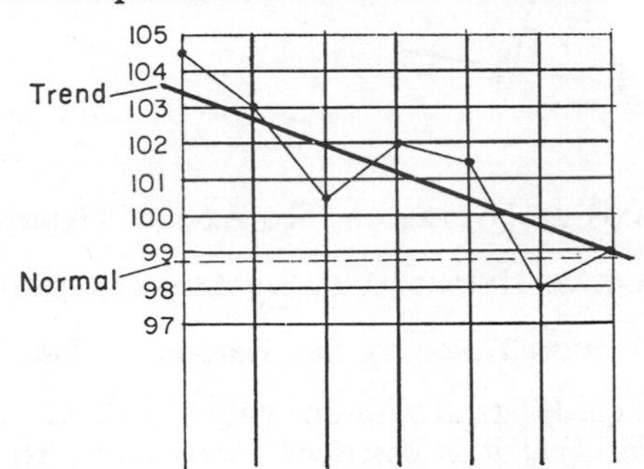

TREND LINE. The line which shows the general **trend** of a set of data. Fluctuations are about the trend. See **Line of Best Fit**.

TRIANGLE (PLANE TRIANGLE). Synonym of *trigon*. In general, a plane figure with three straight sides defining three interior angles of total value 180° or π radians. Six shapes are distinguished as follows:

acute. All angles less than 90°.
equilateral. Three equal sides; three angles of 60°.
isosceles. Two sides equal; two equal acute angles.
obtuse. One angle greater than 90° (this may or may not be isosceles).
right (right-angled). One angle of 90° (this may or may not be isosceles).
scalene. No two sides equal.

See **Spherical Triangle.**

TRIANGLE OF FORCES, VELOCITIES, ETC. See **Triangle of Vectors.**

TRIANGLE OF VECTORS. A graphical method for the **composition of vectors**: a convenient adaption of the **parallelogram of vectors.** If **vector quantities, P** and **Q,** acting at a point O, are represented in magnitude and direction by the line segments $\overline{OA}$ and $\overline{AB}$, then their resultant **R** is similarly represented by $\overline{OB}$, along the third side of triangle OAB. It follows that if three vectors acting at a point can be represented by line segments $\overline{OA}$, $\overline{AB}$, $\overline{BO}$ along the sides of triangle OAB, then the resultant is zero: and such a system is in equilibrium. In the diagram: **R** is the **vector sum** of **P** and **Q**; **P,** the vector difference **R** - **Q**; **Q,** the vector difference **R** - **P.**

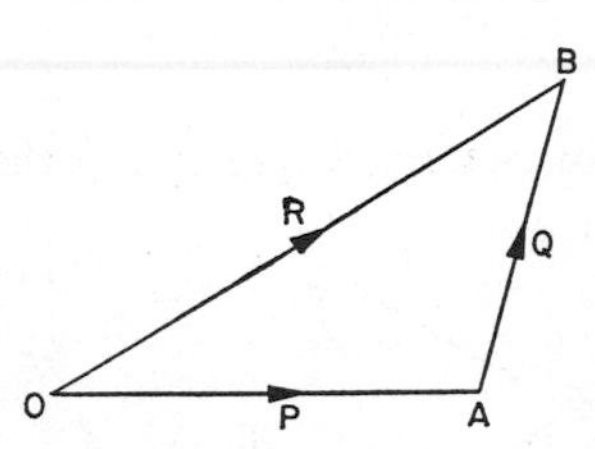

TRIANGULAR NUMBERS. See **Polygonal Numbers.**

TRIANGULATION. A method of surveying a region by surveying a chosen network of triangles. In *pure triangulation*, a base line to one triangle is measured and the rest of the surveying involves the measurement of angles only: in *mixed triangulation* certain sides and angles are measured: in *chain triangulation*, sides only are measured. See **Traversing.**

TRICHOTOMY. See **Order Relations.**

TRIGON. Synonym of **triangle.** A **polygon** with three angles.

TRIGONOMETRIC (TRIGONOMETRICAL). Referring to the measurement of triangles, especially those measurements based on the ratios of the sides of right-angled triangles. See **Trigonometric Ratios.**

TRIGONOMETRIC CURVE. The graph of a **trigonometric function.** See **Sine Curve; Tangent Curve.**

TRIGONOMETRIC EQUATION. An **equation** containing one of the six **trigonometric functions** as a variable.

TRIGONOMETRIC EXPRESSION. An **expression** in which at least one term is a **trigonometric function.**

TRIGONOMETRIC FORM OF COMPLEX NUMBER. Synonym of **polar form of complex number.**

TRIGONOMETRIC FUNCTIONS. Synonym of **circular functions.** Functions of angles defined in terms of **trigonometric ratios.** If point P, (x, y), lies on the circumference of a unit circle, $x^2+y^2=1$, the trigonometric functions of θ, the angle OP makes with the positive direction of the x-axis, are defined as:

sine of θ ($\sin\theta$) $=y$, cosecant of θ ($\csc\theta$) $=1/y$,
cosine of θ ($\cos\theta$) $=x$, secant of θ ($\sec\theta$) $=1/x$,
tangent of θ ($\tan\theta$) $=y/x$, cotangent of θ ($\cot\theta$) $=x/y$.

	$-A$	$90°\pm A$	$180°\pm A$	$270°\pm A$	$360°\pm A$
sin	$-\sin A$	$\cos A$	$\mp\sin A$	$-\cos A$	$\pm\sin A$
cos	$\cos A$	$\mp\sin A$	$-\cos A$	$\pm\sin A$	$\cos A$
tan	$-\tan A$	$\mp\cot A$	$\pm\tan A$	$\mp\cot A$	$\pm\tan A$
cot	$-\cot A$	$\mp\tan A$	$\pm\cot A$	$\mp\tan A$	$\pm\cot A$
sec	$\sec A$	$\mp\csc A$	$-\sec A$	$\pm\csc A$	$\sec A$
csc	$-\csc A$	$\sec A$	$\mp\csc A$	$-\sec A$	$\pm\csc A$

The signs of the coordinates determine the signs of the ratios. If θ is acute, all ratios are positive and are fully tabulated in trigonometric tables. The ratios of angles greater than a right angle can be converted to ratios of acute angles by the **reduction formulae in trigonometry**, summarized above. (No sign implies that both ratios are positive; one negative sign that both are negative.)

TRIGONOMETRIC IDENTITIES. **Identities** which involve **trigonometric functions.** For example, **Pythagoras's Theorem** can be expressed by the identities $\sin^2\theta+\cos^2\theta=1$, $\tan^2\theta+1=\sec^2\theta$, $1+\cot^2\theta=\csc^2\theta$. See **Addition, Subtraction Formulae.**

TRIGONOMETRIC RATIOS. The ratios of the lengths of sides in a right-angled triangle employed in survey or mensuration, or coordinates of a point on the circumference of a unit circle, $x^2+y^2=1$, used to define the **trigonometric functions** in analysis. If θ is an acute angle, the trigonometric

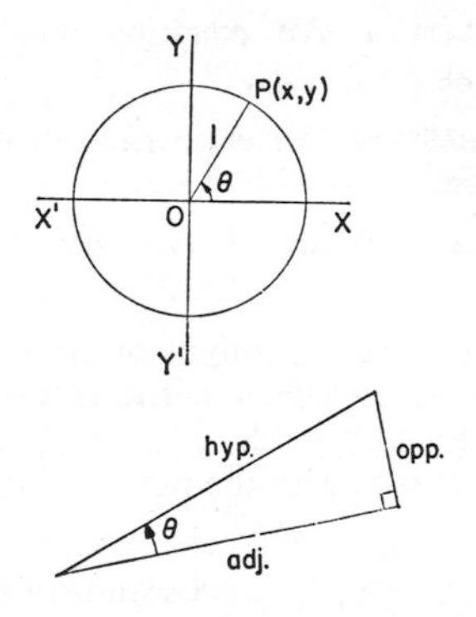

	$\sin\theta$	$\tan\theta$	$\sec\theta$	
$\sin\theta=a$	a	$a/(1-a^2)^{\frac12}$	$1/(1-a^2)^{\frac12}$	$a=\sin\theta$
$\cos\theta=b$	$(1-b^2)^{\frac12}$	$(1-b^2)^{\frac12}/b$	b^{-1}	$b=\cos\theta$
$\tan\theta=c$	$c/(1+c^2)^{\frac12}$	c	$(1+c^2)^{\frac12}$	$c=\tan\theta$
$\cot\theta=d$	$1/(1+d^2)^{\frac12}$	d^{-1}	$(1-d^2)^{\frac12}/d$	$d=\cot\theta$
$\sec\theta=e$	$(e^2-1)^{\frac12}/e$	$(e^2-1)^{\frac12}$	e	$e=\sec\theta$
$\csc\theta=f$	f^{-1}	$1/(f^2-1)^{\frac12}$	$f/(f^2-1)^{\frac12}$	$f=\csc\theta$
	$(\csc\theta)^{-1}$	$(\cot\theta)^{-1}$	$(\cos\theta)^{-1}$	

ratios of θ are conveniently defined as ratios of lengths of the hypotenuse and the sides adjacent to and opposite θ in a right-angled triangle. Thus

sine (sin) θ = opp./hyp.; cosine (cos) θ = adj./hyp.;
tangent (tan) θ = opp./adj.; cosecant (csc, cosec) θ = hyp./opp.;
secant (sec) θ = hyp./adj.; cotangent (cot, ctn) θ = adj./opp.

The *co-ratios* are the ratios of complementary angles, e.g., $\cos\theta = \sin(90° - \theta)$. Every ratio has a reciprocal ratio. The six ratios are related as shown in the table above.
See **Reduction Formulae in Trigonometry.**

TRIGONOMETRIC SERIES. Series involving the **trigonometric functions** as variables. They are widely used in applied mathematics. Fourier's periodic series is an example:

$$f(x) = 2[\sin x - \tfrac{1}{2}\sin 2x + \tfrac{1}{3}\sin 3x - \tfrac{1}{4}\sin 4x + \dots].$$

See **Periodic Function; Fourier Series.**

TRIGONOMETRIC SUBSTITUTIONS. Three are used for the purpose of rationalizing quadratic **surds**: (1) $x = a\sin u$ reduces $\sqrt{(a^2 - x^2)}$ to $a\cos u$; (2) $x = a\tan u$ reduces $\sqrt{(x^2 + a^2)}$ to $a\sec u$; (3) $x = a\sec u$ reduces $\sqrt{(x^2 - a^2)}$ to $a\tan u$. The quadratic surd, $\sqrt{(x^2 + ax + b)}$, can always be reduced to one or other of the three forms if it is written:

$$\sqrt{[(x + \tfrac{1}{2}a)^2 + (b - \tfrac{1}{4}a^2)]}.$$

TRIGONOMETRIC SURVEY. See **Survey; Triangulation.**

TRIGONOMETRY. The measurement of trigons (triangles). It involves all the study outlined in every entry under **trigonometric**. The drawing of triangles on spherical surfaces, arising from the urgencies of navigation, gave rise to **spherical trigonometry**. The **transcendental** nature of the ratios has given rise to branches of the subject dealing with **exponential, elliptic** and **hyperbolic functions.**

TRIHEDRAL. A figure formed by three lines meeting in one point and thus defining three planes.

TRIHEDRAL ANGLE. A **solid angle** as in a triangular pyramid. At each vertex is a **trihedral** figure.

TRIHEXAFLEXAGON. See **Flexagon.**

TRILINEAR COORDINATES. The **coordinates of point** in a plane, which are the numerical values of the lengths of the three perpendiculars from the point to the sides of a *triangle of reference* in the plane.

TRILLION. In the United Kingdom, 10^{18} (1 million)^3; in U.S.A. and France, 10^{12}.

TRINOMIAL EXPRESSION. An **algebraic expression** which contains three terms. See **Multinomial Expression.**

TRIPLE. A set of three elements as in (x, y, z), an ordered triple of coordinates.

TRIPLE INTEGRAL. See **Multiple Integral.**

TRIPLE POINT. See **Multiple Point.**

TRIPLE ROOT. See **Repeated Roots.**

TRIRECTANGULAR SPHERICAL TRIANGLE. A **spherical triangle** with three right angles.

TRISECTION. The process of dividing into three equal parts.

TRISECTION OF ANGLE. The classical problem of trisecting an angle by Euclidean constructions (use of straight lines and circles only) was proved to be insoluble by Wantzel in 1847. The trisection requires the use of cube roots which the circle does not produce. The trisection can be performed theoretically by the use of straight lines and certain other curves, for example, the **conchoid of Nicomedes**, the **limaçon of Pascal** and the **trisectrix of Maclaurin.** The trisection of an angle 3θ depends upon the solution of the cubic trigonometric equation for $\cos\theta$: $4\cos^3\theta - 3\cos\theta + \cos 3\theta = 0$, where $\cos 3\theta$ is known.

TRISECTRIX OF MACLAURIN. The locus of the equation $x^3 + xy^2 + ay^2 - 3ax^2 = 0$. The curve passes through the origin, has $x = -a$ as an **asymptote** and is symmetric with respect to the x-axis. Maclaurin showed how to trisect an angle using straight lines and the trisectrix only. If any line through the point $(2a, 0)$ is inclined at an angle 3θ to the positive direction of the x-axis and cuts the curve at P, then the line through the origin and P is inclined at an angle θ to the positive direction of the x-axis. See **Trisection of Angle.**

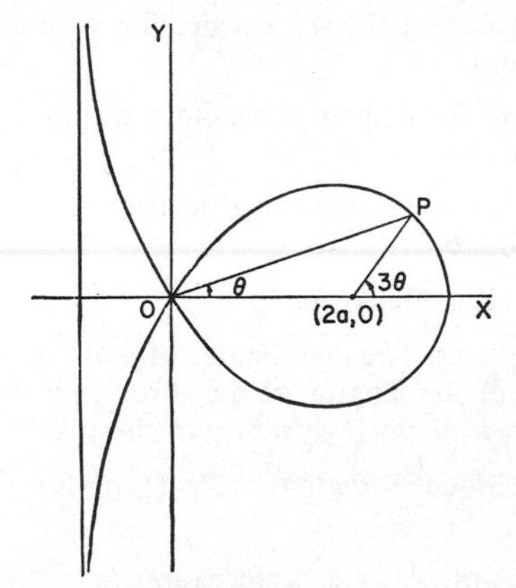

TROMINO. A **polyomino** made of three adjacent squares.

TROPICAL YEAR. The period of rotation of the sun relative to the first point of Aries. The period between two vernal equinoxes. Aries has a retrograde angular motion of 50.22 seconds among the stars. Consequently the tropical year is 20 minutes 23.5 seconds shorter than the **sidereal year** and 25 minutes 7 seconds shorter than the **anomalistic year.** Its length is 365.2422 **mean solar days** or 365 days 5 hours 48 minutes 46 seconds. See **Calendar; Second of Time Interval.**

TROY WEIGHT. A system of measure used mainly in weighing precious metals and gems.

24 grains	=1 pennyweight
20 pennyweight	=1 ounce
12 ounces	=1 pound
3.168 grains	=1 carat

The grain, ounce and pound are the same weights as those of **apothecaries' weight.** See **Avoirdupois.**

TRUE BEARING. See **Bearing.**

TRUE HORIZON. See **Pictorial Projection.**

TRUE SOLAR DAY. Synonym of apparent solar day. See **Apparent Solar Time.**

TRUE SOLAR NOON. Synonym of **apparent noon.**

TRUNCATED CONE, CYLINDER, PRISM, PYRAMID. That part of these **solids** bounded by two non-parallel plane **sections.** When the sections are parallel the truncated solid is a **frustum.**

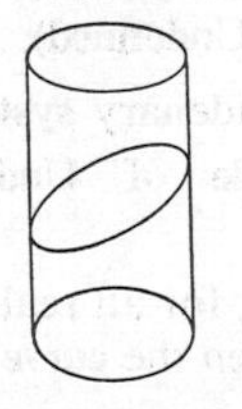

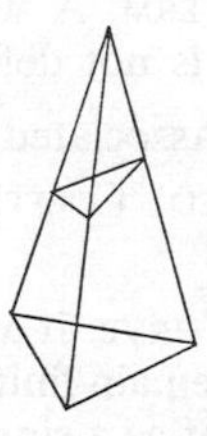

TRUTH SET. In any given problem there are sets of logical possibilities in which a given statement p is true or false. If P is the set of logical possibilities in which p is true, P is referred to as the *truth set* of the statement p. There are two bounds for P, the universal set U of all logical possibilities and the empty set ϕ. For any two given statements p and q with corresponding truth sets P and Q, the universal set U is divided into four disjoint and exhaustive subsets. In the notation of the **algebra of sets** they are (1) $P \cap Q$, (2) $P \cap Q'$, (3) $P' \cap Q$, (4) $P' \cap Q'$. The following relations exist between compound statements (written in the notation of **algebra of propositions**), their truth sets and the four subsets:

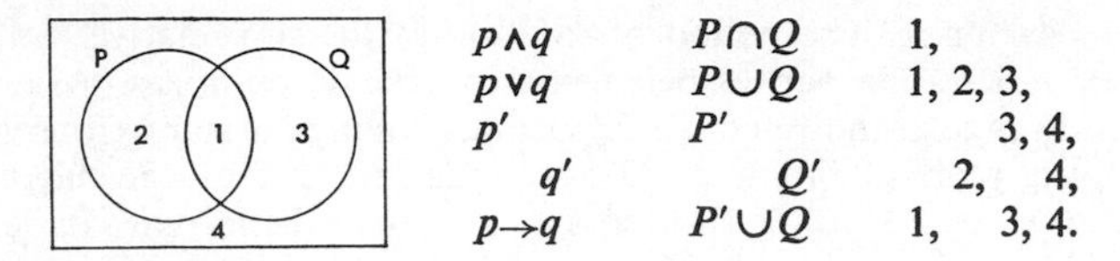

$p \wedge q$	$P \cap Q$	1,
$p \vee q$	$P \cup Q$	1, 2, 3,
p'	P'	3, 4,
q'	Q'	2, 4,
$p \rightarrow q$	$P' \cup Q$	1, 3, 4.

See **Probability in Logic; Set Theory.**

TURNING EFFECT. See **Moment of Forces.**

TURNING POINTS, VALUES. See **Maximum, Minimum Values.**

TWISTED (SKEW) CURVE. A space **curve** which does not lie in a plane.

TWO DIMENSIONAL GEOMETRY. Synonym of **plane geometry**. See **Dimensions of Ordinary Space.**

TWO DIMENSIONAL SPACE. See **Dimension of Space.**

U

U. Universal set. See **Set Theory.**

U, u. Symbols often used instead of V_0, v_0 to express **initial conditions** in **velocity.**

UNBOUNDED. See **Bound of Function; Bound of Set.**

UNCONDITIONAL INEQUALITY. See **Inequality.**

UNDEFINED TERM. A word which, when used, satisfies certain **axioms** but, otherwise, is not defined. See **Primitive (Undefined).**

UNDENARY. Associated with eleven as in undenary **system of notation.**

UNDETERMINED COEFFICIENT. See **Principle of Undetermined Coefficients.**

UNICURSAL CURVE. If $x = u(t)$, $y = v(t)$ and if, for all real finite values of t, $u(t)$ and $v(t)$ remain finite and continuous, then the **curve** in the xy plane can be swept out as a single continuous curve.

UNIFORM CIRCULAR MOTION. The motion of a particle moving at a constant speed, along the circumference of a circle. The restraints cause a constant **acceleration** towards the centre of the circle, equal to v^2/r or $r\omega^2$, where r is the radius of the circle, v the speed of the particle and ω the **angular velocity** of the particle.

UNIFORM (CONSTANT) ACCELERATION. **Acceleration** in which there are equal changes in **velocity** in equal intervals of time.

UNIFORM (CONSTANT) SPEED. Motion over equal distances in equal intervals of time. The motion may be in any direction. See **Speed.**

UNIFORM (CONSTANT) VELOCITY. Rectilinear motion over equal distances in equal intervals of time. See **Velocity.**

UNIFORM GRAVITATIONAL FIELD. A region of space subjected to gravitational forces where the ratio of gravitational **force** to **mass** is constant. In such a field, the **centre of gravity** coincides with the **centre of mass.** See **Force of Gravity.**

UNIFORM SCALE. A **scale** in which equal intervals of length correspond to equal values of a **variable.**

UNILATERAL SURFACE. A surface with only one side. See **Klein Bottle; Möbius Band (Strip).**

UNIMODULAR MATRIX. A square **matrix** with a **determinant** 1.

UNION (SUM) OF SETS. See **Set Theory.**

UNIT CIRCLE. A **circle** of unit radius. Its circumference is 2π and its area π. The circumference subtends 2π radians at the centre. The equation $x^2+y^2=1$ represents a unit circle having its centre at the origin of coordinates.

UNIT CUBE. A **cube in geometry** with an edge of one unit length (inch cube, foot cube). Its **volume** is a cubic unit (cubic inch, cubic foot). It is used in the measure of volume.

UNIT FRACTION. A common fraction of the form $1/n$, in which n is an integer. The reciprocal of n when n is an integer.

UNIT, FUNDAMENTAL. See **Fundamental Units.**

UNIT LINE SEGMENT, SQUARE, CUBE. A line segment of unit linear dimension; a square, cube with unit linear dimensions.

UNIT MEASURE. Any arbitrarily chosen quantity used as a standard for assessing the relative sizes of other quantities of a similar nature. See **Absolute Units; Fundamental Units.**

UNIT NUMBER. The number having unit **place value** in any **system of notation.**

UNIT SET. See **Set Theory.**

UNIT SPHERE. A **sphere** with unit radius. Its area is 4π and its volume $4\pi/3$. The surface subtends 4π **steradians** at the centre. The equation $x^2+y^2+z^2=1$ represents a unit sphere with its centre at the origin of coordinates.

UNIT VECTOR. A **vector quantity** of magnitude 1. See **Basis Vectors.**

UNITY. The symbol 1 can be interpreted in several ways:
1. A **cardinal number** associated with a **unit set.**
2. The first **ordinal number.**

3. An **odd number** in the sequence . . . $-3, -1, +1, +3,$ . . .
4. The **index** which has no effect ($2^1=2$).
5. The **identity element** of multiplication, $a \times 1 = 1 \times a = a$.
6. The **quotient** of any quantity and itself.
7. The implicit denominator of all integers when they are defined as **rational numbers.**
8. The **unit of measure** of length, area, volume, time, etc., when the system of units is not considered.
9. The symbol -1 used as an index forms a reciprocal ($2^{-1} = \frac{1}{2}$).
10. Used as a superscript -1 is the symbol of an **inverse operation** ($y = \sin^{-1}x \Leftrightarrow \sin y = x$).
11. $1 = -i^2$ or $\sqrt{-1} = i$. See **Complex Number.**

UNITY, ROOTS OF. See **Roots of Unity.**

UNIVERSAL SET (UNIVERSE), U, $\mathscr{E}$. See **Set Theory.**

UNKNOWN QUANTITY. The **symbol** entering into an equation or an expression, the finding of the numerical value of which constitutes the solution of the problem involved.

UNLIKE TERMS. Terms which are not **similar (like) terms.**

UNSOLVED PROBLEMS. There are various mathematical problems which have remained unsolved to the present time. Some exist as theorems for which no one, to date, has propounded a rigid **proof.** Some are geometrical constructions which have been proved impossible with the instruments permitted. See **Fermat's Theorems; Four-colour Problem; Goldbach's Conjecture; Levy's Conjecture; Quadrature in Geometry.**

UNSTABLE EQUILIBRIUM. See **Equilibrium.**

UPPER BOUND. See **Bound of Function; Bound of Set.**

V

V. Vector space.

VP. Vertical plane. See **First Angle, Third Angle Projection.**

V, v. Roman symbol for 5. Usual symbol for **velocity.**

VALIDITY. The state or quality of being sufficiently supported by fact or reason. A valid statement is true or provable. The validity of a *logically valid statement* depends upon the use of words and not upon the meaning of the statement.

VALUE, ABSOLUTE. See **Absolute Value.**

VALUE OF EXPRESSION. The result of simplifying an expression when the variables have been replaced by chosen numbers.

VALUE OF FUNCTION. The result of simplifying the expression obtained from replacing the variables by chosen numbers.

VALUE OF ROOTS OF EQUATION. Those numbers which satisfy the equation.

VANISH. To equal zero. Thus, the function $2x-5$ vanishes when $x=2\frac{1}{2}$.

VANISHING POINTS. See **Pictorial Projection.**

VARIABLE. Any symbol for any member of a set of numbers, points, values, etc. An element of the set is then called a value of the variable, and the whole set its **range.** See **Dependent, Independent Variable.**

VARIANCE. A statistical term for the square of the **standard deviation,** or the mean of the squares of the individual deviations. See **Deviation in Statistics.**

VARIATION. A relation between a set of values of one **variable** and a set of values of other variables. It involves the concept of a **function** and is expressed in the form of an **identity.**

combined variation. A relation between one variable and a combination of others, e.g., $F=kmm'/r^2$.

direct variation. A relation between two variables such that their ratio is constant. Thus, for the variation $y=kx$, *y varies directly as x*, or y and x are in *direct proportion.*

inverse or indirect variation. A relation between two variables such that the ratio of one to the reciprocal of the other is constant. Thus, for the variation $xy=k$, *y varies inversely* as x, or y and x are in *inverse proportion.*

joint variation. A direct variation between one variable and the product of others, e.g., $I=Prn/100$.

See **Proportion; Ratio.**

VARIATION IN NAVIGATION. See **Magnetic Declination.**

VARIATION IN STATISTICS. The *deviation* of any term from the mean of the set to which it belongs. See **Deviation in Statistics.**

VECTOR ANALYSIS. The study of **vector quantities,** their relationships and applications.

VECTOR DIFFERENCE. The difference **A** – **B** of **vector quantities A** and **B,** equivalent to the sum of **A** and –**B.** See **Composition of Vectors.**

VECTOR PRODUCT. Synonym of cross product. The product of two vectors resulting in another **vector quantity,** and usually written $\mathbf{A}\times\mathbf{B}$ to distinguish it from a **scalar product, A B.** If **A** is $(x,y,z)_1$ and **B** is $(x,y,z)_2$, then $\mathbf{A}\times\mathbf{B}$ is given by

$$\begin{vmatrix} \mathbf{i} & \mathbf{j} & \mathbf{k} \\ x_1 & y_1 & z_1 \\ x_2 & y_2 & z_2 \end{vmatrix} = \mathbf{i}\begin{vmatrix} y_1 & z_1 \\ y_2 & z_2 \end{vmatrix} + \mathbf{j}\begin{vmatrix} z_1 & x_1 \\ z_2 & x_2 \end{vmatrix} + \mathbf{k}\begin{vmatrix} x_1 & y_1 \\ x_2 & y_2 \end{vmatrix}.$$

Since the **minors** are **scalar quantities** and **i, j, k** are **unit vectors**, the cross product is a vector quantity.

If the vectors are given geometrically, the vector product is a vector through the common origin at right angles to the plane of the given vectors and equal to **A B** (sin θ) **n**, where θ is the angle between **A** and **B** and **n** is a unit vector in the direction of $\mathbf{A} \times \mathbf{B}$. See **Right-hand Rule for Vector Multiplication.**

VECTOR QUANTITY. A quantity which has magnitude and direction as compared with a **scalar quantity** which has magnitude only. A vector quantity can be represented by a **directed line segment** related to a fixed directed line in a plane or three fixed directed lines in three-dimensional space. Vectors may be interpreted as physical quantities such as velocity, force, displacement, which have magnitude and direction or as ordered sets of numbers, represented in terms of a set of coordinates, e.g. (a, b) in a plane, (a, b, c) in three dimensional space. The magnitudes of the vectors will be represented by the lengths of the line from the origin of coordinates to the points (a, b) and (a, b, c) respectively: their absolute (numerical) values will be $\sqrt{(a^2+b^2)}$ and $\sqrt{(a^2+b^2+c^2)}$, respectively. The concept can be extended to spaces of higher dimensions (**vector spaces**). There are special laws governing the **composition of vectors.** See **Absolute Value of Vector; Position Vector; Scalar Multiplication; Scalar Product; Vector Product.**

VECTOR SPACE. A set of **elements** which satisfies the axioms: (1) for any number a, $a\mathbf{V}$ is an element of the set; (2) $1 \cdot \mathbf{V} = \mathbf{V}$ for any $\mathbf{V}$; (3) the set is a **commutative group** under addition; (4) the set obeys the **associative law**; (5) the set obeys the **distributive law**. The concept of vector spaces was developed in the middle of the nineteenth century. They may be two-dimensional (planar), three-dimensional (Euclidean space) or n-dimensional. See **Composition of Vectors.**

VECTOR SUM. If two vectors **A** and **B** are specified by their components (a_x, a_y, a_z) and (b_x, b_y, b_z) their sum is specified by its components $(a_x+b_x,$ $a_y+b_y,$ $a_z+b_z)$. Addition of vectors obeys the **associative law** and the **commutative law.** See **Composition of Vectors.**

VECTOR VELOCITY. See **Velocity.**

VECTORIAL ANGLE. See **Polar Coordinates in Space.**

VELOCITY. The rate of change of displacement with respect to time when the direction of motion is specified: a directed **speed.** When the displacement is along a line (*rectilinear motion*) the **velocity** of a point with displacement x, is the **derivative** dx/dt, or $\dot{x}$ (linear or *rectilinear velocity*). When the displacement is along a curve (*curvilinear motion*) the velocity (*curvilinear velocity*) is in the direction of the tangent to the curve. Since velocity is a **vector quantity**, the velocity of a point can be represented by the sum of the derivatives of its displacements parallel to the coordinate axes. Thus the *vector velocity* of a point (x, y, z) with position vector $\mathbf{R} = \mathbf{i}x + \mathbf{j}y + \mathbf{k}z$ is $d\mathbf{R}/dt$ or $\mathbf{i}\dot{x} + \mathbf{j}\dot{y} + \mathbf{k}\dot{z}$. The speed of a point is equal to the length of the vector velocity, or $\sqrt{(\dot{x}^2 + \dot{y}^2 + \dot{z}^2)}$.

absolute velocity. The velocity of a moving point with respect to a fixed point. The velocity computed relative to a fixed system of coordinates.

average velocity. The difference in position vectors of a point at the beginning and end of a time interval divided by the length of the interval, written $\Delta x/\Delta t$ or $\delta x/\delta t$.

instantaneous velocity. The limit of the *average velocity* when the time interval, Δt, approaches zero, written dx/dt.

See **Angular Velocity**; **Relative Acceleration, Velocity**; **Equations of Motion.**

VELOCITY OF LIGHT (c). 186,326 miles/s: $2.998\,5 \times 10^{10}$ cm/s; approximately 300,000 Km/s. See **Light Year.**

VELOCITY RATIO. See **Machine.**

VENN DIAGRAM. A schematic diagram used for the representation of relation and operation in **set theory.** See **Truth Set.**

VERSED, COVERSED. Versed sine of θ = versine θ = vers $\theta = 1 -$ cosine θ; coversed sine of θ = versed cosine θ = covers $\theta = 1 -$ sine θ; haversine of θ = hav $\theta = \frac{1}{2}$ vers θ. The relations between versed, coversed sine, **exsecant, excosecant** and the **trigonometric ratios** are shown in relation to the coordinates of a point on a **unit circle** with centre at the origin of coordinates.

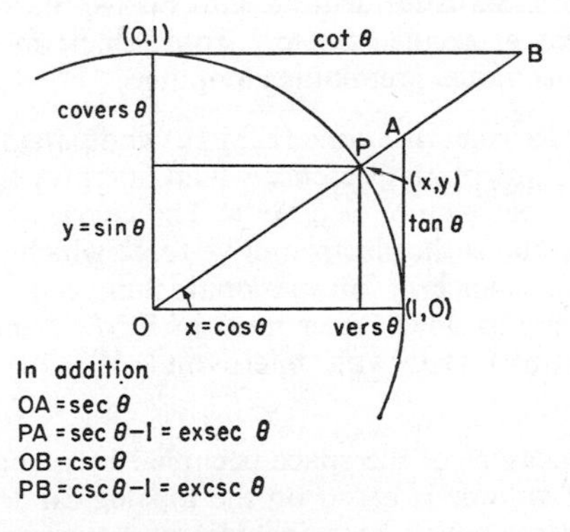

VERSIERA. Synonym of **witch of Agnesi.**

VERTEX, VERTICES. A defining point in various geometrical configurations.

1. The intersection of the arms of an **angle.**
2. The intersection of the sides of a **polygon.**
3. The intersection of the edges of a **polyhedron.**

4. The intersection of a **conic section** with its major axis.
5. The intersection of the **generating lines** of a pyramid or cone.
6. The common point of a **pencil of lines.**

VERTICAL ANGLES. Synonym of **vertically opposite angles.**

VERTICAL INTERVAL. See **Contours.**

VERTICAL LINE. A line in the direction towards or away from the centre of the earth at a given point. This, the plumb-line direction, is perpendicular to the plane of the true horizon, used in **pictorial projection.**

VERTICALLY OPPOSITE ANGLES. (1) A pair of **angles** having a common **vertex** such that the arms of one are the extensions through the vertex of the arms of the other. Two intersecting lines form two pairs of equal vertically opposite angles. (2) A pair of **polyhedral angles** having a common vertex such that the edges of one are the extensions through the vertex of the edges of the other. Three or more concurrent lines (not in the same plane) form two symmetric or oppositely congruent polyhedral angles which are equal **solid angles. See Congruent Figures in Space; Transversal.**

VIBRATION. Synonym of **oscillation.**

VINCULUM. (1) A horizontal line used to separate the numerator from the denominator in a **common fraction.** (2) A form of bracket: a horizontal line placed above terms to be treated collectively, e.g., $2a+\overline{n-1}\cdot d \equiv 2a+nd-d$. See **Aggregation.**

VITAL STATISTICS. Statistics dealing with births, marriages and deaths; incidence of diseases, accidents, etc.; from which mortality tables are constructed and insurance premiums computed.

VOLT. Unit of **electromotive force** (E.M.F.) and **potential difference.** The *absolute volt* is the potential difference which liberates 1 joule of energy in the transfer of one **coulomb** of electricity. The *international volt* (equal to 1.004 3 absolute volts) is the electromotive force which, when applied to a conductor with resistance 1 international **ohm,** causes a current of 1 international **ampere** to flow. Other units of E.M.F. and potential difference are millivolt (mV) $=10^{-3}$ volt, microvolt (μV) $=10^{-6}$ volt and kilovolt (kV) $=1\,000$ volts.

VOLUME. The measure of the space occupied by a geometric solid. The **metric** concept of volume is based on the **topological properties** of a continuous **closed surface** which both includes and excludes a part of space. In the concrete situation, a solid occupies an amount of space called its *external volume* (or volume) and certain solids have an *internal volume* (or *cubic capacity*). (The term **capacity** when not associated with cubic, is usually reserved for the volume of liquids or materials which pour, and special sets of units, e.g., gallon, litre, are used.) The volume of a solid can be considered as the common limit of a set of inscribed and another set of circumscribed **polyhedra**. The volumes of solids and the internal volume of containers are measured in terms of equivalent volumes of **unit cubes**. The volume of any geometric solid is the **least upper bound** of the sum of the volumes of the non-overlapping cubes contained by the solid. If the solids are cuboid (rectangular parallelepipeds), the volume can be computed directly from the formula $V=l\ b\ h$, where l, b and h are the numbers of unit cubes that can be placed along the edges of the cuboid described as length, breadth and height. The volumes of other solids are computed from formulae derived from this basic one, and, where this is not possible, computed by use of the **calculus**. See **Cubic Measure.**

VOLUME OF REVOLUTION. See **Solid of Revolution.**

VOLUMETRIC. Appertaining to the measurement of **volume.**

VULGAR FRACTION. Synonym of **common fraction.**

W

W.C.B. Whole circle bearing. See **Bearing.**

WALLIS'S FORMULAE. The **reduction formulae** in **integration** for the evaluation of the definite integrals from 0 to $\frac{1}{2}\pi$ of the functions $\sin^m x$, $\cos^m x$ and $\sin^m x \cdot \cos^n x$ where m and n are positive integers.

WALLIS'S PRODUCT FOR π. The infinite product

$$\left(\frac{2}{1}\cdot\frac{2}{3}\right)\cdot\left(\frac{4}{3}\cdot\frac{4}{5}\right)\cdot\left(\frac{6}{5}\cdot\frac{6}{7}\right)\cdots\left(\frac{2r}{2r-1}\cdot\frac{2r}{2r+1}\right)\cdots$$

which yields $\frac{1}{2}\pi$.

WATT. A unit of power measured by the rate of **work** done in **joules** per second. It is also the power yielded by a current of 1 **ampere** with an **electromotive force** of 1 **volt.** One kilowatt (1 000 watts) is equivalent to 1.340 **horse-power**; 1 horse-power is equivalent to 746 watts. See **Power in Mechanics.**

WAVE. Any regular periodic fluctuation that travels through a medium in such a way that throughout the process it recognizably retains its identity in some special respect, such as shape. The word has come to be applied to periodic fluctuations in electromagnetic intensity, associated with the phenomena of heat, light, magnetism and electricity.

WAVE LENGTH. The distance between any two successive points on a wave which represent the same phase of disturbance. Alternatively, the distance between two consecutive maxima. It is given by the relation $\lambda=c/f$, where c is the **wave velocity** and f, the **frequency of oscillation.** See **Phase in Simple Harmonic Motion.**

WAVE MOTION. See **Simple Harmonic Motion.**

WAVE VELOCITY. Synonym of **velocity** of propogation. The velocity with which the whole system of **oscillation** travels outwards from the source of oscillation. It is given by the relation $c = \lambda f$, where λ is the **wave length** and f, the **frequency of oscillation.**

WEDGE. The physical form of a **dihedral angle.** Used as a mechanical tool from antiquity as a device for causing a **force** in one direction to be effective in another. The principle involved is the **resolution of forces.** See **Spherical Wedge; Resolution of Vectors.**

WEIGHT. The force of attraction of the earth on any object near its surface. $W = mg$, where W is the *weight*, m the **mass** and g the **acceleration due to gravity.** See **Pound Force; Gramme (Gram) Force.**

WEIGHT, BRITISH UNITS. See **Apothecaries' Weight; Avoirdupois; Troy Weight.**

WEIGHT, METRIC UNITS. See **Metric System.**

WEIGHTED AVERAGE (MEAN). The process by means of which the elements of a set of numbers are given coefficients which establish their relative importance when the average or mean is being computed, is known as weighting. The weighted average is thus the mean of a set of numbers which have been altered in this way. If the numbers of marks allotted to subjects A, B and C are 60, 75 and 90 respectively, the **arithmetic mean** is 75. If in relative importance the subjects are as 1 : 2 : 3, the weighted average (mean) will be $(1 \cdot 60 + 2 \cdot 75 + 3 \cdot 90)/(1 + 2 + 3) = 80$. See **Mean in Statistics.**

WEIGHTING. The method by which statistical ratios, mostly percentages, are combined. If the ratio of a subset of n_1 terms is p/q and the ratio of another subset of n_2 terms is r/s, each ratio is multiplied by a number called a *weight* before they can be combined to give the ratio of the larger group of $n_1 + n_2$ terms. The weights would be $n_1/(n_1 + n_2)$ and $n_2/(n_1 + n_2)$ respectively. The combined ratio is then $(p/q)n_1/(n_1 + n_2) + (r/s)n_2/(n_1 + n_2)$ which reduces to $(n_1ps + n_2qr)/(n_1 + n_2)qs$. See **Weighted Average (Mean).**

WHOLE NUMBER. An integer; one of the natural **numbers,** 1, 2, 3, 4, etc.

WIDTH. Synonym of **breadth.**

WILSON'S THEOREM. The number $(n - 1)! + 1$ is divisible by n if and only if n is a **prime number.** When $n = 3$, a prime, Wilson's number, 3, is divisible by 3; when $n = 6$, a **composite number,** Wilson's number, 121, is *not* divisible by 6. Leibniz proved that this condition was necessary: Lagrange, that it was sufficient.

WITCH OF AGNESI. The curve defined by the **parametric equations** $x = 2a \cot \theta$, $y = 2a \sin^2 \theta$, or **Cartesian equation** $4a^2(2a - y) = x^2y$. Draw a circle to touch two parallel lines, at a distance of $2a$ apart. Take one point of contact as origin of coordinates, O, the tangent at O as the x-axis and the diameter through O as the y-axis. Let any line through O cut the circle at C and the parallel tangent at T. Draw a right-angled triangle CPT with CP parallel to the x-axis and PT parallel to the y-axis. The set of all possible points P is the curve known as the *witch of Agnesi*, named after Maria Agnesi (1718–1799) who in 1748 discussed the curve and referred to as it the *versiera*, which in Italian meant *versed sine* or *witch*. It was studied earlier by Grandi in 1703 and Fermat before 1666. One property of the witch is that the area between the curve and the x-axis is four times that of the generating circle. The *pseudo versiera* is the curve obtained by doubling the ordinates of the versiera. Its equation is $a^2(2a - y) = x^2y$. It was studied by Gregory in 1658 and used by Leibnitz in 1674 in obtaining the series $\pi/4 = 1 - 1/3 + 1/5 - 1/7 + \dots$.

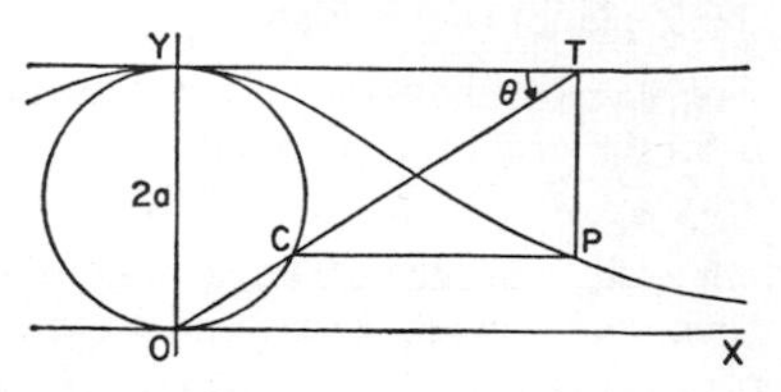

WORK. Work is done when energy is transferred from one body A to another body B via a **force** F exerted by A on B. If the force F produces a displacement x in the direction of the force F, then the work done is measured by the **scalar product** $F\ x$. Measured in **ergs, joules** or **foot-pounds** the relations are:

$$1 \text{ ft lbf} \equiv 1.356 \times 10^7 \text{ ergs} \equiv 1.356 \text{ joules};$$
$$1 \text{ joule} \equiv 10^7 \text{ ergs} \equiv 0.737\,6 \text{ ft lb}$$

See **Kinetic Energy**; **Potential Energy**.

X

X, **x.** Roman symbol for ten (10).

*x-, y-, z-*AXES. The first, second and third axes of a system of **Cartesian coordinates**. The lines $X'OX$, $Y'OY$, $Z'OZ$, through the origin of co-ordinates.

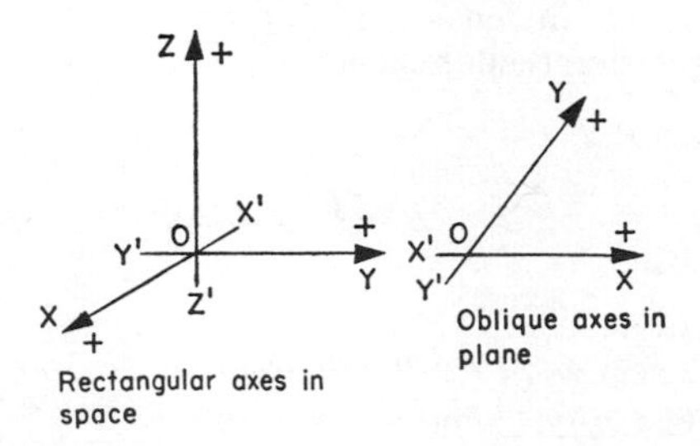

Rectangular axes in space

Oblique axes in plane

x, y, z COORDINATES. In a system of **Cartesian coordinates**, the coordinates measured parallel to the *x*-axis, *y*-axis or *z*-axis from a point in space to the *yz*-plane, *zx*-plane or *xy*-plane (*YOZ* plane, *ZOX* plane or *XOY* plane) respectively.

x, y, z INTERCEPTS. In a system of **Cartesian coordinates**, the distances from the origin to the points of intersection of a curve or surface with the *x*-axis, *y*-axis or *z*-axis respectively.

xy, yz, zx PLANES. In a system of **Cartesian coordinates**, the planes which contain the *x*-axis and *y*-axis, *y*-axis and *z*-axis or *z*-axis and *x*-axis respectively. They are sometimes referred to as the *XOY*, *YOZ*, *ZOX* planes.

Y

y-AXIS. See ***x*-, *y*-, *z*-Axes.**

y COORDINATE. See ***x, y, z* Coordinates.**

y INTERCEPT. See ***x, y, z* Intercepts.**

yz PLANE. See ***xy, yz, zx* Planes.**

YARD. The British Imperial and U.S. Standard unit of length, preserved as the distance between two fine marks on a metal bar under controlled temperature conditions. It is divided into 3 feet or 36 inches. One one-thousand-seven-hundred-and-sixtieth part of the **statute mile**. Its equivalent in the **metric system** is 0.9144 m.

YEAR. The longest unit of time; the period of revolution of the earth about the sun. It was known to the Egyptians as 365 days. Between one vernal equinox and the next there are 365 days, 5 hours, 48 minutes and 46 seconds of solar time. See **Anomalistic Year**; **Calendar**; **Sidereal Year**; **Tropical Year.**

YOUNG'S MODULUS. A constant found by Young in 1807 when studying the behaviour of elastic bodies subjected to external **forces.** If T is the **stress** in the cross-section of a thin rod, and e the extension **(strain)** associated with the stress, then $T = Ee$, where E is *Young's modulus* in tension. The modulus in tension is not necessarily the corresponding modulus in compression (**bulk modulus**).

Z

z-AXIS. See ***x*-, *y*-, *z*-Axes.**

z COORDINATE. See ***x, y, z* Coordinates.**

z INTERCEPT. See ***x, y, z* Intercepts.**

zx PLANE. See *xy*, *yz*, *zx* **Planes.**

ZENITH. The point on the **celestial sphere** vertically above the observer.

ZENITH-NADIR LINE. The line joining the **zenith** to the **nadir.** At any point on the earth it is the extension of a vertical line.

ZENITHAL PROJECTION. See **Map Projection.**

ZERO. The symbol 0 can be interpreted in several ways:
1. A **place holder** which facilitates arithmetical computation.
2. A **cardinal number** associated with a **null set** (an empty set).
3. A point on a number line, the boundary between positive and negative **directed numbers** $(-3, -2, -1, 0, +1, +2, +3, \ldots)$.
4. An **even number** in the sequence $\ldots -4, -2, 0, +2, +4 \ldots$
5. One of the equal coordinates $(0, 0, 0)$ of the origin in **Cartesian coordinates.**
6. The limit of $1/n$ as n approaches **infinity.**
7. The power **index** which transforms any number into **unity.**
8. The **identity element** of addition $a+0=0+a=a$.
9. The only number which can be divided exactly by any other non-zero number, the **quotient** always being zero.
10. The only number which cannot be considered as a divisor; the **ratios** $a/0$, $1/0$, $0/0$ having meaning only in the field of **limits.**

ZONE OF SPHERE. See **Spherical Cap; Spherical Zone.**

ZONE TIME. Synonym of **standard time.**